Problem Solving in General Chemistry

Whitten • Davis • Peck

Fifth Edition

LESLIE N. KINSLAND
University of Southwestern Louisiana

Saunders College Publishing
Harcourt Brace College Publishers

Fort Worth   Philadelphia   San Diego   New York   Orlando   San Antonio   Austin
Montreal   London   Sydney   Tokyo

Copyright ©1996, 1992, 1988, 1984, 1982 by Saunders College Publishing

All rights reserved. No part of this publication may be reproduced or transmitted in any form or by any means, electronic or mechanical, including photocopy, recording, or any information storage and retrieval system, without permission in writing from the publisher.

Requests for permission to make copies of any part of the work should be mailed, to: Permissions Department, Harcourt Brace & Company, 6277 Sea Harbor Drive, Orlando, Florida 32887-6777.

Printed in the United States of America.

Kinsland; Problem Solving in General Chemistry, 5E. Whitten, Davis, Peck.

ISBN 0-03-015698-X

567 021 987654321

Contents

Chapter 1 The Foundations of Chemistry — 1
    Exercises — 14
    Answers to Exercises — 18

Chapter 2 Chemical Formulas and Composition Stoichiometry — 19
    Exercises — 31
    Answers to Exercises — 36

Chapter 3 Chemical Equations and Reaction Stoichiometry — 38
    Exercises — 52
    Answers to Exercises — 60

Chapter 4 Some Types of Chemical Reactions — 63
    Exercises — 76
    Answers to Exercises — 80

Chapter 5 The Structure of Atoms — 85
    Exercises — 97
    Answers to Exercises — 100

Chapter 6 Chemical Periodicity — 103
    Exercises — 110
    Answers to Exercises — 111

Chapter 7 Chemical Bonding — 112
    Exercises — 123
    Answers to Exercises — 125

Chapter 8 Molecular Structure and Covalent Bonding Theories — 133
    Exercises — 143
    Answers to Exercises — 145

Chapter 9 Molecular Orbitals in Chemical Bonding — 149
Exercises — 156
Answers to Exercises — 158

Chapter 10 Reactions in Aqueous Solutions I: Acids, Bases and Salts — 161
Exercises — 169
Answers to Exercises — 171

Chapter 11 Reactions in Aqueous Solutions II: Calculations — 174
Exercises — 193
Answers to Exercises — 200

Chapter 12 Gases and the Kinetic Molecular Theory — 204
Exercises — 220
Answers to Exercises — 226

Chapter 13 Liquids and Solids — 227
Exercises — 240
Answers to Exercises — 243

Chapter 14 Solutions — 244
Exercises — 259
Answers to Exercises — 262

Chapter 15 Chemical Thermodynamics — 264
Exercises — 284
Answers to Exercises — 290

Chapter 16 Chemical Kinetics — 292
Exercises — 305
Answers to Exercises — 310

Chapter 17 Chemical Equilibrium — 311
Exercises — 330
Answers to Exercises — 336

Chapter 18 Ionic Equilibria I: Acids and Bases — 339
  Exercises — 357
  Answers to Exercises — 361

Chapter 19 Ionic Equilibria II: Hydrolysis — 364
  Exercises — 379
  Answers to Exercises — 382

Chapter 20 Ionic Equilibria III: The Solubility Product Principle — 388
  Exercises — 404
  Answers to Exercises — 408

Chapter 21 Electrochemistry — 410
  Exercises — 429
  Answers to Exercises — 435

To the Student

Problem solving is one of the key concerns in general chemistry courses. Chemistry instructors frequently hear from students "I know how to work the problem. I just don't know how to set it up." or "I just didn't have enough time to work the exam." Successfully and efficiently solving chemistry problems requires learning certain techniques, patterns of thought, and ways of inspecting data. To master those techniques, as to become proficient in any other skill, practice is essential. Although general chemistry text books have detailed explanations and numerous exercises, many students find they still require additional material for drill. **Problem Solving in General Chemistry** is designed to fill that need by providing supplemental examples and exercises that are geared to the level of those in the main text.

Problem Solving in General Chemistry consists twenty-one chapters that correspond to the first twenty-one chapters in the texts, **General Chemistry 5/E** and **General Chemistry with Qualitative Analysis 5/E**, by Whitten, Davis, and Peck. This edition of the problem solving book has been extensively modified to reflect the changes that those authors have made in the main text and to provide different exercises for you to practice the techniques you have learned. Although this problem book is designed to be used with the Whitten texts, it may be used with other texts because the coverage corresponds to the common core of most general chemistry courses. Throughout this book, references are made to tables, figures, and appendices in the listed texts. Similar information in other texts can be found by consulting the indices. You may find it useful to photocopy data tables, both for use with this book and when doing homework assignments in the main text, to have vital physical data and conversion factors readily available as you read problems.

This book is designed as a supplement to, **not** as a replacement for, **General Chemistry 5/E**. To gain maximum benefit from this book, you should attend class and pay careful attention to what topics and techniques your professor most emphasizes. Next, you should study the examples in your notes and in the text and work the assigned homework problems. Then, after you have thoroughly explored those primary sources, if you still feel that you need additional help, turn to this text for the extra review and practice you require.

Each chapter in this book consists of:

(1) A **brief** discussion of the appropriate topics. If the material in the brief summary does not seem familiar, you should review the more detailed explanations, discussion, and problem solving tips in the main text.

(2) A variety of examples to illustrate the topics. The examples are different from those in the main text, but are worked out in a similar manner to complement that work's coverage.

(3) A set of exercises coded by topic. The level of the exercises is similar to that in the main text to provide practice on fundamental skills and problem solving techniques.

(4) A set of miscellaneous exercises that review material from all parts of the chapter.
(5) The answers to all exercises.

We hope that this supplement will increase your success in your general chemistry course. If you have any comments or suggestions for future editions of this text, please contact Leslie N. Kinsland, Chemistry Department, The University of Southwestern Louisiana, Lafayette, Louisiana, 70504-4370, or by email at lnk7547@usl.edu.

Thanks for the improved format and readability of this edition of the problem solving book go to Jennifer Bortel, Associate Editor at Saunders, who suggested many of the changes, and Cynthia L. Kinsland, an organic graduate student at Cornell University, who designed the plan to implement the proposed changes.

Chapter 1

The Foundations of Chemistry

1-1 Units of Measurement

Most measurements in the sciences are reported in units of the metric system or its successor, the International System of units (SI, from Système International), which was adopted by the National Bureau of Standards in 1964. The seven fundamental units of the SI system, from which other units are derived, are shown in Table 1-1.

Table 1-1 The Fundamental SI Units

Physical Property	Name of Unit	Symbol
Length	Meter	m
Mass	Kilogram	kg
Time	Second	s
Electric current	Ampere	A
Thermodynamic temperature	Kelvin	K
Luminous intensity	Candela	cd
Quantity of Substance	Mole	mol

The advantages of a metric system of units relative to the English units, which are still widely used in this country, are that the same basic unit is always used to describe the same physical property and that it is a decimal system. In the English system, feet, miles, and inches are all used for measuring lengths, whereas in the metric system the base unit for length is the meter. When different sizes or magnitudes of a property are to be measured, metric systems use a series of prefixes to indicate fractions and multiples of ten. These prefixes have the same meaning regardless of what unit they precede. The relative powers of ten permit easier conversions than the many differing relationships between the English units (5280 feet = 1 mile, but 12 inches = 1 foot, for example). Some of the prefixes used in the SI system are shown in Table 1-2. Table 1-7 in the text summarizes provides some conversion factors that relate length, volume, and mass units. Appendix C in the text includes additional metric prefixes and conversion factors. It also tabulates some of the SI derived units, such as the joule (kg m^2 s^{-2}) for energy.

Table 1-2 Some Common Metric Prefixes

Prefix	Symbol	Meaning	Example
mega-	M	10^6	1 megagram (Mg) = 1 x 10^6 g
kilo-	k	10^3	1 kilometer (km) = 1 x 10^3 m
deci-	d	10^{-1}	1 decisecond (ds) = 1 x 10^{-1} s
centi-	c	10^{-2}	1 centiampere (cA) = 1 x 10^{-2} A
milli-	m	10^{-3}	1 millijoule (mJ) = 1 x 10^{-3} J
micro-	μ	10^{-6}	1 microsecond (μs) = 1 x 10^{-6} s
nano-	n	10^{-9}	1 nanometer (nm) = 1 x 10^{-9} m
pico-	p	10^{-12}	1 picomole (pmol) = 1 x 10^{-12} mol

1-2 Use of Numbers in Scientific Calculations

Scientific notation is used to deal with very large or very small numbers to make their magnitude more apparent. In scientific notation we place one nonzero number to the left of the decimal point. To the right of the decimal point are the rest of the significant figures multiplied by the appropriate power of ten. To determine the power of ten, we must recognize that any number is equal to itself x 10^0, because $10^0 = 1$. To convert a number to scientific notation we increase the power of ten by one for each place we shift the decimal point to the left or we decrease the power of ten by one for each place we shift the decimal point to the right. Appendix A in the text has more details about scientific notation and discusses calculator usage.

Example 1-1 Scientific Notation

Convert these numbers to scientific notation.
(a)   0.0001456
(b)   0.00203
(c)   663,000,000,000

Plan

Decrease the power of ten by one for each place we shift the decimal point to the right or increase the power of ten by one for each place we shift the decimal point to the left.

Solution

(a) $0.0001456 = 0.0001456 \times 10^0 = \boxed{1.456 \times 10^{-4}}$ (decimal point moved four places to the right)

(b) $0.00200 = 0.00200 \times 10^0 = \boxed{2.00 \times 10^{-3}}$ (decimal point moved three places to the right)

(c) $663,000,000,000 = 663,000,000,000 \times 10^0 = \boxed{6.63 \times 10^{11}}$ (decimal point moved eleven places to the left)

Numbers may be either **measured** or **exact**. Most measurements yield **inexact numbers**, whose accuracy is limited by the measuring device used. The number of **significant figures** in a measurement consists of all the numbers that are considered to be correct by a competent person making a measurement. The number of significant figures consists of all digits that can be read directly plus one estimated digit.

For example, the smallest units on most triple beam balances are 0.1 gram units. The first estimated digit in a mass obtained from such a balance would be 0.01 gram. The mass of a penny might be reported as 2.56 grams on a triple beam balance. This number has three significant figures. This indicates that the person who weighed the penny is certain that its mass is between 2.5 and 2.6 grams and that it is most likely 2.56 grams. The digit **6** is considered significant because it is the best estimate of the person who weighed the penny. The result of a measurement should always be reported to the maximum number of significant figures allowed by the measuring device and no more.

Analytical balances are much more accurate and have smaller divisions than triple beam balances. Most have divisions to 0.001 grams and, therefore, can be used to determine mass to the nearest 0.0001 gram. If the same penny were weighed on an analytical balance, its mass might be reported as 2.5622 grams, which has five significant figures. Four of these digits are exactly known and the fifth, **2**, is an estimated digit.

Exact numbers, which are either defined or counted, are known to be absolutely correct. For example, one can count the number of people in a classroom. If there are thirty-seven people, there is no doubt that it is **exactly** thirty-seven, not 37.1 or 36.9. A kilometer is defined to be **exactly** 1000 meters. All metric prefixes and defined quantities (such as inches in a foot) are exact numbers. Such numbers are understood to have an infinite number of significant figures; they will not limit the number of significant figures in a calculation.

If only the results of measurements are reported, the following conventions are used to tell which digits are significant:
1. All non-zero numbers are significant.
2. Zeros used only to position the decimal point are not significant.

These conventions are shown in the following examples.
(a) 0.00225   three significant figures (2.25×10^{-3}, the zeros are place holders)
(b) 1.0085   five significant figures
(c) 51.00   four significant figures
(d) 138,000   ambiguous significant figures
(e) 0.00001   one significant figure (1×10^{-5}, the zeros are place holders)

Three statements cover all cases involving zeros:
1. **Leading zeros** (those to the left of the first nonzero digit) are **never** significant.
2. **Embedded zeros** (those between nonzero digits) are **always** significant.
3. **Trailing zeros** (those to the right of nonzero digits) are significant **only** if the decimal point is shown.

To avoid the ambiguity on very large numbers (and also for clarity on very small numbers), scientific notation should be used. In example "d" above, the number could be written as:
1.38×10^5 to show three significant figures
1.380×10^5 to show four significant figures
1.3800×10^5 to show five significant figures
The appropriate notation would reflect the accuracy of the device used for the measurement.

When calculations are done using the results of measurements, the following rules apply:
1. In **addition** and **subtraction**, the last digit retained in the sum or difference is determined by the position of the **first doubtful digit**.
2. In **multiplication** or **division**, an answer contains the same **number of significant figures** as the number with the least significant figures of those used in the operation.
3. To show the results of calculations to the proper number of significant figures, rounding rules must be used. When the number to be dropped is less than 5, the preceding number is left unchanged (i.e., 2.681 rounds to 2.68). When the number to be dropped is greater than 5, the preceding number is increased by 1 (i.e., 2.687 rounds to 2.69). When the number to be dropped is exactly equal to 5, the preceding number is not changed when it is even, or is increased by 1 when it is odd (2.685 rounds to 2.68, but 2.695 rounds to 2.70).

In the following examples, the first estimated digit is shown in bold.

Example 1-2 Significant Figures (Addition)

Using different balances, a student found the mass of a penny to be 2.56 g, a dime to be 2.2624 g and a quarter to be 5.505 g. What is the total mass of coins?

Plan
The masses are in the same units. If they had not been, they would need to be converted to the same units. Because this is an addition, we use significant figure rule number 1 and round at the first doubtful digit.

Solution

```
    2.56     g
    2.2624   g
+   5.505    g
   ─────────
   10.3274   g      (calculator result)
```

$\boxed{10.33 \text{ g}}$ (correct significant figures)

Example 1-3 Significant Figures (Multiplication)

What is the volume of a carton if its base has a width of 12.5 cm and a length of 44.23 cm? The height of the carton is 24 cm.

Plan
The volume is equal to the height times the width times the length. Because all measurements are expressed in the same units, we can multiply directly, then apply significant figure rule number 2. The answer must have the same number of significant figures as the component with the fewest.

Solution

```
     12.5     cm
     44.23    cm
  x  24       cm
   ─────────
  13269       cm³      (calculator answer)
```

$\boxed{1.3 \times 10^4 \text{ cm}^3}$ (correct significant figures)

Problem Solving in General Chemistry

In the rest of the examples in this chapter, the calculator answer will be given in parentheses and the answer that has correct number of significant figures will be boxed. On multi-step problems, answers may vary slightly if rounding is done after each step or at the end of the process. Even when intermediate answers are shown rounded to the correct number of significant figures, final results in this book are reported using the latter approach.

Example 1-4 Significant Figures

Perform the following operations.
(a) 5.58 x 3.617

(b) 438.6 + 0.0102

(c) 24.38 - 0.00219

(d) 25.78 + 88.63 + 33.841

Plan

We will use a calculator to perform the calculations. The appropriate significant figure conventions will be used and the answers will be rounded.

Solution

(a) $\boxed{20.2}$ (20.18286)

(b) $\boxed{4.30 \times 10^4}$ (43000)

(c) $\boxed{24.38}$ (24.37781)

(d) $\boxed{3.3808}$ (3.38081026)

Note that in (d) the addition gives 25.78 + 88.63 = 114.41; when we add (or subtract) we can gain (or lose) a counted number of significant figures.

1-3 Unit Factors and Dimensional Analysis

Units are an integral part of any measurement and should always be included. They can facilitate the solving of many problems in an approach called **dimensional analysis**, the **factor-label method**, or the **unit factor** method. In this method, a given quantity is converted into the desired quantity by multiplying the given number and unit by **unit factors**, which are constructed from equalities. In these factors, the denominator and the numerator are equal to the same quantity, so the factor equals one. Multiplying by one does not change the value of an expression. Consider the equality: 2.54 cm = 1.00 inch. Division of both sides by 1.00 inch yields one unit factor. Division of both sides by 2.54 cm yields a second unit factor.

$$\frac{2.54 \text{ cm}}{1.00 \text{ in}} = \frac{1.00 \text{ in}}{1.00 \text{ in}} = 1 \qquad\qquad \frac{2.54 \text{ cm}}{2.54 \text{ cm}} = \frac{1.00 \text{ in}}{2.54 \text{ cm}} = 1$$

Chapter 1 The Foundations of Chemistry

Two reciprocal unit factors such as these can be constructed from any equality. Several examples follow using relationships found in Appendix C of the text book.

Example 1-5 Length Conversion

How many centimeters are there in 56.3 inches?

Plan
First we write down the units we wish to know, preceded by a question mark. Then we set the desired units equal to what is given.

$$? \text{ cm} = 56.3 \text{ inches}$$

Finally, we multiply by the appropriate unit factor that will convert the given units of inches to the desired unit of centimeters. The appropriate factor has inches in the denominator and centimeters in the numerator, so that, after multiplication, the "inches cancel" and centimeters remain as the desired unit.

Solution

$$? \text{ cm} = 56.3 \text{ inches} \times \frac{2.54 \text{ cm}}{1.00 \text{ in}} = \boxed{143 \text{ cm}} \qquad (143.002)$$

Example 1-6 Mass Conversion

How many kilograms are in 175 pounds?

Plan
Use the equalities 454 g = 1 lb and 10^3 g = 1 kg to construct unit factors to do the conversion.

Solution

$$? \text{ kg} = 175 \text{ lb} \times \frac{454 \text{ g}}{1 \text{ lb}} \times \frac{1 \text{ kg}}{10^3 \text{ g}} = \boxed{79.4 \text{ kg}} \qquad (79.45)$$

Example 1-7 Area Conversion

How many square millimeters, mm², are in 0.368 square inches, in²?

Plan

If both sides of an equality are squared, they are still equal. By squaring both sides of the previous equality, it is found that $(2.54 \text{ cm})^2 = (1 \text{ inch})^2$. Also, 10 mm = 1 cm, an exact relationship, so $(10 \text{ mm})^2 = (1 \text{ cm})^2$.

Solution

$$? \text{ mm}^2 = 0.368 \text{ in}^2 \times \left[\frac{2.54 \text{ cm}}{1.00 \text{ in}}\right]^2 \times \left[\frac{10 \text{ mm}}{1 \text{ cm}}\right]^2 = \boxed{237 \text{ mm}^2} \quad (237.41888)$$

Example 1-8 Rate Conversion

A small plane can travel at 355 miles per hour. Convert this rate to meters per second.

Plan

Use the equalities 1 mile = 1.61 kilometer, 1 kilometer = 10^3 meters, 60 seconds = 1 minute, and 60 minutes = 1 hour to write the required unit factors. Note that "per" means that the quantity is in the denominator.

Solution

$$? \frac{m}{s} = \frac{355 \text{ mi}}{\text{hr}} \times \frac{1.61 \text{ km}}{\text{mi}} \times \frac{10^3 \text{ m}}{\text{km}} \times \frac{1 \text{ hr}}{60 \text{ min}} \times \frac{1 \text{ min}}{60 \text{ s}} = \boxed{159 \text{ m/s}} \quad (158.7638889)$$

Percentages are also used as unit factors. The percentage of A, %A by mass, is defined as:

$$\%A \text{ by mass} = \frac{\text{parts A by mass}}{100 \text{ parts mixture by mass}}$$

Example 1-9 Percentage

When intravenous medication delivery is desired, the drug is often introduced in a 0.90% salt (sodium chloride) solution. How many grams of this solution are needed to provide 0.250 grams of sodium chloride?

Chapter 1 The Foundations of Chemistry   9

Plan

We know that there are: $\dfrac{0.90 \text{ g salt}}{100 \text{ g solution}}$ or $\dfrac{100 \text{ g solution}}{0.90 \text{ g salt}}$

We will use the factor that cancels out the mass of salt and gives the mass of solution.

Solution

$$? \text{ g solution} = 0.250 \text{ g salt} \times \dfrac{100 \text{ g solution}}{0.90 \text{ g salt}} = \boxed{28 \text{ g solution}} \quad (27.7777777)$$

1-4 Density and Specific Gravity

The **density** of a substance is its mass per unit volume. Density usually has units of g/cm^3 or g/mL ($1 \text{ mL} = 1 \text{ cm}^3$) for solids and liquids, but g/L for gases. Density is an **intensive** property, which is independent of the size of the sample, whereas mass and volume are **extensive** properties, which do depend on sample size. Densities of several substances appear in Table 1-8 in the text. Density problems may be performed algebraically with a formula:

$$\text{Density} = \dfrac{\text{mass}}{\text{volume}} \quad \text{or} \quad D = \dfrac{m}{V}$$

or by the dimensional analysis approach.

Example 1-10 Density, Mass, Volume

A bar of chromium having a mass of 84.876 g displaces 11.8 mL of water when placed in water in a graduated cylinder. What is the density of chromium?

Plan

We use the definition of density directly for this problem.

Solution

$$D = \dfrac{m}{V} = \dfrac{9.959 \text{ g}}{1.1 \text{ mL}} = \boxed{9.1 \text{ g/mL}} \quad (9.053636364)$$

Example 1-11 Density, Mass, Volume

What is the volume of 125 g of sugar (sucrose)? The density of sucrose is 1.59 g/cm³.

Plan

We can rearrange the formula for density to solve for volume or, alternatively, we can use the unit factor method.

Solution

$$D = \frac{m}{V}, \text{ so } V = \frac{m}{D} = \frac{125 \text{ g}}{1.59 \text{ g/cm}^3} = \boxed{78.6 \text{ cm}^3} \quad (78.6163522)$$

or, $? \text{ cm}^3 = 125 \text{ g} \times \dfrac{1 \text{ cm}^3}{1.59 \text{ g}} = \boxed{78.6 \text{ cm}^3}$

In Example 1-11, the volume and mass each must have been known to at least three significant figures to report the density to three significant figures. The numerator in the unit factor is understood to contain at least three significant figures. The **1** does not limit the number of significant figures in the calculation.

Specific gravity (S.G.) is the density of a substance compared to a standard at the same temperature. For solids and liquids, the standard is water, which has a density of 1.00 g/mL (1.00 g/cm³) at 4°C. Therefore, in the metric system, specific gravity is numerically equal to density, but has no units.

$$\text{specific gravity} = \frac{\text{density of substance}}{\text{density of water}} = \frac{\text{density of substance}}{1.00 \text{ g/cm}^3} = \frac{\text{density of substance}}{1.00 \text{ g/mL}}$$

Example 1-12 Specific Gravity

Sulfur has a density of 2.07 g/cm³. What is the specific gravity of sulfur?

Plan

We use the definition of specific gravity.

Solution

$$? \text{ Specific gravity} = \frac{\text{density of sulfur}}{\text{density of water}} = \frac{2.07 \text{ g/cm}^3}{1.00 \text{ g/cm}^3} = \boxed{2.07}$$

Example 1-13 Specific Gravity, Percent by Mass, Volume

Muriatic acid, sold for adjusting the pH of swimming pools and for cleaning freshly poured concrete, is a concentrated solution of hydrochloric acid in water. It has a specific gravity of 1.18 and is typically 36.5% hydrochloric acid by mass. How many grams of hydrochloric acid are in a quart (three significant figures) of muriatic acid?

Plan

We can use the specific gravity to find the density of the muriatic acid. We must convert the volume to mL to agree with the units of the density. The text gives the conversion factor that 1 quart = 0.9463 liters and we know that 10^{-3} L = 1 mL. The density can then be used to find the mass of the solution. Finally, we use the mass percentage to find the mass of acid.

Solution

Because the density of water is 1.00 g/mL, the density of the solution is 1.18 g/mL

Find the volume of the solution in mL.

$$? \text{ mL solution} = 1.00 \text{ quart} \times \frac{0.9463 \text{ L}}{1 \text{ quart}} \times \frac{1 \text{ mL}}{10^{-3} \text{ L}} = 946 \text{ mL solution } (946.3)$$

Now determine the mass of the solution.

$$? \text{ g solution} = 946 \text{ mL solution} \times \frac{1.18 \text{ g solution}}{1 \text{ mL solution}} = 1.12 \times 10^3 \text{ g solution}$$

Because the solution is 36.5% by mass, the factor needed is $\frac{36.5 \text{ g acid}}{100 \text{ g solution}}$.

$$? \text{ g acid} = 1.12 \times 10^3 \text{ g solution} \times \frac{36.5 \text{ g acid}}{100 \text{ g solution}} = \boxed{408 \text{ g acid}} \quad (407.57141)$$

1-5 Heat and Temperature Scales

Heat is a form of energy and can be expressed in any energy unit (see Appendix C). The SI unit of energy is the joule, J, (1 joule = 1 kg·m²/s² or 1 newton-meter). Another commonly used unit of heat is the calorie (1 calorie = 4.184 joules). **Temperature** is a measure of the intensity of heat. If two objects made from the same material are at the same temperature, the larger object will have higher heat content. The three most often used temperature scales are Fahrenheit (°F), Celsius (°C) and absolute or Kelvin (K). By convention, no degree sign is used with Kelvin. The reference points for the scales are the boiling and freezing points of water at one atmosphere pressure (Figure 1-1.) The Celsius degree is 180/100 or 9/5 as large as the Fahrenheit degree and there is a 32° difference between the scales at the freezing point of water. From this information, the following relationships can be derived:

$$? \,°F = \left(y\,°C \times \frac{1.8°F}{1.0°C}\right) + 32°F \quad \text{or} \quad ? \,°C = \frac{1.0°C}{1.8°F}(y°F - 32°F)$$

The absolute temperature scale is based on the observed behavior of matter. A temperature of 0 K is theoretically the lowest temperature obtainable, the temperature of zero point vibrational energy. Absolute zero, 0 K, corresponds to -273.15°C and, therefore, the relation between °C and K is as follows:

$$? \,K = °C + 273.15 \quad \text{or} \quad ? \,°C = K - 273.15$$

Table 1-3 Comparison of Temperature Scales

	°F	°C	K
Boiling point of water	212°	100°	373.15
Freezing point of water.	32°	0°	273.15

Example 1-14 Temperature Conversion

Oxygen is a gas at room temperature and atmospheric pressure. When cooled, it liquefies at -297° F. (This is the same as the boiling point of liquid oxygen.) What is the liquefaction temperature (boiling point) of oxygen in °C and in K?

Plan
We use the relationships given earlier in turn to carry out the desired conversions.

Solution

$$? \,°C = \frac{1.0°C}{1.8°F}(°F - 32°) = \frac{1.0°C}{1.8°F}(-297°F - 32) = \boxed{-183°C} \quad (-182.77778)$$

$$? \,K = °C + 273.15 = -183\,°C + 273.15 = 90.\,K \quad (90.15)$$

1-6 Heat Transfer and the Measurement of Heat

The **specific heat** of a substance is an intensive property defined as the amount of heat required to raise the temperature of 1 gram of a substance 1°C **with no change in state**:

$$\text{specific heat} = \frac{\text{(amount of heat in J)}}{\text{(mass of substance in g)(temperature in °C)}}$$

A table of specific heats is provided in Appendix E in the text.

Example 1-15 Specific Heat

The specific heat of lead is 0.159 J/g·°C. How many joules are required to raise the temperature of 80.0 grams of lead from 13.5°C to 40.0°C?

Plan
We can find the temperature change required and then use dimensional analysis.

Solution
The temperature change, Δt, is 40.0°C - 13.5°C = 26.5°C.

$$\text{amount of heat} = (\text{mass}) \times (\text{specific heat}) \times (\Delta t)$$

$$?J = 80.0 \text{ g} \times 0.159 \frac{J}{g \cdot °C} \times 26.5° \text{ C} = \boxed{337 \text{ J}} \;(337.08)$$

EXERCISES

On multi-step conversions, final answers may vary to a small degree, depending upon whether rounding is done at the end of the problem or at the end of each of the intermediate steps. Consult your text book for appropriate conversion factors or data.

Significant Figures and Scientific Notation

1. Use scientific notation to express the following numbers to the number of significant figures indicated in parentheses.

    (a) 374,628 (2)   (b) 115,332,000 (3)   (c) 0.00034221 (4)
    (d) 0.005456 (3)   (e) 33,598,776,332 (1)   (f) 9887 (2)

2. How many significant figures do the following numbers contain?

    (a) 0.005543   (b) 121.8   (c) 0.000012   (d) 5.2210×10^{11}
    (e) 120,001   (f) 1445   (g) 0.00000077   (h) 25.00

3. Perform the following operations and express the results to the proper number of significant figures. Assume that these numbers represent measurements.

    (a) 816 + 97
    (b) (56.85 + 74.328) × 1.0881
    (c) 12.6 - 0.03
    (d) $3.058 \times 10^4 + 695$
    (e) $\dfrac{22.45 - 3.98}{6.14 \times 10^{-3}}$
    (f) $\dfrac{3.78 \times 10^{12}}{6.022 \times 10^{23}}$
    (g) 18.002 + 97.55 - 2.4
    (h) $8.66 \times 10^3 + 9.36 \times 10^2$

Unit Factors and Dimensional Analysis

4. How many meters are in 95.6 inches?

5. Many soft drinks are sold in 12 ounce cans. How many deciliters are in 12 ounces?

6. The earth is 93 million miles from the sun. Express this distance in km and in Mm.

7. A salt water aquarium has a volume of 157 gallons. Express this volume in cubic feet and in cubic meters.

8. A Nordic Track® skiing machine is supposed to burn 1000 Calories per hour. A "food" Calorie is a kilocalorie. Convert this value to kJ/s.

Chapter 1 The Foundations of Chemistry

9. A 6 ounce serving of a flavored coffee beverage has an energy value of 50 Calories. Express this caloric value in kJ/mL to 2 significant figures.

10. The volume of a cube is 226 mm^3. How long is each side in mm, cm and in inches?

11. An acre has an area of 43,560 square feet. What is the area of a 3550 acre wheat farm in square miles and in square kilometers?

12. Precious metals are sold by the "troy ounce" instead of by the standard "avoirdupois" ounce with which we measure most staples. A troy ounce is equal to 1.097 avoirdupois ounces. If platinum costs $421 per troy ounce, what would be the cost of the metal (to the nearest dollar) required to produce a platinum medallion with a mass of 46.0 g?

13. A sulfur ion has a radius of about 0.184 nm. Assuming the ion is spherical, calculate the volume of a sulfide ion in nm^3 and in in^3.

14. One brand of toaster pastry has a mass of 52 grams and contains 220 mg sodium. What is the percent by mass sodium in one of these pastries?

15. Chromium steel permanent magnets are made from an alloy that is 3.5% chromium, 0.9% carbon, and 0.3% manganese, with the remainder of the mass being iron. How many tons of chromium steel alloy can be made from 56 kilograms of chromium? (1 ton = 2000 pounds)

Density and Specific Gravity

16. Ethyl butanoate is a colorless liquid with a pineapple odor and a density of 0.879 g/mL. It is used for manufacturing artificial rum flavoring.

    (a) What is the mass of 50.0 mL ethylbutanoate?

    (b) How many milliliters of ethylbutanoate have a mass of 50.0 g?

17. Carbon disulfide is an extremely toxic liquid that is a commercially valuable solvent. When a 15.00 mL sample of carbon disulfide is added to a beaker with a mass of 34.023 g, the mass of the liquid and beaker is 53.414 g. What are the (a) density and, (b) the specific gravity of carbon disulfide?

18. A block of chromium, which is used for chrome plating on appliances and automobiles, has a mass of 234.56 g, a length of 4.86 cm, and a height of 5.56 cm. If the density of chromium is 7.19 g/mL, what is the width of the block?

19. Which has the larger:

    (a) volume--15.5 grams of lead or 2.3 mL of lead?

    (b) mass--1.50 pounds of gold or 0.0305 liters of gold?

20. The "lift" of a balloon is the difference between the weight of the gas in the balloon and the weight of an equal volume of air. How many pounds, including its own weight, can be lifted

by a spherical balloon, filled with helium, that has a diameter of 150 feet? The density of air is 1.205 g/L and that of helium is 0.178 g/L at 25 °C.

Heat and Temperature Scales

21. The melting point of iodine, which is an important nutrient for regulating thyroid function, is 113.5 °C. What is this temperature on both the Kelvin and the Fahrenheit scales?

22. The high temperature on a winter's day in northern Alaska might be -26 °F. What is this temperature in °C and Kelvin?

23. Convert 0 Kelvin to Fahrenheit.

24. Convert 7.82×10^3 calories to (a) joules, (b) kilocalories, (c) ergs, and (d) liter-atmospheres.

Heat Transfer

25. The specific heat of mercury is 0.138 J/g•°C.
    (a) How many joules are required to raise the temperature of 15.0 grams of mercury from 20.0°C to 90.0°C?
    (b) Repeat the calculation in (a) for water.
    (c) What is the specific heat of mercury in cal/lb•°F?

26. The specific heat of nickel is 0.444 J/g•°C. When 338 joules of heat is applied to a sample of nickel, the temperature changes from 22.5 °C to 47.8°C. What is the mass of the sample?

27. When 8.95 kilojoules of heat are applied to a 834 g sample of osmium at 16.7°C, the temperature of the sample rises to 98.6°C. What is the specific heat of osmium in J/g•°C?

28. How many kilojoules are needed to raise the temperature of a block of copper with a mass of 75.0 grams from 5.0°C to 125.0°C? The specific heat of copper is 0.385 J/g•°C.

29. What is the specific heat of ethanol ("drinking alcohol") if 4.65 kilojoules are required to raise the temperature of 250.0 grams of ethanol from 22.00°C to 29.56°C?

Miscellaneous Exercises

30. A gallon of latex paint is supposed to cover about 450 square feet. If we could transfer all the paint to the walls in a perfectly even coat, how thick (in mm to 2 significant figures) would the paint layer be?

31. Urea, sometimes used as a fertilizer, is 46% by mass nitrogen. It takes about 2.1 pounds of nitrogen to produce a bushel of corn. How many pounds of this fertilizer would be needed to produce 600 bushels of corn?

32. The density of bromine is 3.12 g/mL and its specific heat is 0.473 J/g·°C. How many kilojoules are required to raise the temperature of 2.50 quarts of bromine from 10.5°C to 37.8°C?

33. Concentrated hydrochloric acid, commercially sold as muriatic acid, is 37.4% by mass hydrochloric acid and has a specific gravity of 1.18. How many grams of hydrochloric acid are in 500.0 mL of muriatic acid?

34. The specific heat of diethyl ether, used as an anesthetic, is 3.74 J/g·°C and its density is 0.741 g/mL. What is the volume, in liters, of a sample of diethyl ether if 86.6 kilojoules applied to the sample raises its temperature from 13.4°C to 24.9°C

35. The LD_{50} (lethal dosage) rating of a substance is the amount of that substance needed to kill 50 % of the animals tested. For male rats, the LD_{50} rating of ibuprofen, a common pain-relieving substance found in Advil® and Nuprin®, is 1.05 g/kg body weight. How many 200 mg ibuprofen tablets would be required to kill a 1.50 pound rat?

36. The diameter of wire is still usually measured in "mils", where a mil is defined as 0.001 inch. What is the mass of a piece of copper wire that is 35.0 mils in diameter and has a length of 75.0 feet? The density of copper is 8.92 g/mL.

37. A normal adult human is about 1.5% calcium by mass. The density of calcium is 1.54 g/mL. If we could remove all the calcium from a 169 pound person, how many liters of pure calcium would we have?

38. Many consumers use solutions of ammonia in water for cleaning glass and other household articles. A commercially available concentrated solution is 29% by mass ammonia and has a solution density of 0.90 g/mL. How many grams of ammonia are in a 20.0 ounce bottle of this solution?

39. A container holds 22.60 mL of water. When a 39.9 gram sample of bismuth is place in the water, the final volume of the water and metal is 26.70 mL. What is the density of bismuth in pounds per cubic foot?

40. The furlong, which is equal to 660 feet, is still used to measure the length of some horse races. How many kilometers are in a 16.5 furlong steeplechase race?

ANSWERS TO EXERCISES

1. (a) 3.7×10^5 (b) 1.15×10^8 (c) 3.422×10^{-4} (d) 5.46×10^{-3}
   (e) 3×10^{10} (f) 9.9×10^3

2. (a) 4 (b) 4 (c) 2 (d) 5 (e) 6 (f) 4 (g) 2 (h) 2

3. (a) 8.4 (b) 142.73 (c) 12.6 (d) 44.0 (e) 3.01×10^3
   (f) 6.28×10^{-12} (g) 113.2 (h) 9.60×10^3

4. 2.43 m 5. 355 dL 6. 1.5×10^8 km; 1.5×10^2 Mm

7. 21.0 ft³; 0.595 m³ 8. 1.16 J/s 9. 1.2 kJ/mL

10. 3.40 mm; 0.340 cm; 0.134 in 11. 5.55 mi²; 14.4 km² 12. $623

13. 0.0261 nm³; 1.59×10^{-24} in³ 14. 0.42% sodium 15. 1.8 tons

16. (a) 44.0 g (b) 56.9 mL 17. 1.293 g/mL, 1.293 18. 1.21 cm

19. (a) 2.3 mL of lead (15.5 grams is only 1.37 mL)
    (b) 1.50 pounds (681 grams versus 589 grams in 0.0305 L)

20. 1.13×10^5 pounds 21. -236.3°F; 284.6 K 22. -32°C; 241 K 23. -459.67 °F

24. (a) 3.27×10^4 J (b) 7.82 kcal (c) 3.27×10^{11} ergs
    (d) 3.23×10^2 liter-atmospheres

25. (a) 145 J (b) 4.39×10^3 J (c) 8.32 cal/lb·°F

26. 30.1 g 27. 0.131 J/g·°C 28. 3.46 kJ 29. 2.46 J/g·°C

30. 0.092 mm 31. 2.7×10^3 lb 32. 95.4 kJ 33. 221 g 34. 2.72 L

35. 4 tablets (3.57) 36. 127 g 37. 0.75 L 38. 1.5×10^2 g

39. 81.2 lb/gal 40. 3.32 km

Chapter 2: Chemical Formulas and Composition Stoichiometry

2-1 Symbols and Formulas

The symbols of the elements consist of either a capital letter or a capital letter followed by a lower case letter. For example, O represents oxygen and Na represents sodium. Elements may exist as monatomic species, such as Ne, diatomic molecules, such as O_2, or polyatomic molecules, such as P_4.

Formulas indicate the elements present in a compound and their relative proportions. Each covalent molecule of methane, CH_4, a component in natural gas, consists of one atom of carbon and four atoms of hydrogen. Each formula unit of ionic Na_2CO_3, sodium carbonate or washing soda, consists of two sodium atoms, one carbon atom, and three oxygen atoms. Parentheses or dots (•) set off groups of atoms. Each formula unit of barium hydroxide octahydrate, $Ba(OH)_2 \cdot 8H_2O$, represents one barium atom, eighteen hydrogen atoms, and ten oxygen atoms (two OH groups and eight H_2O groups).

Ionic compounds are composed of ions in the proper ratios such that the compound has no net charge. See **Table 2-3** in the text for an initial list of common ions. Ionic compounds are named by naming the ions from left to right. Conversely, the formula of an ionic compound is written by choosing subscripts that make the sum of the positive charges equal to the sum of the negative charges for a net charge of zero.

Example 2-1 Names of Ionic Compounds

Name the following compounds.

(a) NaBr   (b) $AgCH_3COO$ (c) Cu_2CO_3 (d) Fe_2S_3 (e) $(NH_4)_3PO_4$

Plan

Consult Table 2-3 in the text, checking the charges of the ions that have more than one possible charge.

Solution

(a) sodium bromide
(b) silver acetate
(c) Copper can have a charge of either 1+ or 2+. Because carbonate is CO_3^{2-}, the compound contains Cu^+ and is named copper(I) carbonate.
(d) Iron can be either 2+ or 3+. Because sulfide is S^{2-}, the compound contains Fe^{3+} and is named iron(III) sulfide.
(e) ammonium phosphate

Example 2-2 Formulas for Ionic Compounds

Write the formulas of the ionic compounds below.

(a) aluminum fluoride
(b) copper(II) nitrate
(c) zinc sulfate
(d) calcium hydroxide
(e) iron(III) sulfite
(f) ammonium phosphate

Plan

When writing names of ionic compounds from formulas, ions must be present in the proper ratios to give a net charge of zero.

Solution

(a) Al^{3+} and F^-, so AlF_3
(b) Cu^{2+} and NO_3^-, so $Cu(NO_3)_2$
(c) Zn^{2+} and SO_4^{2-}, so $ZnSO_4$
(d) Ca^{2+} and OH^-, so $Ca(OH)_2$
(e) Fe^{3+} and SO_3^{2-}, so $Fe_2(SO_3)_3$
(f) NH_4^+ and PO_4^{3-}, so $(NH_4)_3PO_4$

2-2 Atomic Weights and the Mole

The relative masses of atoms are expressed in terms of atomic units on an **atomic weight** scale. In this system, an **atomic mass unit (amu)** is defined as exactly 1/12 the mass of a specific atom of carbon called a carbon-12 or ^{12}C atom. A ^{12}C atom has a mass of exactly 12 amu. Atomic weights are found below the symbol of an element on a periodic table, which can be found inside the front cover of this book. They are also tabulated inside the back cover. The atomic weights of other atoms are found by comparing them to the carbon-12 atom. An average calcium, Ca, atom has an atomic weight of 40.078 amu, or about three and a third times the mass of a ^{12}C atom. An average helium, He, atom has an atomic weight of 4.00260 amu, or about one third times the mass of a ^{12}C atom. When atomic weights are used, they are often rounded to give at least three significant figures. Atomic weights should never be rounded such that they would be the limiting factor for the significant figures in the final result of a calculation. In this book, atomic weights will be rounded to two decimal places for calculations.

The atomic mass unit is a convenient, arbitrary unit for discussing the relative masses of atoms, but is not practical for laboratory purposes because atoms are so small. Instead, a system in which masses are measured in grams is used. The quantity unit **mole (mol)** is defined as the amount of a substance that has the same number of particles (atoms, molecules, or formula units) as there are atoms present in exactly 0.012 kg (12 g) of carbon-12 atoms. This number of particles is **6.022 x 10^{23}** and is also called **Avogadro's number** of particles. A mole can also be interpreted as the number of particles that has a mass in grams that is numerically equal to the mass of one particle in atomic mass units. When we look at the Periodic Table, we can interpret the atomic weight as the number of amu in one atom of the element or as the number of grams in one mole, which is 6.022 x 10^{23} atoms, of the element. When the atomic weight is expressed in grams, it is called the **molar mass**, with the units expressed as grams/mole. A mole is analogous to a dozen. Just as one dozen is always twelve "things", a mole is always 6.022 x 10^{23} "things". There are 6.022 x 10^{23} calcium atoms (one mole of calcium atoms) in 40.078 grams of calcium. There are 6.022 x 10^{23} helium atoms (one mole of helium atoms) in 4.00260 grams of helium.

Example 2-3 Number of Moles and Number of Atoms

Calculate the number of moles and the number of atoms in 0.00426 grams of neon atoms.

Plan

The atomic weight of neon is 20.18. This means that an average atom of neon has a mass of 20.18 amu or that one mole of neon atoms has a mass of 20.18 g. This can be expressed as two unit factors:

$$\frac{1 \text{ mol Ne}}{20.18 \text{ g Ne}} \quad \text{or} \quad \frac{20.18 \text{ g Ne}}{1 \text{ mol Ne}}$$

We will use the factor that converts grams to moles and then use Avogadro's number to convert the number of moles to the number of atoms.

Solution

$$? \text{ mol Ne} = 0.00426 \text{ g Ne} \times \frac{1 \text{ mol Ne atoms}}{20.18 \text{ g Ne}} = \boxed{2.11 \times 10^{-4} \text{ mol Ne}}$$

$$? \text{ atoms Ne} = 2.11 \times 10^{-4} \text{ mol Ne} \times \frac{6.022 \times 10^{23} \text{ atoms Ne}}{1 \text{ mol Ne}} = \boxed{1.27 \times 10^{20} \text{ atoms Ne}}$$

Example 2-4 Atomic Weight and Element Identification

A sample of an element with a mass of 8.00 grams contains 3.96×10^{22} atoms. What is the atomic weight of this element and what element is it?

Plan

We know that the molar mass of an element should have units of g/mol and that the molar mass is numerically equivalent to the atomic weight. We must use Avogadro's number to find the number of moles of the element and then divide its mass in grams by its number of moles.

Solution

$$? \text{ mol} = 3.96 \times 10^{22} \text{ atoms} \times \frac{1 \text{ mol}}{6.022 \times 10^{23} \text{ atoms}} = 0.0658 \text{ mol}$$

$$? \text{ g/mol} = \frac{8.00 \text{ g}}{0.0658 \text{ mol}} = 122 \text{ g/mol or } 122 \text{ amu/atom}.$$

The atomic weight is $\boxed{122 \text{ amu}}$, so the element must be antimony, $\boxed{Sb}$.

2-3 Formula Weights and Molecular Weights

The mass of one formula unit of an ionic compound is its **formula weight**. The formula weight of a compound is the sum of the masses of the individual atoms that comprise the compound. For compounds that are not ionic, the term **molecular weight** is often used. The formula (or molecular) weight in amu is numerically equal to the mass of one mole of formula units (or molecules) in grams. As in the case of elements, the term **molar mass**, with units of grams/mole, is used for the mass of one mole of a compound.

Example 2-5 Formula Weight

What are the formula weight and mass of one mole of ammonium phosphate, $(NH_4)_3PO_4$, an ionic compound that is used as a fertilizer when hydrated?

Plan

We can find the formula weight by multiplying the atomic weight of each element by the number of its occurrences in the compound and summing the contributions of the elements.

Solution

kind of atoms	number of atoms per formula unit	x	atomic weight	=	contribution
N	3	x	14.01 amu	=	42.03 amu
H	12	x	1.01 amu	=	12.12 amu
P	1	x	30.97 amu	=	30.97 amu
O	4	x	16.00 amu	=	64.00 amu
			formula weight	=	149.12 amu
			1 mole	=	149.12 grams

We can also find the molar mass of a compound directly by recognizing that there will be the same number of moles of atoms in one mole of a compound as there are atoms in one formula unit or molecule of the compound.

Example 2-6 Molar Mass of a Compound

What is the mass of a mole of acetic acid, CH_3CO_2H? Acetic acid is the substance responsible for the characteristic taste of vinegar.

Plan

One mole of this compound contains 2 moles of carbon, 4 moles of hydrogen, and 2 moles of oxygen. We can calculate the mass in grams of each element within the compound and sum them to find the molar mass.

Solution

kind of atoms	number of moles of atoms per mole of compound	x	molar mass of element	=	contribution
H	4	x	1.01 g	=	4.04 g
C	2	x	12.01 g	=	24.02 g
O	2	x	16.00 g	=	32.00 g
			1 mole CH_3CO_2H	=	60.06 grams

The following table should put the previous concepts into perspective.

Table 2-1 One Mole of Some Common Substances

Substance	Mass of 1 Mole of Substance	Contains
iron	55.85 grams Fe	6.022×10^{23} Fe atoms
uranium	238.03 grams U	6.022×10^{23} U atoms
iodine	253.80 grams I_2	6.022×10^{23} I_2 molecules [$2(6.022 \times 10^{23}$ I atoms)]
water, H_2O	18.02 grams H_2O	6.022×10^{23} H_2O molecules [$2(6.022 \times 10^{23}$ H atoms)] [6.022×10^{23} O atoms]
calcium iodide, CaI_2	293.88 grams CaI_2	6.022×10^{23} formula units [6.022×10^{23} Ca^{2+} ions] [$2(6.022 \times 10^{23}$ I^- ions)]

Example 2-7 Number of Moles

Calculate the number of moles of molecules in 4.00 grams of acetic acid.

Plan

We use the molar mass calculated in Example 2-6 as a unit factor to find the number of moles.

Solution

$$? \text{ mol } CH_3CO_2H = 4.00 \text{ g } CH_3CO_2H \times \frac{1 \text{ mol } CH_3CO_2H \text{ molecules}}{60.06 \text{ g } CH_3CO_2H}$$

$$= \boxed{0.0666 \text{ mol } CH_3CO_2H \text{ molecules}}$$

The 4.00 g sample of acetic acid in Example 2.7 contains $0.0666(6.022 \times 10^{23}) = 4.01 \times 10^{22}$ molecules of CH_3CO_2H. These molecules would be composed of $4(4.01 \times 10^{22}) = 1.60 \times 10^{23}$ atoms of hydrogen, $2(4.01 \times 10^{22}) = 8.02 \times 10^{22}$ atoms of carbon and $2(4.01 \times 10^{22}) = 8.02 \times 10^{22}$ atoms of oxygen.

Chapter 2 Chemical Formulas and Composition Stoichiometry

Example 2-8 Number of Atoms in a Compound

How many atoms of nitrogen are in a 2.375 g sample of $(NH_4)_3P$?

Plan

By inspecting the formula we can see that there are 3 moles of N atoms in one mole of $(NH_4)_3P$. If we find the molar mass of the compound, we can find the number of moles of $(NH_4)_3P$, and the number of moles of nitrogen. We can then use Avogadro's number to find the number of nitrogen atoms.

Solution

kind of atoms	number of moles of atoms per mole of compound	×	molar mass of element	=	contribution
N	3		14.01	=	42.03 g
H	12		1.01	=	12.12 g
P	1		30.97	=	30.97 g
			1 mol $(NH_4)_3P$	=	85.12 g

$$? \text{ atoms N} = 2.375 \text{ g }(NH_4)_3P \times \frac{1 \text{ mol }(NH_4)_3P}{85.12 \text{ g }(NH_4)_3P} \times \frac{3 \text{ mol N}}{1 \text{ mol }(NH_4)_3P}$$

$$\times \frac{6.022 \times 10^{23} \text{ atoms N}}{1 \text{ mol N}} = \boxed{5.041 \times 10^{22} \text{ atoms N}}$$

When we find the molar mass of a compound, we determined an extensive number of other mass relationships. These can be used to calculate the relative amounts of each element in any mass of a compound and the amount of the compound needed to provide a given amount of an element.

Example 2-9 Composition of Compounds

Calculate the mass of carbon in 12.6 grams of aniline, C_5H_5N. This compound is the starting material for the synthesis of many dyes.

Plan

One mole of this compound contains 2 moles of carbon, 4 moles of hydrogen, and 2 moles of oxygen. We can calculate the mass in grams of each element within the compound and sum them to find the molar mass. We can then use mass relationships as a unit factor to determine the amount of carbon in the given mass of aniline.

Solution

kind of atoms	number of moles of atoms per mole of compound	×	molar mass of element	=	contribution
H	5	×	1.01 g	=	5.05 g
C	5	×	12.01 g	=	60.05 g
N	1	×	14.01 g	=	14.01 g
			1 mole C_5H_5N	=	79.11 grams

$$? \text{ g C} = 12.6 \text{ g } C_5H_5N \times \frac{60.0 \text{ g C}}{79.11 \text{ g } C_5H_5N} = \boxed{9.56 \text{ g C}}$$

2-4 Percent Composition and Formulas of Compounds

If the formula for a compound is known, its composition may be expressed in terms of the percent by mass of its constituent elements. **Percent by mass** is the mass of the element in the compound divided by the mass of the whole compound times 100. The masses may be reported in either amu or grams. Percentages should add to 100% within slight rounding errors.

Example 2-10 Percent Composition

Calculate the percent composition of the elements in acetic acid, CH_3CO_2H.

Plan

In Example 2-6, we found the mass of each element in one mole of acetic acid. We can use these masses directly to find the percent composition.

Solution

$$? \% \text{ H} = \frac{4.04 \text{ g H}}{60.06 \text{ g } CH_3CO_2H} \times 100 = \boxed{6.73 \% \text{ H}}$$

$$? \% \text{ C} = \frac{24.02 \text{ g C}}{60.06 \text{ g CH}_3\text{CO}_2\text{H}} \times 100 = \boxed{39.99 \% \text{ C}}$$

$$? \% \text{ O} = \frac{32.00 \text{ g O}}{60.06 \text{ g CH}_3\text{CO}_2\text{H}} \times 100 = \boxed{53.28 \% \text{ O}}$$

2-5 Derivation of Formulas from Elemental Composition

If the elemental percentages or elemental composition by mass of a compound are known, the procedure in the above problems can be reversed and the **simplest** or **empirical formula** of a compound can be found. The empirical formula represents the lowest whole number ratio of atoms present. If the molecular weight of the compound is also known, the **true** or **molecular formula** can be determined. The molecular formula is the actual number of atoms per molecule.

The simplest formula of benzene is CH. This indicates that hydrogen and carbon are present in a 1:1 ratio. However, the molecular weight of benzene is 78 amu. Since one atom of carbon plus one atom of hydrogen only accounts for 13 amu, the true formula must be some multiple of CH (i.e., [CH]$_n$, where n is the number of repeating units). Because 78/13 = 6, the true formula must be C_6H_6. Glucose, $C_6H_{12}O_6$, would have a simplest formula of CH_2O and a chlorine, Cl_2, molecule would have a simplest formula of Cl. For some substances, the true and empirical formulas are identical. Examples of this are water, H_2O, sulfuric acid, H_2SO_4, and nitrobenzene, $C_6H_5NO_2$.

Example 2-11 Empirical Formula

Hydrogen peroxide, used as an antiseptic and a bleaching agent, is 5.94% hydrogen and 94.06% oxygen by mass. What is its empirical formula?

Plan
The ratio of moles of atoms in a sample of a compound is the same as the ratio of atoms in a sample. When using percents, it is useful to assume that 100.00 grams of the compound are present. In this case, there would then be 5.94 g of H and 94.06 g of O. We can carry this out in two steps. In the first step, we calculate the number of moles of each element and, in the second, we find the lowest whole number of moles.

Solution

$$? \text{ mol H} = 5.94 \text{ g H} \times \frac{1 \text{ mol H}}{1.01 \text{ g H}} = 5.88 \text{ mol H}$$

$$? \text{ mol O} = 94.06 \text{ g O} \times \frac{1 \text{ mol O}}{16.00 \text{ g O}} = 5.879 \text{ mol O}$$

Divide each by the smaller number of moles.

$$\frac{5.88 \text{ mol H}}{5.879 \text{ mol O}} = 1 \quad \text{and} \quad \frac{5.879 \text{ mol O}}{5.879 \text{ mol O}} = 1$$

Because these are whole numbers, the empirical formula for hydrogen peroxide is $\boxed{HO}$ (The true formula is H_2O_2.)

Example 2-12 Molecular Formula

A sample is found to be 43.7% phosphorus and 56.3% oxygen. The molecular weight is about 284 amu. What is the true formula?

Plan

We can find the empirical formula as in Example 2-10. We can then find the mass of an empirical unit and compare it to the molecular mass to see how many empirical units are in one molecule.

Solution

$$? \text{ mol P} = 43.7 \text{ g P} \times \frac{1 \text{ mol P}}{30.97 \text{ g P}} = 1.41 \text{ mol P}$$

$$? \text{ mol O} = 56.3 \text{ g O} \times \frac{1 \text{ mol O}}{16.00 \text{ g O}} = 3.52 \text{ mol O}$$

$$\frac{3.52 \text{ mol O}}{1.41 \text{ mol P}} = 2.50 \quad \frac{1.41 \text{ mol P}}{1.41 \text{ mol P}} = 1$$

This is still not a whole number, so the ratio is multiplied by small whole numbers in turn until it is converted to a whole number. In this case, the number 2 accomplishes this and the simplest formula is P_2O_5.

The mass of a P_2O_5 empirical unit is: $2(30.97) + 5(16.00) = 141.94$ amu.

$$\frac{\text{molecular mass}}{\text{empirical mass}} = \frac{284 \text{ amu}}{141.94 \text{ amu}} = 2$$

The molecular or true formula is $\boxed{P_4O_{10}}$.

Example 2-13 Combustion Analysis to Determine Formula

Isopropyl alcohol, which contains only hydrogen, oxygen and carbon, is sold commercially as rubbing alcohol. When an 11.63 g sample of isopropyl alcohol is burned for combustion analysis, 25.2 g CO_2 and 14.0 g of H_2O are formed. What is the empirical formula of isopropyl alcohol?

Plan

We can use molar mass determinations of CO_2 and H_2O to find the masses of carbon and hydrogen in these compounds and, therefore, in the original sample. Any remaining mass must be due to the oxygen. Once these masses have been determined, the number of moles of each element can be found. The same procedure that was used in previous examples can then be employed.

Solution

$$? \text{ molar mass of } CO_2 = 1(12.01 \text{ g C}) + 2(16.00 \text{ g O}) = 44.01 \text{ g } CO_2$$

$$? \text{ molar mass of } H_2O = 2(1.01 \text{ g H}) + 1(16.00 \text{ g O}) = 18.02 \text{ g } H_2O$$

$$? \text{ g C} = 25.5 \text{ g } CO_2 \times \frac{12.01 \text{ g C}}{44.01 \text{ g } CO_2} = 6.96 \text{ g C}$$

$$? \text{ g H} = 14.0 \text{ g } H_2O \times \frac{2.02 \text{ g H}}{18.02 \text{ g } H_2O} = 1.57 \text{ g H}$$

$$? \text{ g O} = 11.63 \text{ g compound} - 6.96 \text{ g C} - 1.57 \text{ g H} = 3.10 \text{ g O}$$

$$? \text{ mol C} = 6.96 \text{ g C} \times \frac{1 \text{ mol C}}{12.01 \text{ g C}} = 0.580 \text{ mol C}$$

$$? \text{ mol H} = 1.57 \text{ g H} \times \frac{1 \text{ mol H}}{1.01 \text{ g H}} = 1.55 \text{ mol H}$$

$$? \text{ mol O} = 3.10 \text{ g O} \times \frac{1 \text{ mol O}}{16.00 \text{ g O}} = 0.194 \text{ mol O}$$

$$\frac{0.580 \text{ mol C}}{0.194 \text{ mol O}} = 2.99 \quad \frac{1.55 \text{ mol H}}{0.194 \text{ mol O}} = 7.99 \quad \frac{0.194 \text{ mol O}}{0.194 \text{ mol O}} = 1$$

Within data accuracy, the empirical formula is : $\boxed{C_3H_8O}$

This is also the molecular formula of isopropyl alcohol.

2-6 Purity of Samples

Often samples are contaminated with impurities. If impurities are present, correction must be made for the contaminants in the sample when doing calculations. Percent purities are often listed on reagent bottles. Dimensional analysis can be used for calculations in the same mannr as for other percentages in Chapter 1.

$$\text{Percent Purity} = \frac{\text{mass of substance of interest}}{\text{mass of sample}} \times 100$$

Example 2-14 Percent Purity

How many grams of sodium are in a 4.50 g sample that is 25.0 % NaCl? There is no sodium in the impurities.

Plan

We need to use the percent purity to find the mass of NaCl in the sample. We can find the molar mass of NaCl and use the relationship between the mass of Na and the mass of the NaCl to determine the amount of Na.

Solution

? molar mass of NaCl = 1(22.99 g Na) + 1(35.45 g Cl) = 58.44 g NaCl

$$?g\ Na = 4.50\ g\ sample \times \frac{25.0\ g\ NaCl}{100\ g\ sample} \times \frac{22.99\ g\ Na}{58.44\ g\ NaCl} = \boxed{0.443\ g\ Na}$$

EXERCISES

Ionic Nomenclature

1. Provide the correct names of the following compounds.
   - (a) $Fe(NO_3)_3$
   - (b) $CaCO_3$
   - (c) $Cu_3(PO_4)_2$
   - (d) ZnS
   - (e) Al_2O_3
   - (f) $(NH_4)_2SO_4$
   - (g) $AgCl$
   - (h) KOH

2. Write the correct formulas for the following compounds.
   - (a) sodium sulfite
   - (b) magnesium bromide
   - (c) zinc phosphate
   - (d) iron(II) acetate
   - (e) copper(I) sulfide
   - (f) calcium fluoride

Atomic Weights and the Mole

3. Calculate:
   - (a) the number of moles of argon in 52.3 grams of argon.
   - (b) the number of grams in 2.68 moles of potassium.
   - (c) the number of copper atoms in 0.0452 grams of copper.
   - (d) the number of grams of tin in 7.821×10^{25} Sn atoms.
   - (e) the number of atomic mass units in 1.00 gram of silicon.
   - (f) the atomic weight of an element if a 1.08×10^{-8} g sample contains 2.87×10^{13} atoms.

Formula Weights and Molecular Weights

4. Calculate the formula weight of:
   - (a) aluminum chlorate, $Al(ClO_3)_3$ (a disinfectant).
   - (b) ammonium sulfide, $(NH_4)_2S$ (used in photographic developers).
   - (c) potassium phosphate, K_3PO_4 (which makes strongly basic solutions).
   - (d) copper(I) selenide, Cu_2Se (used in semiconductors).
   - (e) barium hydroxide octahydrate, $Ba(OH)_2 \cdot 8H_2O$ (for drilling fluids in the oil industry).
   - (f) parathion (an insecticide), $C_{10}H_{14}SNO_5P$.

5. Calculate the molecular weight of:
   (a) butane, C_4H_{10} (a component of natural gas).
   (b) phosphine, PH_3 (a toxic gas).
   (c) citric acid, $C_6H_8O_7$ (used for pleasantly sour taste and as an antioxidant).
   (d) picric acid, $C_6H_3O(NO_2)_3$ (an explosive compound).
   (e) fungichromin (an antibiotic), $C_{35}H_{58}O_{12}$.
   (f) fluchloralin (an herbicide), $C_{12}H_{13}ClF_3N(NO_2)_2$

6. Calculate:
   (a) the number of moles of N_2O_3 molecules in 33.7 grams of N_2O_3.
   (b) the number of moles of oxygen atoms in 33.7 grams of N_2O_3.
   (c) the number of N_2O_3 molecules in 33.7 grams of N_2O_3.
   (d) the number of nitrogen atoms in 33.7 grams of N_2O_3

7. Calculate:
   (a) the number of moles of magnesium iodide, MgI_2, in 4.778 grams of MgI_2.
   (b) the number of moles of calcium chloride, $CaCl_2$, in 10.0 grams of $CaCl_2$.
   (c) the number of grams of tetraphosphorus decoxide, P_4O_{10}, in 0.0881 moles of P_4O_{10}.
   (d) the number of dinitrogen oxide, N_2O (laughing gas), molecules in 6.50 grams of N_2O.
   (e) the number of formula units of $Ca(CN)_2$ in 17.22 grams of $Ca(CN)_2$.
   (f) the number of curine, $C_{36}H_{38}N_2O_6$, (a toxin) molecules in 2.61 micrograms of curine.

8. Calculate:
   (a) the number of bromide, Br^-, ions in 6.00 moles of sodium bromide, NaBr.
   (b) the number of aluminum, Al^{3+}, ions in 2.04 grams of aluminum thiosulfate, $Al_2(S_2O_3)_3$.
   (c) the number of grams of lithium carbonate, Li_2CO_3, that contain 10.7 grams of lithium.
   (d) the number of moles of ammonium ions, NH_4^+, in 1.50 grams of ammonium sulfate, $(NH_4)_2SO_4$.
   (e) the number of moles of cyanide, CN^-, ions in 0.135 moles of iron(III) cyanide, $Fe(CN)_3$.

(f) the number of moles of isopropyl alcohol ("rubbing alcohol"), C_3H_8O, molecules in 423 grams of C_3H_8O

9. How many grams of $(CH_3)_3N$, trimethylamine, an insect attractant, contain 4.88×10^{26} atoms of carbon?

10. How many grams of sulfur are in 675 grams of $Pt(SO_4)_2 \cdot 4H_2O$?

11. When a hydrate is heated, the water may be removed. How many grams of water can be released when 350.0 grams of $Al_2(SO_4)_3 \cdot 18H_2O$, used in antiperspirants, is heated?

12. How many grams of zirconium can be obtained from 76.8 grams of $Zr(CN)_2$?

Percent Composition

13. Calculate the elemental percent composition of the following compounds.

    (a) $CO(NH_2)_2$
    (b) N_2O_3
    (c) $Fe(CO)_5$
    (d) $(NH_4)_2CrO_7$
    (e) IF_7
    (f) $Ga_2(C_2O_4)_3$
    (g) $C_4H_7Cl_2O_4P$
    (h) $C_6H_5NO_2$

14. How many grams of potassium, manganese, and oxygen are present in 6.216 grams of potassium permanganate, $KMnO_4$?

Derivation of Formulas from Elemental Composition

15. Styrene, the material from which Styrofoam is made, has a molar mass of about 104 g/mol. Styrene is 92.6% carbon and 7.74% hydrogen. What are the empirical and molecular formulas of this compound?

16. A compound composed of chromium and sulfate ions is found to be 26.52% chromium, 24.52% sulfur, and 48.95% oxygen. What is the formula of the compound and what is the charge on the chromium ion?

17. Rubies and sapphires are composed primarily of a compound that is 52.92% aluminum and 47.08% oxygen. Impurities cause the different colors seen in these gems. What is the empirical formula of the major component in these precious stones?

18. An 18.40 gram sample of oil of wintergreen is found to contain 11.60 grams of carbon, 0.97 grams of hydrogen and the rest oxygen. The molecular weight of the compound is 152 amu. What is the molecular formula of oil of wintergreen?

19. Kethoxal, a substance used to combat viruses, is 48.6% carbon, 8.17% hydrogen, and 43.2% oxygen. It has a molar mass of 148 g/mol. What is the true formula of kethoxal?

20. A 2.65 gram sample of isooctane (the basis for the octane rating of gasoline) is found to contain only carbon and hydrogen. When it is burned in air, 8.180 grams of CO_2 and 3.77 grams of H_2O are formed. The molar mass of isooctane is 114 g/mol. What is its molecular formula?

21. Ethylbutanoate is responsible for the smell of pineapple. The compound contains only carbon, hydrogen and oxygen. When a 50.00 gram sample of ethylbutanoate is burned in air, it yields 114 grams of CO_2 and 46.6 grams of H_2O. The mass of one mole of the compound is about 116 grams. What is the true formula of ethylbutanoate?

Purity of Samples

22. How many grams of lithium are in a 356 gram sample of lithium phosphate, Li_3PO_4, which is found to be only 73.8% pure?

23. What is the percent purity of potassium bromide, KBr, in a 13.987 gram sample if sodium chloride, NaCl, is found to account for 0.848 grams of that mass?

24. How many grams of bismuth are in a 50.0 gram ore sample which is 3.85% Bi_2S_3?

Miscellaneous Exercises

25. Aspartame is an artificial sweetener sold under the name Nutrasweet®. A sample of aspartame is found to be 57.14% carbon, 6.16 % hydrogen, 9.52% nitrogen, and 27.18% oxygen. What is the simplest formula (also the molecular formula) of this compound?

26. Buckminsterfullerene is a molecule named for the characteristic shape of the architect's geodesic domes. Buckminsterfullerene is composed solely of carbon atoms. If one molecule has a mass of 1.197×10^{-21}g, what is the formula of the molecule?

27. The compound n-butyl phthalate is used to treat fabrics to repel insects. It has a molar mass of 278 g/mol. A 22.6 gram sample of n-butyl phthalate is found to contain 15.6 grams of carbon, and 1.8 grams of hydrogen. The remainder of the mass is due to oxygen. What is the molecular formula of this compound?

28. Acyclovir is administered orally to inhibit several herpes viruses. It has a formula of $C_8H_{11}N_5O_3$. What is the elemental percent composition of this compound?

29. "Epsom salt" is a magnesium sulfate hydrate. When a 2.922 gram sample of this compound is heated, 1.420 grams of H_2O are released. What is the formula of the hydrate?

30. The common alum used in pickling is a hydrate with formula $KAl(SO_4)_2 \cdot 12H_2O$. How many grams of aluminum would be needed to make 650.0 g of this alum?

Chapter 2 Chemical Formulas and Composition Stoichiometry   35

31. Both anhydrous ammonia, NH_3, and urea, $CO(NH_2)_2$, are used as nitrogen fertilizers. How many kilograms of NH_3 would be needed to provide the same mass of nitrogen as does 475 kilograms of urea?

32. What mass of sodium has the same number of atoms as 25.0 grams of silver?

33. A gaseous compound has the formula PX_3. If a 54.0 gram sample of the compound is found to contain 19.0 grams of phosphorus, what element is X?

34. What is the empirical formula of the ionic compound that is composed of 405.9 grams of gold and 219.1 grams of chlorine?

35. Viridin, an agent used against fungus, has a molar mass of 352 g/mol. When a 103.1 gram sample of this compound, which contains only carbon, hydrogen, and oxygen, is burned in oxygen, 258 grams of carbon dioxide and 42.2 grams of water vapor are formed. What are the empirical and molecular formulas of viridin?

36. Neomycin B, an antibacterial agent, has the formula $C_{23}H_{46}N_6O_{13}$. How many molecules of $C_{23}H_{46}N_6O_{13}$ are in 12.6 milligrams of this substance?

37. The density of carbon tetrachloride, CCl_4, is 1.589 g/mL. How many moles and how many molecules of carbon tetrachloride are in a 25.00 mL sample of this extremely toxic solvent? How many atoms are in this sample?

ANSWERS TO EXERCISES

1. (a) iron(III) nitrate
   (b) calcium carbonate
   (c) copper(II) phosphate
   (d) zinc sulfide
   (e) aluminum oxide
   (f) ammonium sulfate
   (g) silver chloride
   (h) potassium hydroxide

2. (a) Na_2SO_3
   (b) $MgBr_2$
   (c) $Zn_3(PO_4)_2$
   (d) $Fe(CH_3COO)_2$
   (e) Cu_2S
   (f) CaF_2

3. (a) 1.31 mol Ar
   (b) 105 g K
   (c) 4.28×10^{20} Cu atoms
   (d) 1.542×10^4 g Sn
   (e) 6.02×10^{23} amu Si
   (f) 227 g/mol (element = Ac)

4. (a) 277.33 amu
   (b) 68.17 amu
   (c) 212.27 amu
   (d) 206.06 amu
   (e) 315.52 amu
   (f) 291.28 amu

5. (a) 58.18 amu
   (b) 34.00 amu
   (c) 192.14 amu
   (d) 229.12 amu
   (e) 670.93 amu
   (f) 355.73 amu

6. (a) 0.443 mol N_2O_3 molecules
   (b) 1.33 mol O atoms
   (c) 2.67×10^{23} N_2O_3 molecules
   (d) 5.34×10^{23} N atoms

7. (a) 0.01718 mol MgI_2
   (b) 0.0901 mol $CaCl_2$
   (c) 25.0 g P_4O_{10}
   (d) 8.89×10^{22} N_2O molecules
   (e) 1.126×10^{23} $Ca(CN)_2$ formula units
   (f) 2.64×10^{15} $C_{36}H_{38}N_2O_6$ molecules

8. (a) 3.61×10^{24} Br^- ions
   (b) 6.29×10^{21} Al^{3+} ions
   (c) 56.9 g Li_2CO_3
   (d) 0.0227 mol NH_4^+ ions
   (e) 0.405 mol CN^- ions
   (f) 7.04 mol C_3H_8O

9. 1.60×10^4 g $(CH_3)_3N$ 10. 94.2 g S   11. 175.1 g H_2O 12. 48.9 g Zr

13. (a) 19.99% C, 26.64% O, 46.65% N, 6.73% H
    (b) 36.86 % N, 63.14% O
    (c) 28.51% Fe, 30.65% C, 40.84% O
    (d) 11.11% N, 3.20% H, 41.26% Cr, 44.42% O
    (e) 48.83% I, 51.17% F
    (f) 34.56% Ga, 17.86% C, 47.58% O
    (g) 21.74% C, 3.19% H, 32.09% Cl, 28.96% O, 14.01% P
    (h) 58.53% C, 4.10% H, 11.38% N, 25.99% O

14. 1.538 g K, 2.161 g Mn, 2.517 g O   15. C_2H_3Cl 16. $Cr_2(SO_4)_3$, Cr^{3+}

17. Al_2O_3 18. $C_8H_8O_3$ 19. $C_6H_{12}O_4$ 20. C_8H_{18} 21. $C_6H_{12}O_2$

22. 47.2 g Li   23. 93.94% KBr   24. 1.56 g Bi   25. $C_{14}H_{18}N_2O_5$

26. C_{60} 27. $C_{16}H_{22}O_4$ 28. 42.66% C, 4.92% H, 31.10% N, 21.31% O

29. $MgSO_4 \cdot 7H_2O$ 30. 36.97 g Al   31. 270 kg NH_3 32. 5.33 g Na

33. F   34. $AuCl_3$ 35. $C_{10}H_8O_3$ (empirical formula), $C_{20}H_{16}O_6$ (molecular formula)

36. 1.23×10^{19} $C_{23}H_{46}N_6O_{13}$ molecules

37. 0.2583 mol CCl_4 molecules, 1.555×10^{23} CCl_4 molecules, 7.777×10^{23} atoms

Chemical Equations and Reaction Stoichiometry

3-1 Chemical Equations

Chemical equations provide a shorthand description of chemical reactions. In an equation, the correct formulas for the **reactants** are shown on the left and the correct formulas for the **products** are shown on the right. Once written, formulas cannot be changed, because different subscripts indicate different compounds, and, therefore, a different reaction would be being described. According to the **Law of Conservation of Matter,** there is no gain or loss of mass during ordinary chemical reactions. Coefficients are used in front of a formula to indicate relative amounts of the substances in the equation. The coefficients are adjusted until the same number and type of atoms appear on both the left and right sides of a chemical equation. When this process is complete, the equation is said to be balanced. In general, the element that occurs in the fewest different species is adjusted first and the one in the most different species last. Additionally, if a free element is present, ordinarily it is balanced last.

Example 3-1 Balancing Equations

Balance the reaction in which sodium metal, Na, reacts with bromine molecules, Br_2, to make ionic sodium bromide, NaBr.

Unbalanced: $Na + Br_2 \rightarrow NaBr$

Plan

Since there must be the same number and kind of atoms on both sides of the equation, a coefficient of 2 must precede Na and NaBr. The coefficients indicate the relative numbers of molecules, formula units, atoms and/or ions involved in a reaction and the relative numbers of moles of each.

Solution

$2Na$	$+$	Br_2	$\rightarrow$	$2NaBr$
2 atoms		1 molecule		2 formula units
$2(6.022 \times 10^{23})$ atoms		6.022×10^{23} molecules		$2(6.022 \times 10^{23})$ formula units
or 2 moles		1 mole		2 moles

Example 3-2 Balancing Equations

Balance the following equation for the complete combustion of butane, C_4H_{10}.

$$C_4H_{10} + O_2 \rightarrow CO_2 + H_2O$$

Plan

C and H appear in just one substance on each side, so begin by balancing them. Each molecule of C_4H_{10} contains 4 C atoms and 10 H atoms. Therefore, choose 4 and 5 as the respective coefficients of the CO_2 and the H_2O.

$$C_4H_{10} + O_2 \rightarrow 4CO_2 + 5H_2O$$

Now there are $8 + 5 = 13$ O atoms on the right side. We could use a coefficient of 13/2 on the left to balance the O.

$$C_4H_{10} + 13/2\,O_2 \rightarrow 4CO_2 + 5H_2O$$

This has atoms balanced, but does not have lowest whole number coefficients. We can multiply all coefficients by 2 to give the balanced equation.

Solution

$$2C_4H_{10} + 13O_2 \rightarrow 8CO_2 + 10H_2O$$

3-2 Calculations Based on Chemical Equations

The reaction of gaseous ammonia with red-hot copper(II) oxide is described by the balanced equation given below. The reaction ratio on the mole level is the same as on the molecular level. Molar masses are given in parentheses.

$$2\text{ NH}_3 + 3\text{ CuO} \rightarrow 3\text{ H}_2\text{O} + \text{N}_2 + 3\text{ Cu}$$

2 molecules	3 formula units	3 molecules	1 molecule	3 atoms
2(17.04 g)	3(79.55 g)	3(18.02 g)	(28.02 g)	3(63.55 g)
34.08 g	238.65 g	54.06 g	28.02 g	190.65 g

Many different unit factors can be constructed from the information given here using any pair of equivalent quantities to create each factor and its reciprocal. Some of the many different possibilities are illustrated in the following examples.

Example 3-3 Number of Moles Required

How many moles of CuO are required to react with 0.445 moles of NH_3?

Plan

The balanced equation shows that 3 mol of CuO are needed to react with 2 mol NH_3. We can write two unit factors:

$$\frac{2\text{ mol NH}_3}{3\text{ mol CuO}} \quad \text{and} \quad \frac{3\text{ mol CuO}}{2\text{ mol NH}_3}$$

and use the second in this case.

Solution

$$?\text{ mol CuO} = 0.445\text{ mol NH}_3 \times \frac{3\text{ mol CuO}}{2\text{ mol NH}_3} = \boxed{0.668\text{ mol CuO}}$$

Example 3-4 Mass of Product Formed

How many grams of N_2 can be formed by reacting 3.18 moles of CuO in the presence of excess NH_3?

Plan

We can use unit factors as in Example 3-4, however, in this case, we must then convert the number of moles formed to grams using the molar mass of N_2.

$$\text{mol CuO} \rightarrow \text{mol N}_2 \rightarrow \text{g N}_2$$

Solution

$$?\text{ g N}_2 = 3.18\text{ mol CuO} \times \frac{3\text{ mol CuO}}{1\text{ mol N}_2} \times \frac{28.02\text{ g N}_2}{1\text{ mol N}_2} = \boxed{267\text{ g N}_2}$$

Chapter 3 Chemical Equations and Reaction Stoichiometry

Example 3-5 Mass of Reactant Required

How many grams of NH_3 must react in the presence of excess CuO to produce 55.0 grams of copper?

Plan
We must convert the mass of Cu to moles and then proceed as in previous examples.

$$g\ Cu \rightarrow mol\ Cu \rightarrow mol\ NH_3 \rightarrow g\ NH_3$$

Solution

$$?\ g\ NH_3 = 55.0\ g\ Cu \times \frac{1\ mol\ Cu}{63.55\ g\ Cu} \times \frac{2\ mol\ NH_3}{3\ mol\ Cu} \times \frac{17.04\ g\ NH_3}{1\ mol\ NH_3} = \boxed{9.83\ g\ NH_3}$$

Alternatively, we can employ a one step approach using the mass relationships shown below the balanced equation as a unit factor.

$$?\ g\ NH_3 = 55.0\ g\ Cu \times \frac{2(17.04\ g\ NH_3)}{3(63.55\ g\ Cu)} = \boxed{9.83\ g\ NH_3}$$

3-3 The Limiting Reactant Concept

The coefficients of the species in a balanced equation always show the ratios in which substances react and are produced, but not necessarily the ratios of reactants that are mixed together. In some of the previous examples, the minimum amount of one reactant required to react with the other was calculated. In the other cases, one reactant was present in excess and the other (the **limiting reactant**) reacted completely. When nonstoichiometric quantities of reactants are mixed, it is necessary to determine which reactant is the limiting reactant to determine how much product can be made. If a reactant goes to completion, the limiting reactant is totally depleted. By being completely consumed, the limiting reactant restricts the extent to which the reaction will occur and, therefore, the amount of product that can be produced. The text shows several techniques for determining limiting reactants. We can calculate how much of one reactant is needed to react with the other. If there is not enough of that reactant, it is the limiting reactant and will the limit the amount of product that can be made. If there is more than enough of the reactant present, it is the reactant present in excess and the first reactant is the limiting reactant. The limiting reactant can also be determined by comparing ratios of the numbers of moles required to the number of moles of each reactant available. An alternate approach of predicting the amount of product that could be made if each reactant were to be totally consumed is shown here.

Example 3-6 Limiting Reactants

If 50.0 grams of CS_2 and 60.0 grams of O_2 are reacted, what is the maximum mass of SO_2 that could be produced? The balanced equation is interpreted below:

$$CS_2 \quad + \quad 3\,O_2 \quad \rightarrow \quad CO_2 \quad + \quad 2\,SO_2$$

CS_2	$3\,O_2$	CO_2	$2\,SO_2$
1 mol	3 mol	1 mol	2 mol
(76.13 g)	3(32.00 g)	(44.01 g)	2(64.06 g)

Plan

First, we calculate the amount of SO_2 that could be made if all the CS_2 were able to react using techniques seen in Example 3-5. We repeat the process by calculating the amount of SO_2 that could be made if all the O_2 were to react. The smaller amount of product is the amount that can be produced.

Solution

$$?\,g\,SO_2 = 50.0\,g\,CS_2 \times \frac{1\,mol\,CS_2}{76.13\,g\,CS_2} \times \frac{2\,mol\,SO_2}{1\,mol\,CS_2} \times \frac{64.06\,g\,SO_2}{1\,mol\,SO_2} = 84.1\,g\,SO_2$$

$$?\,g\,SO_2 = 60.0\,g\,O_2 \times \frac{1\,mol\,O_2}{32.00\,g\,O_2} \times \frac{2\,mol\,SO_2}{3\,mol\,O_2} \times \frac{64.06\,g\,SO_2}{1\,mol\,SO_2} = \boxed{80.1\,g\,SO_2}$$

From these two calculations, it can be seen that the O_2 limits the amount of product that can be made. Therefore, oxygen is the limiting reactant and 80.1 grams of SO_2 is the maximum amount that can be made.

3-4 Percent Yields from Chemical Reactions

In the previous examples, it has been assumed that all of at least one of the reactants is completely converted to products. In practice, however, it may not be possible to obtain all the predicted product because: (1) the reaction may not go to completion, i.e., one or more reactants may not react completely; (2) there may be competing side reactions that consume some of the reactants; and, (3) reaction conditions may make it impossible to isolate the product completely from the reaction mixture.

The **theoretical yield** of a product in a reaction is defined as the amount of product that would be obtained if 100% of at least one reactant were converted to products in the desired reaction and if 100% of the desired pure products could be obtained. The **actual yield** is the amount of desired pure product obtained when the experiment is done. The percent yield is defined in terms of actual and theoretical yields.

$$\% \text{ yield} = \frac{\text{actual yield}}{\text{theoretical yield}} \times 100$$

Example 3-7 Percent Yield

Inhaled oxygen, O_2, which is expired as water vapor, H_2O, can be regenerated by reaction with potassium superoxide, KO_2. What are the theoretical and percent yields of KOH in the following reaction if 15.3 grams of KOH are obtained from the reaction of 25.0 grams of H_2O with 25.0 grams of KO_2? The chemical balance chemical equation, interpreted below, is:

$$\begin{array}{cccc}
2H_2O & + & 4KO_2 & \rightarrow & 4KOH & + & 3O_2 \\
2 \text{ mol} & & 4 \text{ mol} & & 4 \text{ mol} & & 3 \text{ mol} \\
2(18.02 \text{ g}) & & 4(71.1 \text{ g}) & & 4(56.11 \text{ g}) & & 3(32.00 \text{ g})
\end{array}$$

Plan

First, we find appropriate molar masses and calculate the theoretical yield by predicting the amount of product that could be made if each reactant could be completely converted to product as we did in Example 3.6. The smaller mass of product is the theoretical yield. We divide the actual yield by the theoretical yield and multiply by 100 to find the percent yield.

Solution

$$? \text{ g KOH} = 25.0 \text{ g } H_2O \times \frac{1 \text{ mol } H_2O}{18.02 \text{ g } H_2O} \times \frac{4 \text{ mol KOH}}{2 \text{ mol } H_2O} \times \frac{56.11 \text{ g KOH}}{1 \text{ mol KOH}} = 156 \text{ g KOH}$$

$$? \text{ g KOH} = 25.0 \text{ g } KO_2 \times \frac{1 \text{ mol } KO_2}{71.10 \text{ g } KO_2} \times \frac{4 \text{ mol KOH}}{4 \text{ mol } KO_2} \times \frac{56.11 \text{ g KOH}}{1 \text{ mol KOH}} = 19.7 \text{ g KOH}$$

The 19.7 g of KOH is the smaller predicted amount of product and is the theoretical yield. The percent yield is:

$$? \% \text{ yield} = \frac{15.3 \text{ g KOH}}{19.7 \text{ g KOH}} \times 100 = \boxed{77.7\%}$$

3-5 Sequential Reactions

Many times it is not possible to make a desired product in one step. Instead a series of reactions is performed in turn to produce the substance of interest. These sequential processes can be interpreted in a similar manner to reactions which occur in a one step process.

Example 3-8 Sequential Reactions

Carbon, in a form called coke, is used in smelting to convert metal oxide ores such as hematite, Fe_2O_3, to the metal in a two-step process involving carbon monoxide, CO, as an intermediate product.

$$2 C + O_2 \rightarrow 2 CO$$
$$Fe_2O_3 + 3 CO \rightarrow 2 Fe + 3 CO_2$$

How many grams of carbon would be needed to produce 750.0 grams of iron?

Plan

We can calculate molar masses and interpret both equations and solve the problem in two steps. The equations tell us that 2 moles of C produce 2 moles of CO and that 3 moles of CO produce 2 moles of Fe.
1. We determine the amount of CO required, leaving the answer in moles.
2. We determine the mass of C needed to make that number of moles of CO.

Solution

1) $\text{? mol CO} = 750.0 \text{ g Fe} \times \dfrac{1 \text{ mol Fe}}{55.85 \text{ g Fe}} \times \dfrac{3 \text{ mol CO}}{2 \text{ mol Fe}} = 20.14 \text{ mol CO}$

2) $\text{? g C} = 20.14 \text{ mol CO} \times \dfrac{2 \text{ mol C}}{2 \text{ mol CO}} \times \dfrac{12.01 \text{ g C}}{1 \text{ mol C}} = \boxed{241.9 \text{ g C}}$

Alternatively, this could have been done in one step, combining unit factors.

3-6 Concentration of Solutions

Solutions are homogeneous mixtures consisting of a **solvent** (the dissolving phase) and a **solute** (the dissolved phase). The term concentration refers to the amount of solute or solvent in a given amount of solvent or solution. Two methods of expressing concentration are percent by mass of solute and molarity (*M*).

The **percent by mass** of solute is defined as mass of solute divided by mass of the entire solution times 100.

$$\text{mass percent} = \dfrac{\text{mass of solute}}{\text{mass of solution}} \times 100$$

For example, 100 grams of 12.0% NaCl solution contains 12.0 grams of NaCl and 88.0 grams of water. (Unless otherwise specified in this book, the solutions may be assumed to be aqueous solutions where the solvent is water and percent is the mass percent.) From this information alone, six factors can be constructed:

$$\frac{12.0 \text{ g NaCl}}{100 \text{ g solution}} \qquad \frac{88.0 \text{ g H}_2\text{O}}{100 \text{ g solution}} \qquad \frac{12.0 \text{ g NaCl}}{88.0 \text{ g H}_2\text{O}}$$

and the reciprocals of these factors.

Example 3-9 Mass of Solute

How many grams of sodium chloride, NaCl, are present in 125.0 grams of solution that is 12.0% by mass NaCl?

Plan
We will use the unit factor that relates the mass of NaCl to the solution mass.

Solution

$$? \text{ g NaCl} = 125.0 \text{ g solution} \times \frac{12.0 \text{ g NaCl}}{100 \text{ g solution}} = \boxed{15.0 \text{ g NaCl}}$$

Example 3-10 Mass of Solvent

What mass of water must be added to 350.0 g NaCl to make a 12.0% solution of NaCl?

Plan
We will use the unit factor that relates mass of H_2O to mass of NaCl.

Solution

$$? \text{ g H}_2\text{O} = 350.0 \text{ g NaCl} \times \frac{88.0 \text{ g H}_2\text{O}}{12.0 \text{ g NaCl}} = \boxed{2.57 \times 10^3 \text{ g H}_2\text{O}}$$

Amounts of solutions are more often measured by volume than by mass. In these cases, density relates measured volumes of solutions having a known percent by mass. For example, if a solution density is known to be 1.24 g/mL, two unit factors could be made. They would be:

$$\frac{1.24 \text{ g solution}}{1 \text{ mL solution}} \qquad \frac{1 \text{ mL solution}}{1.24 \text{ g solution}}$$

Example 3-11 Percent Solute and Density

The density of a 15.0% solution of NaCl is 1.108 g/mL. How many milliliters of this solution are needed to provide 75.0 grams of NaCl?

Plan

We can use an appropriate combination of unit factors to first find the mass of the solution from the mass of solute and then to find the volume of solution from the density.

Solution

$$\text{? mL solution} = 75.0 \text{ g NaCl} \times \frac{100 \text{ g solution}}{15.0 \text{ g NaCl}} \times \frac{1 \text{ mL solution}}{1.108 \text{ g solution}} = \boxed{451 \text{ mL solution}}$$

The molarity or molar concentration, M, of a solute is the number of moles of solute per liter of solution.

$$\text{Molarity} = M = \frac{\text{number of moles of solute}}{\text{number of liters of solution}}$$

Molarity may also be expressed as millimoles/milliliter or moles/1000 milliliters. It is the most frequently employed concentration unit because it combines a number of moles with volume.

Example 3-12 Number of Moles of Solute

How many moles of "battery acid" (sulfuric acid, H_2SO_4) are present in a 125.00 milliliters of 0.446 M H_2SO_4?

Plan

We see that the molarity multiplied by the volume in liters will yield the number of moles. We convert the volume to liters and perform the multiplication

Solution

$$\text{? mol } H_2SO_4 = 125.0 \text{ mL solution} \times \frac{1 \text{ L solution}}{1000 \text{ mL solution}} \times \frac{0.446 \text{ mol } H_2SO_4}{1 \text{ L solution}}$$

$$= \boxed{0.0558 \text{ mol } H_2SO_4}$$

Chapter 3 Chemical Equations and Reaction Stoichiometry

Example 3-13 Molarity and Mass of Solute

How many grams of calcium hydroxide, $Ca(OH)_2$, must be dissolved in water to give 250.0 milliliters of 0.0200 M $Ca(OH)_2$ solution?

Plan

We need to convert the volume to liters because the volume in liters multiplied by the molarity will give us moles. We can then use the molar mass of calcium hydroxide to find the mass of solute required.

Solution

$$? \text{ g } Ca(OH)_2 = 250.0 \text{ mL solution} \times \frac{1 \text{ L solution}}{1000 \text{ mL solution}} \times \frac{0.0200 \text{ mol } Ca(OH)_2}{1 \text{ L solution}}$$

$$\times \frac{74.10 \text{ g } Ca(OH)_2}{1 \text{ mol } Ca(OH)_2} = \boxed{0.370 \text{ g } Ca(OH)_2}$$

Example 3-14 Molarity

What is the molarity of potassium nitrate, KNO_3, in a solution prepared by dissolving 3.765 grams of KNO_3 in enough water to make 375 milliliters of solution?

Plan

We can find the number of moles using the mass and the molar mass. We then use the definition of molarity.

Solution

$$? \text{ mol } KNO_3 = 3.765 \text{ g } KNO_3 \times \frac{1 \text{ mol } KNO_3}{101.10 \text{ g } KNO_3} = 0.03724 \text{ mol } KNO_3$$

$$? M = \frac{0.03724 \text{ mol } KNO_3}{300.0 \text{ mL solution}} \times \frac{1000 \text{ mL solution}}{1 \text{ L solution}} = \boxed{0.1241 \, M \, KNO_3}$$

Example 3-15 Molarity and Mass Percent

What is the molarity of a 20.00% solution of sodium nitrate, $NaNO_3$? The solution density is 1.143 g/mL.

Plan

We know that we want to find the number of moles of solute in 1 liter of solution and that the solution contains 20.00 grams of solute in 100.0 grams of solution. We can assume that we have 100.0 grams of solution. We can use the molar mass to find the number of moles of $NaNO_3$ and the solution density to determine the volume of solution. We divide the number of moles by the volume to yield the molarity.

Solution

$$? \text{ mol } NaNO_3 = 20.00 \text{ g } NaNO_3 \times \frac{1 \text{ mol } NaNO_3}{85.00 \text{ g } NaNO_3} = 0.2353 \text{ mol } NaNO_3$$

$$? \text{ L solution} = 100.0 \text{ g solution} \times \frac{1 \text{ mL solution}}{1.143 \text{ g solution}} \times \frac{1 \text{ L}}{1000 \text{ mL}} = 0.08749 \text{ L solution}$$

$$? M \text{ } NaNO_3 = \frac{0.2353 \text{ mol } NaNO_3}{0.08749 \text{ L solution}} = \boxed{2.689 \text{ } M \text{ } NaNO_3}$$

Alternatively, we can use a series of unit factors to solve this problem.

$$? M \text{ } NaNO_3 = \frac{20.00 \text{ g } NaNO_3}{100 \text{ g solution}} \times \frac{1.143 \text{ g solution}}{1 \text{ mL solution}} \times \frac{1000 \text{ mL solution}}{1 \text{ L solution}} \times \frac{1 \text{ mol } NaNO_3}{85.00 \text{ g } NaNO_3}$$

$$= \boxed{2.689 \text{ } M \text{ } NaNO_3}$$

3-7 Dilution of Solutions

Dilution is the process of making a solution less concentrated by the addition of pure solvent or by the mixing of two solutions. In all dilutions in which only solvent is added, the total number of moles of solute remains the same. The number of moles of solute can always be found by multiplying the number of liters of solution by the molarity of the solution. After dilution, the new molarity is the number of moles of solute present divided by the new total volume.

Example 3-16 Dilution

What is the molarity of potassium sulfate, K_2SO_4, in the solution made when 14.5 milliliters of 2.00 M K_2SO_4 solution are diluted with water to a final volume of 300.0 milliliters?

Plan

We need to find the number of moles of K_2SO_4 initially present and then divide it by the new total volume.

Solution

$$? \text{ mol } K_2SO_4 = 0.0145 \text{ L} \times \frac{2.00 \text{ mol } K_2SO_4}{1 \text{ L}} = 0.0290 \text{ mol } K_2SO_4$$

$$? M \, K_2SO_4 = \frac{0.0290 \text{ mol } K_2SO_4}{0.300 \text{ L solution}} = \boxed{0.0967 \, M \, K_2SO_4}$$

In the previous example and in any dilution process in which a solute is present from only one source, the following equation applies:

$$V_1 \times M_1 = V_2 \times M_2$$

where V stands for the volume in either milliliters or liters and M stands for the molarity. The subscript "1" indicates the initial conditions before dilution and the subscript "2" indicates the conditions after dilution.

Example 3-17 Dilution

What is the molarity of hydrochloric acid, HCl, in the solution made when 5.00 mL of 12.0 M HCl is added to 95.0 mL of water? Assume that the volumes are additive.

Plan

We know the initial molarity, M_1, is 12.0 M, the initial volume, V_1 is 5.00 mL and the final volume, V_2, is 5.00 mL + 95.0 mL or 100.0 mL. The equation must be solved for the final molarity, M_2.

Solution

$$? M_2 = \frac{V_1 \times M_1}{V_2} = \frac{5.00 \text{ mL} \times 12.0 M}{100.0 \text{ mL}} = \boxed{0.600 \, M \, HCl}$$

3-8 Using Solutions in Chemical Reactions

Whenever a chemical reaction is to be done in solution, molarities can be used as factors in stoichiometric calculations. In this way, the volumes of reactants of known molarities that are needed for a reaction can be calculated. Stoichiometry also provides a method for determining unknown molarities in a process called **titration**. In a titration, a measured amount of a solution is added to a known mass or measured volume of another substance until an equivalence point is reached. The equivalence point may be detected by an indicator that changes colors, or by some other technique.

Example 3-18 Solution Stoichiometry

How many grams of silver nitrate, $AgNO_3$, are required to react with 75.00 milliliters of 0.420 M potassium chromate, K_2CrO_4? The chemical equation is:

$$2AgNO_3 + K_2CrO_4 \rightarrow Ag_2CrO_4 + 2KNO_3$$

Plan

We can first calculate the number of moles of $AgNO_3$ present. We then find the number of moles of K_2CrO_4 needed to react with it. Finally, we use the molarity as a unit factor to find the volume of interest. These steps can be combined into one expression.

$$M \text{ and L } K_2CrO_4 \text{ solution} \rightarrow \text{mol } K_2CrO_4 \rightarrow \text{mol } AgNO_3 \rightarrow \text{g } AgNO_3$$

Solution

$$? \text{ g } AgNO_3 = 75.00 \text{ mL } K_2CrO_4 \text{ solution} \times \frac{0.420 \text{ mol } K_2CrO_4}{1000 \text{ mL solution}} \times \frac{2 \text{ mol } AgNO_3}{1 \text{ mol } K_2CrO_4}$$

$$\times \frac{169.98 \text{ g } AgNO_3}{1 \text{ mol } AgNO_3} = \boxed{10.71 \text{ g } AgNO_3}$$

Example 3-19 Volume of Solution Required

How many milliliters of 0.0585 M AgNO$_3$ solution are required to react with 16.28 milliliters of 0.0321 M K$_2$CrO$_4$ solution? The equation of the reaction is given in Example 3-18.

Plan

We can first find the number of moles of K$_2$CrO$_4$ that are available by multiplying the molarity by the volume. We can then find the number of moles of AgNO$_3$ needed by using the mole ratios. Finally, we can determine the volume of AgNO$_3$ solution that is required by using molarity as a unit factor, in this case expressing the molarity in mol/1000 mL. These steps can be combined into one expression as in Example 3-18.

Solution

$$? \text{ mL AgNO}_3 \text{ solution} = 16.28 \text{ mL K}_2\text{CrO}_4 \text{ solution} \times \frac{0.0321 \text{ mol K}_2\text{CrO}_4}{1000 \text{ mL K}_2\text{CrO}_4 \text{ solution}}$$

$$\times \frac{2 \text{ mol AgNO}_3}{1 \text{ mol K}_2\text{CrO}_4} \times \frac{1000 \text{ mL AgNO}_3 \text{ solution}}{0.0585 \text{ mol AgNO}_3 \text{ solution}} = \boxed{17.9 \text{ mL AgNO}_3 \text{ solution}}$$

Example 3-20 Titration

What is the molarity of a barium hydroxide, Ba(OH)$_2$, solution if 15.48 milliliters of it are required to react with 25.00 milliliters of 0.3026 M HCl solution? The appropriate equation is:

$$2\text{HCl} + \text{Ba(OH)}_2 \rightarrow \text{BaCl}_2 + 2 \text{ H}_2\text{O}$$

Plan

We determine the number of moles of HCl available by multiplying the molarity by the volume. We then use mole ratios to determine the number of moles of Ba(OH)$_2$ that were required to react. Finally, we use the definition of the molarity.

Solution

$$? \, M \text{ Ba(OH)}_2 = 25.00 \text{ mL HCl solution} \times \frac{0.3026 \text{ mol HCl}}{1000 \text{ mL HCl solution}} \times \frac{1 \text{ mol Ba(OH)}_2}{2 \text{ mol HCl}}$$

$$\times \frac{1}{15.48 \text{ mL Ba(OH)}_2 \text{ solution}} \times \frac{1000 \text{ mL Ba(OH)}_2 \text{ solution}}{1 \text{ L Ba(OH)}_2 \text{ solution}}$$

$$= \boxed{0.2443 \, M \text{ Ba(OH)}_2 \text{ solution}}$$

EXERCISES

Balancing Chemical Equations

1. Balance the following equations with lowest whole number coefficients.

   (a) $Ga + S_8 \rightarrow Ga_2S_3$

   (b) $Al + HBr \rightarrow AlBr_3 + H_2$

   (c) $SiO_2 + BrF_3 \rightarrow SiF_4 + Br_2 + O_2$

   (d) $Mg(OH)_2 + H_3AsO_4 \rightarrow Mg_3(AsO_4)_2 + H_2O$

   (e) $P_4S_3 + O_2 \rightarrow P_4O_{10} + SO_2$

   (f) $NH_4NO_3 \rightarrow N_2O + H_2O$

   (g) $P_4 + NO \rightarrow P_4O_6 + N_2$

   (h) $AsCl_3 + H_2O \rightarrow H_3AsO_3 + HCl$

   (i) $C_{12}H_{22} + O_2 \rightarrow CO_2 + H_2O$

   (j) $C_4H_{10}O_2 + O_2 \rightarrow CO_2 + H_2O$

   (k) $B_{10}H_{18} + O_2 \rightarrow B_2O_3 + H_2O$

   (l) $C_3H_6 + NH_3 + O_2 \rightarrow C_3H_3N + H_2O$

   (m) $NaBF_4 + H_2O \rightarrow H_3BO_3 + NaF + HF$

   (n) $BF_3 + LiAlH_4 \rightarrow B_2H_6 + LiF + AlF_3$

   (o) $SiO_2 + Na_2CO_3 \rightarrow Na_2SiO_3 + CO_2$

   (p) $SiF_4 + H_2O \rightarrow H_4SiO_4 + H_2SiF_6$

   (q) $N_2H_4 + H_2O_2 \rightarrow N_2 + H_2O$

   (r) $BCl_3 + NH_4Cl \rightarrow HCl + (ClB\text{-}NH)_3$

   (s) $SO_2 + O_2 + H_2O \rightarrow H_2SO_4$

   (t) $Zn + HNO_3 \rightarrow Zn(NO_3)_2 + NH_4NO_3 + H_2O$

   (u) $Cu + HNO_3 \rightarrow Cu(NO_3)_2 + NO_2 + H_2O$

   (v) $Cu + HNO_3 \rightarrow Cu(NO_3)_2 + NO + H_2O$

   (w) $(NH_4)_2Cr_2O_7 \rightarrow N_2 + Cr_2O_3 + H_2O$

(x) $Fe + K_2Cr_2O_7 + H_2SO_4 \rightarrow FeSO_4 + Cr_2(SO_4)_3 + K_2SO_4 + H_2O$

(y) $P_4O_{10} + H_2O \rightarrow H_3PO_4$

(z) $C_5H_{12}S + O_2 \rightarrow CO_2 + H_2O + SO_3$

(aa) $As + NaOH \rightarrow Na_3AsO_3 + H_2$

(bb) $N_2H_4 + N_2O_4 \rightarrow N_2 + H_2O$

(cc) $Na_2S + Cr(NO_3)_3 \rightarrow Cr_2S_3 + NaNO_3$

(dd) $C_3H_5N_3O_9 \rightarrow CO_2 + H_2O + N_2 + O_2$

(ee) $NaOH + CS_2 \rightarrow Na_2CS_3 + Na_2CO_3 + H_2O$

(ff) $Al(OH)_3 + H_2SO_4 \rightarrow Al_2(SO_4)_3 + H_2O$

(gg) $S_2Cl_2 + NH_3 \rightarrow S_4N_4 + S_8 + NH_4Cl$

Interpretation of Chemical Equations

2. Chlorine trifluoride can be made from elemental chlorine and fluorine according to the reaction below:

$$Cl_2 + 3F_2 \rightarrow 2ClF_3$$

(a) How many moles of Cl_2 are needed to react with 3.44 moles of F_2?

(b) How many grams of ClF_3 can be produced when 0.204 moles of F_2 react in the presence of excess Cl_2?

(c) How many grams of ClF_3 can be made from 130.0 grams of Cl_2 in the presence of excess F_2?

(d) How many kilograms of Cl_2 are required to produce 55.0 kilograms of ClF_3?

(e) How many grams of F_2 are required to react with 3.50 grams of Cl_2?

3. Boric acid, H_3BO_3, a substance used to kill cockroaches and also as a mild eyewash when dissolved in water, can be prepared by treating tetraborane, B_4O_{10}, with water according to the equation below.

$$B_4O_{10} + 12H_2O \rightarrow 4 H_3BO_3 + 11H_2$$

(a) How many moles of H_2 are produced when 25.0 moles of water react?

(b) How many moles of B_4O_{10} must react to produce 0.440 moles of H_3BO_3?

(c) How many moles of H_3BO_3 can be produced when 75.00 grams of B_4O_{10} react?

(d) How many grams of H_3BO_3 can be produced from 230.0 grams of water in the presence of excess B_4O_{10}?

(e) How many grams of H_2 can be produced when 12.89 grams of B_4O_{10} react?

(f) How many grams of H_2 can be produced when 685 grams of H_3BO_3 are made?

4. Pure bismuth metal can be produced by the reacting the oxide at high temperatures with carbon in an impure form called coke according to the reaction below.

$$2 Bi_2O_3 + 3 C \rightarrow 4Bi + 3 CO_2$$

(a) How many moles of Bi_2O_3 must have reacted if 11.8 moles of CO_2 are produced?

(b) How many grams of Bi can be produced from the reaction of 125 grams of Bi_2O_3?

(c) How many tons of bismuth can be produced if 7.96 tons of coke are available?

(d) How many kilograms of CO_2 are produced in the process of making 35.0 kilograms of bismuth?

Limiting Reactant, Theoretical Yield and Percent Yield

5. Tetraphosphorus decoxide and phosphorus pentachloride react to produce phosphorus oxychloride, used as a chlorinating agent for many organic reactions, according to the equation below.

$$P_4O_{10} + 6PCl_5 \rightarrow 10POCl_3$$

(a) How many grams of $POCl_3$ can theoretically be produced by the reaction of 225.0 grams of P_4O_{10} and 675.0 grams of PCl_5?

(b) How many grams of the excess reactant from part (b) are left over?

(c) A student reacts 42.66 grams of PCl_5 with excess P_4O_{10}. She isolates 47.22 grams of purified $POCl_3$. What is her percent yield?

6. The poisonous gas hydrogen cyanide can be produced by treating cyanide compounds with acid. How many grams of HCN could be produced from the reaction of 28.0 grams of $Ca(CN)_2$ and 28.0 grams of HCl?

$$Ca(CN)_2 + 2HCl \rightarrow CaCl_2 + 2HCN$$

7. If 12.3 grams of carbon tetrachloride, CCl_4, are produced from the reaction of 18.0 grams of carbon disulfide, CS_2, with 22.0 grams of Cl_2, what is the percent yield of CCl_4?

$$CS_2 + 3Cl_2 \rightarrow CCl_4 + S_2Cl_2$$

8. If 2.82 grams of CCl_4 are produced from the reaction of 2.25 grams of CS_2 with 14.60 grams of Cl_2, what percent yield of CCl_4 is obtained? Use the equation in Exercise 7.

9. Pure iron can be liberated from a mixture of iron oxides (FeO and Fe_2O_3, shown as Fe_3O_4) by the action for aluminum metal according to the reaction below.

$$8Al + 3Fe_3O_4 \rightarrow 4Al_2O_3 + 9Fe$$

   (a) How many grams of iron are released by the reaction of 225.0 grams of Al and 225.0 grams of Fe_3O_4?

   (b) How many grams of Al_2O_3 would also be produced in this process?

   (c) Which reactant is in excess and how many grams of it are left over?

10. Calcium sulfite, $CaSO_3$, undergoes thermal decomposition to yield calcium oxide, CaO, and sulfur dioxide, SO_2.

$$CaSO_3 \rightarrow CaO + SO_2$$

   (a) If 50.0 grams of $CaSO_3$ undergo the reaction, how many grams of SO_2 can be produced?

   (b) If an **impure** 50.0 gram sample containing $CaSO_3$ undergoes thermal decomposition, only 17.2 grams of SO_2 are produced. What is the percent purity of the sample? Assume that the impurities will not produce SO_2.

Sequential Reactions

11. Sulfides are often converted to free metal in a two-step process such as the one below for the production of zinc. How many kilograms of zinc can be produced from 425.0 kilograms of an ore that is 3.55% by mass ZnS?

$$2ZnS + 3O_2 \rightarrow 2ZnO + 2SO_2$$

$$ZnO + CO \rightarrow Zn + CO_2$$

12. The poisonous gas phosphine, PH_3, can be produced by the reaction of 5.42 grams of calcium phosphide, Ca_3P_2, with excess water. What is the maximum mass of phosphoric acid, H_3PO_4, that could be prepared by reacting the phosphine produced with excess O_2?

$$Ca_3P_2 + 6H_2O \rightarrow 2PH_3 + 3Ca(OH)_2$$

$$PH_3 + 2O_2 \rightarrow H_3PO_4$$

13. Consider the reaction in Exercise 12. If the first reaction produces only an 87.4% yield of phosphine, and the second gives only a 74.2% yield, what mass of H_3PO_4 is actually be produced?

14. Consider the reaction in Exercise 12. If only 3.44 grams of phosphoric acid are obtained from 5.42 grams of Ca_3P_2, what is the overall percent yield?

15. Acetylene, (C_2H_2) used in welding torches, is made commercially from calcium carbide in the following reaction.

$$CaC_2 + 2H_2O \rightarrow C_2H_2 + Ca(OH)_2$$

The acetylene then burns with a bright flame in the reaction below.

$$2C_2H_2 + 5O_2 \rightarrow 4 CO_2 + 2 H_2O$$

What is the maximum amount of CO_2 that would be produced from these reactions if we began with 135.0 grams of CaC_2?

Concentrations of Solutions

16. Solutions that are 5.50% by mass glucose ($C_6H_{12}O_6$, also known as "blood sugar") are often administered intravenously to provide patients nourishment.

    (a) How many grams of glucose are in 450.0 grams of a 5.50% solution of glucose?

    (b) How many grams of water are contained in 325 grams of the solution?

    (c) How many grams of solution must be administered to provide 26.5 grams of glucose?

    (d) How many grams of water must be added to 6.78 grams of glucose to make a solution that is 5.50% by mass glucose?

17. A 24.0% by mass solution of potassium iodide, KI, dissolved in water has a solution density of 1.208 g/mL.

    (a) How many grams of KI are contained in 75.0 milliliters of the solution?

    (b) How many grams of water are contained in 50.0 grams of the solution?

    (c) How many grams of solution contain 35.0 grams of KI?

    (d) How many milliliters of the solution contain 10.0 grams of KI?

    (e) How many grams of KI are in 125.0 milliliters of this solution?

18. What is the molarity of each solution described below?

    (a) 4.778 grams of MgI_2 dissolved to make 375.0 milliliters of solution.

    (b) 9.66 grams of Na_2S dissolved to make 450.0 milliliters of solution.

    (c) $CuSO_4$ solution prepared by dissolving 35.0 grams of $CuSO_4 \cdot 5H_2O$ in enough water to give 800.0 milliliters of solution.

    (d) 5.50 grams of KOH dissolved to make 250.0 milliliters of solution.

    (e) 2.476 grams of CaF_2 dissolved to make 375.0 milliliters of solution.

19. How many moles of solute are contained in the following solutions?

    (a) 14.25 milliliters of 2.00 M $CaCl_2$

    (b) 50.0 milliliters of 0.0500 M $Ba(OH)_2$

    (c) 14.0 milliliters of 12.0 M HCl

20. How many grams of solute are contained in the following solutions?

    (a) 131 milliliters of 0.235 M K_2SO_3

    (b) 47.5 milliliters of 0.0337 M $CaCl_2$

    (c) 54.4 milliliters of 0.550 M KOH

    (d) 750.0 milliliters of 0.400 M $HC_2H_3O_2$

    (e) 355.0 milliliters of 0.2500 M $(NH_4)_3PO_4$

21. What is the molarity of 15.00% aqueous sodium chloride, NaCl, solution? The solution density is 1.111 g/mL.

Dilution

22. How many milliliters of 3.66 M NaOH solution are required to make 75.0 milliliters of 0.425 M NaOH solution?

23. How many milliliters of 0.200 M KBr solution can be made from 13.5 milliliters of 6.00 M KBr solution?

24. What is the molarity of the solute in the solution that results when the following mixtures are prepared?

    (a) 600.0 mL of water is added to 20.0 mL of 0.480 M NaOH solution?

    (b) 200.0 mL of water is added to 30.0 mL of 6.00 M NaOH?

    (c) 30.0 mL of 0.800 M KBr is added to 80.0 mL of 0.350M KBr?

    (d) 300.0 mL of 12.0 M HNO_3 is added to 100.0 mL of 3.00 M HNO_3?

Using Solutions in Chemical Reactions and Titrations

25. What volume of 0.407 M KOH solution is just sufficient to react completely with 3.16 grams of $CuSO_4 \cdot 5H_2O$ according to the equation below? The $CuSO_4 \cdot 5H_2O$ dissolves as the KOH is added to form aqueous $CuSO_4$.

$$CuSO_4 + 2KOH \rightarrow Cu(OH)_2 + K_2SO_4$$

26. How many mL of 0.3682 M HCl solution are required to react with 0.4198 grams of Al(CN)$_3$ according to the reaction below?

$$2Al(CN)_3 + 3 H_2SO_4 \rightarrow 6HCN + Al_2(SO_4)_3$$

27. Sodium hydroxide, NaOH, can neutralize phosphoric acid, H$_3$PO$_4$, according to the equation below.

$$3NaOH + H_3PO_4 \rightarrow Na_3PO_4 + 3H_2O$$

(a) How many milliliters of 0.225 M NaOH will neutralize 4.568 grams of H$_3$PO$_4$?

(b) How many milliliters of 0.385 M H$_3$PO$_4$ solution can be neutralized by 50.0 milliliters of 0.404 M NaOH?

(c) What is the molarity of a phosphoric acid solution if 20.00 milliliters of it require 48.19 milliliters of 0.1026 M NaOH for neutralization?

(d) What is the maximum mass of Na$_3$PO$_4$ that could be made from the reaction of 25.00 milliliters of 0.1050 M NaOH and 15.00 milliliters of 0.08650 M H$_3$PO$_4$?

(e) What is the molarity of Na$_3$PO$_4$ in the solution prepared in part (d)?

Miscellaneous Exercises

28. What is the percent by mass of nitric acid, HNO$_3$, in an aqueous 7.911 M HNO$_3$ solution if the solution has a specific gravity of 1.249?

29. (a) Balance the equation below.

$$Na_2C_2O_4 + La(NO_3)_3 \rightarrow La_2(C_2O_4)_3 + NaNO_3$$

(b) How many grams of La$_2$(C$_2$O$_4$)$_3$ can be made from the reaction of 125.0 grams of Na$_2$C$_2$O$_4$ and 200.0 grams of La(NO$_3$)$_3$?

(c) How many grams of the excess reactant would remain after the reaction in part (a)?

(d) How many milliliters of 0.4000 M Na$_2$C$_2$O$_4$ solution would be needed to react with 50.00 milliliters of 0.2500 M La(NO$_3$)$_3$ solution?

(e) What is the molarity of NaNO$_3$ in the solution prepared in part (d)?

30. How many grams of NaCl are needed to make 675 grams of 2.55% by mass NaCl solution?

Chapter 3 Chemical Equations and Reaction Stoichiometry

31. The Ostwald process for production of nitric acid, HNO_3, involves starting with ammonia, NH_3, and carrying out this series of reactions.

$$4NH_3 + 5O_2 \rightarrow 4NO + 6H_2O$$

$$2NO + O_2 \rightarrow 2NO_2$$

$$3NO_2 + H_2O \rightarrow NO + 2HNO_3$$

   (a) How many kilograms of HNO_3 could be made from 365 kilograms of NH_3?
   (b) If only 844 kilograms of HNO_3 are recovered when the reaction in part (a) is performed, what is the percent yield?
   (c) How many kilograms of O_2 are needed for the process in part (a)?

32. Hydrogen sulfide, H_2S, is an unwanted contaminant in natural gas and other petroleum products. It is sometimes removed by conversion to elemental sulfur in a two-step process.

$$2H_2S + 3O_2 \rightarrow 2SO_2 + 2H_2O$$

$$2H_2S + SO_2 \rightarrow 3S + 2H_2O$$

   (a) How many grams of sulfur are produced when 675 grams of H_2S are consumed?
   (b) How many grams of SO_2 and H_2O are also produced in this process?
   (c) How many grams of O_2 are used in this process?

33. An antacid tablet contains aluminum hydroxide, $Al(OH)_3$, and is designed to react with the hydrochloric acid, HCl, in the stomach. The reaction is shown below.

$$Al(OH)_3 + 3HCl \rightarrow AlCl_3 + 3H_2O$$

   (a) How many grams of $Al(OH)_3$ would be needed to react with 25.00 mL of 0.1038 M HCl solution?
   (b) What is the molarity of an HCl solution if 71.28 milliliters of it are required to react with a 0.1035 gram sample of $Al(OH)_3$?
   (c) What is the limiting reactant when 425 milligrams of $Al(OH)_3$ are mixed with 25.0 milliliters of 4.00 M HCl solution?
   (d) How many grams of $AlCl_3$ could be produced from the mixture in part (c).
   (e) A sample that contains 0.5887 grams of impure $Al(OH)_3$ requires 85.32 milliliters of 0.2054 M HCl for reaction. If none of the impurities in the sample react with hydrochloric acid, what is the percent $Al(OH)_3$ in the sample?

ANSWERS TO EXERCISES

1. (a) $16Ga + 3S_8 \rightarrow 8Ga_2S_3$

   (b) $2Al + 6HBr \rightarrow 2AlBr_3 + 3H_2$

   (c) $3SiO_2 + 4BrF_3 \rightarrow 3SiF_4 + 2Br_2 + 3O_2$

   (d) $3Mg(OH)_2 + 2H_3AsO_4 \rightarrow Mg_3(AsO_4)_2 + 6H_2O$

   (e) $P_4S_3 + 3O_2 \rightarrow P_4O_{10} + 3SO_2$

   (f) $NH_4NO_3 \rightarrow N_2O + 2H_2O$

   (g) $P_4 + 6NO \rightarrow P_4O_6 + 3N_2$

   (h) $AsCl_3 + 3H_2O \rightarrow H_3AsO_3 + 3HCl$

   (i) $2C_{12}H_{22} + 35O_2 \rightarrow 24CO_2 + 22H_2O$

   (j) $2C_4H_{10}O_2 + 11O_2 \rightarrow 8CO_2 + 10H_2O$

   (k) $B_{10}H_{18} + 12O_2 \rightarrow 5B_2O_3 + 9H_2O$

   (l) $2C_3H_6 + 2NH_3 + 3O_2 \rightarrow 2C_3H_3N + 6H_2O$

   (m) $NaBF_4 + 3H_2O \rightarrow H_3BO_3 + NaF + 3HF$

   (n) $4BF_3 + 3LiAlH_4 \rightarrow 2B_2H_6 + 3LiF + 3AlF_3$

   (o) $SiO_2 + Na_2CO_3 \rightarrow Na_2SiO_3 + CO_2$

   (p) $3SiF_4 + 4H_2O \rightarrow H_4SiO_4 + 2H_2SiF_6$

   (q) $N_2H_4 + 2H_2O_2 \rightarrow N_2 + 4H_2O$

   (r) $3BCl_3 + 3NH_4Cl \rightarrow 9HCl + (ClBNH)_3$

   (s) $2SO_2 + O_2 + 2H_2O \rightarrow 2H_2SO_4$

   (t) $4Zn + 10HNO_3 \rightarrow 4Zn(NO_3)_2 + NH_4NO_3 + 3H_2O$

   (u) $Cu + 4HNO_3 \rightarrow Cu(NO_3)_2 + 2NO_2 + 2H_2O$

   (v) $3Cu + 8HNO_3 \rightarrow 3Cu(NO_3)_2 + 2NO + 4H_2O$

   (w) $(NH_4)_2Cr_2O_7 \rightarrow N_2 + Cr_2O_3 + 2H_2O$

   (x) $3Fe + K_2Cr_2O_7 + 7H_2SO_4 \rightarrow 3FeSO_4 + Cr_2(SO_4)_3 + K_2SO_4 + 7H_2O$

   (y) $P_4O_{10} + 6H_2O \rightarrow 4H_3PO_4$

(z) $2C_5H_{12}S + 19O_2 \rightarrow 10CO_2 + 12H_2O + 2SO_3$

(aa) $2As + 6NaOH \rightarrow 2Na_3AsO_3 + 3H_2$

(bb) $2N_2H_4 + N_2O_4 \rightarrow 3N_2 + 4H_2O$

(cc) $3Na_2S + 2Cr(NO_3)_3 \rightarrow Cr_2S_3 + 6NaNO_3$

(dd) $4C_3H_5N_3O_9 \rightarrow 12CO_2 + 10H_2O + 6N_2 + O_2$

(ee) $6NaOH + 3CS_2 \rightarrow 2Na_2CS_3 + Na_2CO_3 + 3H_2O$

(ff) $2Al(OH)_3 + 3H_2SO_4 \rightarrow Al_2(SO_4)_3 + 6H_2O$

(gg) $2S_2Cl_2 + 16NH_3 \rightarrow S_4N_4 + S_8 + 12NH_4Cl$

2. (a) 1.15 mol Cl_2 (b) 12.6 g ClF_3 (c) 339 g ClF_3 (d) 21.1 kg Cl_2 (e) 5.63 g F_2

3. (a) 22.9 mol H_2 (b) 0.110 mol B_4O_{10} (c) 5.624 mol H_3BO_3
   (d) 263.1 g H_3BO_3 (e) 5.37 g H_2 (f) 61.5 g H_2

4. (a) 7.87 mol Bi_2O_3 (b) 112 g Bi (c) 185 tons Bi (d) 5.53 kg CO_2

5. (a) 828.4 g $POCl_3$ (PCl_5 limiting reactant) (b) 71.6 g P_4O_{10} left over (c) 90.19% yield

6. 16.4 g HCN ($Ca(CN)_2$ is limiting)   7. Cl_2 is limiting reactant; 77.3% yield

8. CS_2 is limiting reactant; 62.0% yield

9. (a) 162.9 g Fe (Fe_3O_4 limiting) (b) 132.1 g Al_2O_3 (c) 155.1 g Al left over

10. (a) 26.6 g SO_2 (b) 64.4% $CaSO_4$ 11. 10.1 kg Zn   12. 5.83 g H_3PO_4

13. 3.78 g H_3PO_4 14. 59.0%   15. 185 g CO_2

16. (a) 24.8 g $C_6H_{12}O_6$ (b) 307 g H_2O (c) 482 g solution (d) 116 g H_2O

17. (a) 21.7 g KI (b) 38.0 g H_2O (c) 146 g solution (d) 34.5 mL solution (e) 36.2 g KI

18. (a) 0.04581 M MgI_2 (b) 0.275 M Na_2S (c) 0.175 M $CuSO_4$ (d) 0.392M KOH
    (e) 0.08456 M CaF_2

19. (a) 0.0285 mol $CaCl_2$ (b) 0.00250 mol $Ba(OH)_2$ (c) 0.0168 mol HCl

20. (a) 4.88 g K$_2$SO$_3$ (b) 0.178 g CaCl$_2$ (c) 1.68 g KOH (d) 18.0 g HC$_2$H$_3$O$_2$
    (e) 13.23 g (NH$_4$)$_3$PO$_4$

21. 2.85 M NaCl   22. 8.71 mL NaOH   23. 405 mL KBr

24. (a) 0.0155 M NaOH (b) 0.783 M NaOH (c) 0.473 M KBr (d) 9.75 M HNO$_3$

25. 62.2 mL KOH   26. 16.28 mL HCl

27. (a) 621 mL NaOH (b) 17.5 mL H$_3$PO$_4$ (c) 0.08240 M H$_3$PO$_4$
    (d) NaOH is limiting reactant; 0.141 g Na$_3$PO$_4$ (e) 0.0215 M Na$_2$PO$_4$

28. 40.00% HNO$_3$ by mass

29. (a) 3Na$_2$C$_2$O$_4$ + 2La(NO$_3$)$_3$ → La$_2$(C$_2$O$_4$)$_3$ + 6 NaNO$_3$
    (b) 166.8 g La$_2$(C$_2$O$_4$)$_3$ (c) 1.3 g Na$_2$C$_2$O$_4$ left over (d) 46.88 mL Na$_2$C$_2$O$_4$ solution
    (e) 0.3871 M NaNO$_3$

30. 17.2 g NaCl

31. (a) 899 kg HNO$_3$ (b) 93.8% (c) 1.03 x 10^3 kg O$_2$

32. (a) 476 g H$_2$S (b) 357 g H$_2$O and 317 g SO$_2$ (c) 475 g O$_2$

33. (a) 0.0675 g Al(OH)$_3$ (b) 0.05584 M HCl (c) Al(OH)$_3$ is the limiting reactant
    (d) 0.727 g AlCl$_3$ (e) 77.40% Al(OH)$_3$

Some Types of Chemical Reactions

4-1 The Periodic Table

Elements are arranged in the periodic table according to increasing atomic number, which is the number of protons in the nucleus. **Periodic law** states that chemical and physical properties are periodic functions of the atomic numbers. Similar properties are a characteristic of elements in the same **group** or **family** (vertical column). Gradual changes in the properties occur across a **period** (horizontal row).

The "A" Groups, are known as **representative elements**. The IA elements are usually referred to as the **alkali metals**, while the IIA elements are the **alkaline earth metals,** and the VIIA elements are the **halogens**. The **noble gases** are in group 0 or group VIIIA, depending on the chart consulted.

A stepwise division on the periodic chart separates the non-metals (above and to the right) from the metals (below and to the left). Adjacent to the line are the metalloids, which have properties intermediate to those of the nonmetals and the metals. Properties of metals and nonmetals are summarized in Tables 4-3 and 4-4 in the text. In general, metallic character increases from right to left and from top to bottom, while nonmetallic character shows the opposite trend.

4-2 An Introduction to Aqueous Solutions

Many common reactions take place in aqueous (water) solutions. Substances that are soluble in water are either electrolytes or nonelectrolytes. A substance is an **electrolyte** if a dilute aqueous solution of it conducts electricity and a nonelectrolyte if it does not. Electrolytes may be either **strong** (nearly 100% ionized, if covalent, or nearly 100% dissociated, if ionic) or **weak** (only sparingly ionized). **Acids** produce H^+ in aqueous solution, while **bases** produce OH^-. **Salts** contain a cation other than H^+ and an anion other than OH^- or O^{2-}. The strong acids and strong, soluble bases are tabulated in Table 4-1.

Table 4-1 Common Strong Acids and Strong, Soluble Bases

Strong Acids		Strong Soluble Bases	
HCl	HBr	LiOH	NaOH
HI	HNO_3	KOH	$Ca(OH)_2$
H_2SO_4	$HClO_4$	RbOH	$Sr(OH)_2$
$HClO_3$		CsOH	$Ba(OH)_2$

You may assume that other common acids are weak. Other metal hydroxides are generally insoluble in water. Most common soluble ionic salts are strong electrolytes. A convention used to express the state of a substance in a chemical reaction is to show the state in parentheses after the formula of the substance. Abbreviations commonly used are: (s) = solid, (ℓ) liquid, (g) = gas, and (aq) = aqueous solution (dissolved in water).

The strong bases or salts totally dissociate in aqueous solution. For example:

Base $Ca(OH)_2(aq) \rightarrow Ca^{2+}(aq) + 2\ OH^-(aq)$

Salt $K_2SO_4(aq) \rightarrow 2K^+(aq) + SO_4^{2-}(aq)$

The strong acids totally ionize in aqueous solution:

$HNO_3(aq) \rightarrow H^+(aq) + NO_3^-(aq)$

In this latter case, the hydration of H^+ is sometime emphasized by representing it as the hydronium ion, H_3O^+. The ionization reaction is then shown as:

$HNO_3(aq) + H_2O(\ell) \rightarrow H_3O^+(aq) + NO_3^-(aq)$

Weak acids and bases only partially ionize, achieving a dynamic equilibrium with their covalent molecules. This equilibrium condition is expressed with double arrows which indicate that the forward and back reactions each occur at the same rate and that the reaction does not go to completion.

weak acid $HF(aq) + H_2O(\ell) \rightleftharpoons H_3O^+(aq) + F^-(aq)$

weak base $NH_3(aq) + H_2O(\ell) \rightleftharpoons NH_4^+(aq) + OH^-(aq)$

In addition to acids and bases, many other compounds can dissolve in water. A set of solubility rules is given in Section 4-2.5 and summarized in Table 4-8 in the text. These solubility rules are useful for describing many reactions that occur in aqueous solution.

4-3 Reactions in Aqueous Solutions

The reactions that were considered earlier in the text were written as **formula unit** (molecular) equations in which complete formulas for all species were used without regard for the actual characteristics of the substance. A formula unit equation is:

$$Ba(NO_3)_2(aq) + K_2SO_4(aq) \rightarrow BaSO_4(s) + 2KNO_3(aq)$$

A **total ionic** equation shows all soluble strong electrolytes in their ionic form, with square brackets [ ] used to indicate the source of the ions. The total ionic equation for the reaction above is:

$$[Ba^{2+}(aq) + 2NO_3^-(aq)] + [2K^+(aq) + SO_4^{2-}(aq)] \rightarrow BaSO_4(s) + 2[K^+(aq) + NO_3^-(aq)]$$

Notice that the K^+ and NO_3^- ions are in the same form on both the reactant and product sides of the equation. They are not involved in the reaction and are called **spectator ions**. Cancellation of any spectator ions yields the **net ionic** equation:

$$Ba^{2+}(aq) + SO_4^{2-}(aq) \rightarrow BaSO_4(s)$$

When writing ionic equations, two factors must be considered:

1. Is the substance soluble in water?
2. If it is soluble, is the substance mostly ionized or dissociated in water?

If the answer to **both** questions is yes, the formula is written in ionic form. If the answer to **either** of these questions is no, the formula is written in unionized form.

4-4 Precipitation Reactions

The driving force of a **precipitation** reaction is the formation of an insoluble substance called a **precipitate**, which settles out of a mixture after two or more solutions containing soluble substances are mixed. A precipitation occurs if some combination of ions results in an insoluble material. We must consider ion charges to write the proper formulas of the compounds formed.

Example 4-1 Solubility Rules

Will a precipitate form when aqueous solutions of $AgNO_3$ and Na_2S are mixed?

Plan
These substances yield the ions Ag^+, NO_3^-, Na^+ and S^{2-}. By consulting the tables, we see that the only insoluble combination will be that of Ag^+ with S^{2-}.

Solution
The precipitate Ag_2S will form.

Example 4-2 Writing Chemical Equations

Write the formula unit, total, and net ionic equations for any reaction that occurs when aqueous solutions of $Sr(OH)_2$ and $FeCl_2$ are mixed.

Plan

We know that $Sr(OH)_2$ is a stong, soluble base, while $FeCl_2$ is a salt. Consulting the solubility tables we see that one possible combination, $SrCl_2$, is a soluble salt, while the other, $Fe(OH)_2$, is an insoluble base. We look at the total ionic equation and delete the spectator ions.

Solution

Formula unit equation: $\boxed{Sr(OH)_2(aq) + FeCl_2(aq) \rightarrow SrCl_2(aq) + Fe(OH)_2(s)}$

Total ionic equation:

$\boxed{[Sr^{2+}(aq) + 2OH^-(aq)] + [Fe^{2+}(aq) + 2Cl^-(aq)] \rightarrow [Sr^{2+}(aq) + 2Cl^-(aq)] + [Fe(OH)_2(s)]}$

Net ionic equation: $\boxed{2OH^-(aq) + Fe^{2+}(aq) \rightarrow Fe(OH)_2(s)}$

(Sr^{2+} and Cl^- are spectator ions.)

Example 4-3 Writing Chemical Equations

Write formula unit, total, and net ionic equations for any reaction that occurs when aqueous solutions of KNO_3 and $NaCl$ are mixed.

Plan

We consult the solubility table and find that both possible combinations of product ions, $NaNO_3$ and KCl, are soluble.

Solution

Formula unit equation: $\boxed{NaCl(aq) + KNO_3(aq) \rightarrow NaNO_3(aq) + KCl(aq)}$

Total ionic equation:

$\boxed{[Na^+(aq) + Cl^-(aq)] + [K^+(aq) + NO_3^-(aq)] \rightarrow [Na^+(aq) + NO_3^-(aq)] + [K^+(aq) + Cl^-(aq)]}$

Net ionic equation: $\boxed{\text{None–all of the ions are spectator ions.}}$

4-5 Acid-Base Reactions

The reaction of an acid with a metal hydroxide base produces a salt and water. Such reactions are called **neutralization** reactions because the characteristic properties of acids and bases are cancelled or neutralized. The driving force of these reactions is the formation of unionized water molecules. When a strong acid reacts with a strong soluble base, the net ionic equation is **always** $H^+(aq) + OH^-(aq) \rightarrow H_2O(\ell)$. When a weak acid is present, it will appear in the net ionic equation because it will exist predominantly in the unionized form.

Example 4-4 Neutralization Reactions

Write the formula unit, total, and net ionic equations for the neutralization reaction of $Ba(OH)_2$ and $HClO_4$.

Plan
We have a strong acid and a strong soluble base. $Ba(ClO_4)_2$ is a soluble salt.

Solution
Formula unit equation: $\boxed{Ba(OH)_2(aq) + 2HClO_4(aq) \rightarrow Ba(ClO_4)_2(aq) + 2H_2O(\ell)}$

Total ionic equation:
$\boxed{[Ba^{2+}(aq) + 2OH^-(aq)] + 2[H^+(aq) + ClO_4^-(aq)] \rightarrow [Ba^{2+}(aq) + 2ClO_4^-(aq)] + 2H_2O(\ell)}$

Net ionic equation: $\boxed{OH^-(aq) + H^+(aq) \rightarrow H_2O(\ell)}$

Example 4-5 Neutralization Reactions

Write the formula unit, total ionic, and net ionic equations for the neutralization of H_3PO_4 and LiOH.

Plan
LiOH is a strong soluble base and will dissociate totally; however, H_3PO_4 is a weak acid and appears predominantly in the unionized form. Li_3PO_4 is a soluble salt.

Solution
Formula unit equation: $\boxed{H_3PO_4(aq) + 3LiOH(aq) \rightarrow Li_3PO_4(aq) + 3H_2O(\ell)}$

Total ionic equation:

$H_3PO_4(aq) + 3[Li^+(aq) + OH^-(aq)] \rightarrow [3Li^+(aq) + PO_4^{3-}(aq)] + 3H_2O(\ell)$

Net ionic equation: $H_3PO_4(aq) + 3OH^-(aq) \rightarrow PO_4^{3-}(aq) + 3H_2O(\ell)$

4-6 Oxidation Numbers

The **oxidation number** or **oxidation state** of an element is the real or assigned charge on that element. When the element forms binary ionic compounds, its oxidation number corresponds to the charge on its ion in that compound. In molecular species, oxidation numbers are assigned to each element using a special set of rules. Elements may have different oxidation numbers in different compounds, depending on the other atoms involved in bonding. Three general rules for assigning oxidation numbers are:

1. The oxidation number of any free element is zero.
2. In any ion, the algebraic sum of the oxidation numbers of the constituent elements equals the charge on the ion.
3. In any compound, the algebraic sum of the oxidation numbers of all constituent elements equals zero.

The most common oxidation states for the elements are shown in the text in Tables 4-11 and 4-12. Oxidation numbers are always expressed **per atom**. In this book, ion charges will be shown with magnitude first (n+ or n-) and oxidation numbers will be shown with the charge first (+n or -n) above the symbol of the element.

Example 4-6 Oxidation Numbers

What is the oxidation number of nitrogen in the following species?
(a) $RbNO_3$ (b) N_2 (c) $Ca(NO_2)_2$ (d) NH_3 (e) NF_3 (f) N_2O (g) NO_2 (h) NO_3^-
(i) NH_4^+

Plan
We will apply the rules above.

Solution
(a) In $RbNO_3$, Rb and O exhibit oxidation numbers of +1 and -2, respectively, so:
$+1 + x + 3(-2) = 0$
$x = \boxed{+5}$ = the oxidation number of N

(b) In N_2, the nitrogen is a free element and has an oxidation number of $\boxed{0}$.

(c) In Ca(NO$_2$)$_2$, the Ca and O have oxidation numbers of +2 and -2 respectively, so:
$+2 + 2x + 4(-2) = 0$
$2x = 6$ and $x = \boxed{+3}$ = the oxidation number of N

(d) H is assigned a +1 charge when it is bonded to nonmetals.
$x + 3(+1) = 0$
$x = \boxed{-3}$ = the oxidation number of N

(e) In NF$_3$, the F is always -1 in compounds so:
$x + 3(-1) = 0$
$x = \boxed{+3}$ is the oxidation number of N

(f) In N$_2$O, the O is -2 so:
$2x + (-2) = 0$
$2x = (+2)$ and $x = \boxed{+1}$ is the oxidation number of N

(g) In NO$_2$, the O is -2 so:
$x + 2(-2) = 0$
$x = \boxed{+4}$ is the oxidation number of N

(h) In NO$_3^-$, the O is -2 so:
$x + 3(-2) = -1$
$x = \boxed{+5}$ is the oxidation number of N

(i) In NH$_4^+$, the H is +1 so:
$x + 4(+1) = +1$
$x = \boxed{-3}$ is the oxidation number of N

4-7 Oxidation-Reduction Reactions-An Introduction

Oxidation-reduction reactions (redox or electron transfer reactions) are those in which changes in oxidation number are involved. **Oxidation** is defined as an increase in oxidation number (the loss or apparent loss of electrons) and **reduction** is defined as a decrease in oxidation number (gain or apparent gain of electrons). The species that is reduced causes the oxidation and is called the **oxidizing agent**. The species that is oxidized causes the reduction and is called the **reducing agent**. For example, the following is a redox reaction:

$$\overset{0}{2Na(s)} + \overset{0}{Cl_2(g)} \rightarrow \overset{+1\ -1}{2NaCl(s)}$$

Na is oxidized because its oxidation number increases from 0 to +1; it is the reducing agent. Cl is reduced because its oxidation number decreases from 0 to -1; it is the oxidizing agent.

Example 4-7 Redox Reactions

Assign oxidation numbers to each element in the equation below. Write the net ionic equation for this oxidation-reduction equation.

$$Br_2(\ell) + Na_2SO_3(aq) + H_2O(\ell) \rightarrow 2HBr(aq) + Na_2SO_4(aq)$$

Plan

We will follow the rules for assigning oxidation numbers. Consulting previous tables, we see that Na_2SO_3 and Na_2SO_4 are soluble salts while HBr is a strong acid.

Solution

$$\overset{0}{Br_2}(\ell) + \overset{+1\ +4\ -2}{Na_2SO_3}(aq) + \overset{+1\ -2}{H_2O}(\ell) \rightarrow 2\overset{+1\ -1}{HBr}(aq) + \overset{+1\ +6\ -2}{Na_2SO_4}(aq)$$

Br is reduced and S is oxidized. Br_2 is the oxidizing agent, while Na_2SO_3 is the reducing agent.

The total ionic equation is:

$$Br_2(\ell) + [2Na^+(aq) + SO_3^{2-}(aq)] + H_2O(\ell) \rightarrow 2[H^+(aq) + Br^-(aq)] + [2Na^+(aq) + SO_4^{2-}(aq)]$$

Net ionic equation: $\boxed{Br_2(\ell) + SO_3^{2-}(aq) + H_2O(\ell) \rightarrow 2H^+(aq) + 2Br^-(aq) + SO_4^{2-}(aq)}$

Example 4-8 Redox Reactions

Write the following formula unit equations as net ionic equations. For the redox reactions identify the oxidizing and reducing agents.

(a) $SO_2(s) + O_2(g) \rightarrow 2SO_3(g)$
(b) $CO_2(g) + H_2O(\ell) \rightarrow H_2CO_3(aq)$
(c) $2Cs(s) + 2H_2O(\ell) \rightarrow 2CsOH(aq) + H_2(g)$
(d) $10\ FeSO_4(aq) + 8H_2SO_4(aq) + 2KMnO_4(aq) \rightarrow$
    $5Fe_2(SO_4)_3(aq) + 2MnSO_4(aq) + 8H_2O(\ell) + K_2SO_4(aq)$
(e) $BaSO_3(s) \rightarrow BaO(s) + SO_2(g)$
(f) $2C_2H_6(g) + 7O_2(g) \rightarrow 4CO_2(g) + 6H_2O(g)$

Plan

We will assign oxidation numbers according to the above rules and compare oxidation numbers in the products and reactants. We will also consult solubility tables to determine if a substance is soluble. Oxidation numbers are indicated above each element. We will write the ionic equations where appropriate.

Solution

(a) $\overset{+4\ -2}{2\,SO_2(g)} + \overset{0}{O_2(g)} \rightarrow \overset{+6\ -2}{2\,SO_3(g)}$

The reaction does not occur in aqueous solution, so the molecular and ionic equations are identical. The oxidation number of the sulfur increases from +4 to +6; therefore, it is oxidized and is the reducing agent. The oxidation number of the oxygen decreases from 0 to -2; therefore, it is reduced and is the oxidizing agent.

(b) $\overset{+4\ -2}{CO_2(g)} + \overset{+1\ -2}{H_2O(\ell)} \rightarrow \overset{+1\ +4\ -2}{H_2CO_3(aq)}$

This is not a redox reaction since no oxidation numbers change. All species are primarily molecular, so the ionic equations are the same as the molecular equation.

(c) $\overset{0}{2\,K(s)} + \overset{+1\ -2}{2\,H_2O(\ell)} \rightarrow 2[\overset{+1}{K^+(aq)} + \overset{-2\ +1}{OH^-(aq)}] + \overset{0}{H_2(g)}$ (total ionic equation)

$\boxed{2K(s) + 2H_2O(\ell) \rightarrow 2K^+(aq) + 2OH^-(aq) + H_2(g)}$ (net ionic equation)

Potassium hydroxide is a strong soluble base, so it is written in ionized form. The oxidation number of potassium increases from 0 to +1; it is oxidized and is the reducing agent. The oxidation number of some of the hydrogen in water decreases from +1 to 0. Part of the hydrogen is reduced and the water is the oxidizing agent.

(d) $10[\overset{+2}{Fe^{2+}(aq)} + \overset{+6\ -2}{SO_4^{2-}(aq)}] + 8[\overset{+1}{2H^+(aq)} + \overset{+6\ -2}{SO_4^{2-}(aq)}] + 2[\overset{+1}{K^+(aq)} + \overset{+7\ -2}{MnO_4^-(aq)}] \rightarrow$

$5[\overset{+3}{2Fe^{3+}(aq)} + \overset{+6\ -2}{3SO_4^{2-}(aq)}] + 2[\overset{+2}{Mn^{2+}(aq)} + \overset{+6\ -2}{SO_4^{2-}(aq)}] + \overset{+1\ -2}{8H_2O(\ell)}$

$+ [\overset{+1}{2K^+(aq)} + \overset{+6\ -2}{SO_4^{2-}(aq)}]$ (total ionic equation)

$\boxed{5Fe^{2+}(aq) + 8H^+(aq) + MnO_4^-(aq) \rightarrow Mn^{2+}(aq) + 5Fe^{3+}(aq) + 4H_2O(\ell)}$
(net ionic equation)

The permanganate ion (MnO_4^-) is the oxidizing agent because the manganese is reduced with its oxidation number decreasing from +7 to +2. Iron(II) ion is the reducing agent because the iron is oxidized with its oxidation number increasing from +2 to +3.

(e) $\overset{+2\ +4\ -2}{BaSO_3(s)} \rightarrow \overset{+2\ -2}{BaO(s)} + \overset{+4\ -2}{SO_2(g)}$ (not redox)

(f) $2\overset{-3\ +1}{C_2H_6(g)} + 7\overset{0}{O_2(g)} \rightarrow 4\overset{+4\ -2}{CO_2(g)} + 6\overset{+1\ -2}{H_2O(\ell)}$

C_2H_6 is the reducing agent because its carbon is oxidized from the -3 to the +4 oxidation state. Oxygen is the oxidizing agent because it is reduced from the 0 to the -2 oxidation state. Because this reaction is not in aqueous solution, total and net ionic equations do not apply.

4-8 Displacement Reactions

Displacement reactions are those in which one element displaces another from a compound. A more **active** metal will displace a less active metal from a compound. An **activity series** of the metals (and hydrogen) is presented in Table 4-13 in the text. Zinc is more active than cobalt and aluminum is more active than hydrogen, so the following reactions are observed:

$CoSO_4(aq) + Zn(s) \rightarrow Co(s) + ZnSO_4(aq)$

$2Al(s) + 6HCl(aq) \rightarrow 2AlCl_3(aq) + 3H_2(g)$

A halogen with a higher molecular mass is more active, therefore the following reactions occur:

$2\ NaCl(aq) + F_2(g) \rightarrow 2\ NaF(aq) + Cl_2(g)$

$2\ KI(aq) + Br_2(\ell) \rightarrow 2\ KBr(aq) + I_2(s)$

To decide if a displacement reaction will occur, consult the activity table to see which element is more active. Be sure to consider the ion charges and choose subscripts to make the compound electrically neutral. Consult the solubility table to write the total and net ionic equations

Example 4-9 Displacement Reactions

Will aluminum displace nickel from an aqueous solution of $NiCl_2$? If so write the formula unit, total ionic, and net ionic equations.

Plan

Aluminum is above nickel on the activity chart, so it will displace nickel. $AlCl_3$ is soluble.

Solution

Formula unit equation: $\boxed{2Al(s) + 3NiCl_2(aq) \rightarrow 2AlCl_3(aq) + 3Ni(s)}$

Total ionic equation: $\boxed{2Al(s) + 3[Ni^{2+}(aq) + 2Cl^-(aq)] \rightarrow 2[Al^{3+}(aq) + 3Cl^-(aq)] + 3Ni(s)}$

Net ionic equation: $\boxed{2Al(s) + 3Ni^{2+}(aq) \rightarrow 2Al^{3+}(aq) + 3Ni(s)}$

4-9 Naming Inorganic Compounds

For naming purposes, simple inorganic compounds are classified into two different categories. They may be **binary** compounds, which consist of two elements, or **ternary** compounds, which consist of three elements.

Binary compounds may be either ionic or covalent. In both cases, the more metallic element is named first and the less metallic element second. The more metallic element's normal name is used, while the less metallic element is named by adding an **ide** ending to the element's characteristic stem. The stems to which the suffixes are added are given in Section 4-9 of the text.

Binary ionic compounds contain metal cations and nonmetal anions. The cation is named first and the anion second.

Formula	Name	Formula	Name
LiF	lithium fluoride	Al_2O_3	aluminum oxide
$MgBr_2$	magnesium bromide	Sr_3P_2	strontium phosphide
NaH	sodium hydride	Rb_2Se	rubidium selenide

This method is adequate for naming representative (A group) metals that exhibit just one oxidation state. Most transition elements and many of the less metallic representative metals exhibit more than one oxidation state. These metals can form more than one compound with nonmetals. Therefore, the oxidation number of the metal is indicated by a Roman numeral in parentheses following the metal name. Roman numerals are never used for metals that exhibit only one nonzero oxidation state.

Formula	Name	Formula	Name
$CuCl_2$	copper(II) chloride	CuCl	copper(I) chloride
SnO	tin(II) oxide	SnO_2	tin(IV) oxide
Fe_3N_2	iron(II) nitride	FeN	iron(III) nitride
In_2S	indium(I) sulfide	In_2S_3	indium(III) sulfide

An older method, still in use, but not recommended, involves the use of **ous** and **ic** suffixes to indicate the lower and higher of two oxidation states, respectively. This system is capable of distinguishing between only two oxidation states and is also ambiguous. For example, CuO is copper(II) oxide or cupric oxide, while FeO is iron(II) oxide but ferrous oxide.

Pseudobinary ionic compounds are those whose names also end in **ide**, but contain more than two elements. In these compounds, one or more of the ions consist of more than one element, but behave as if they were simple ions. Three common examples of such ions are the ammonium ion, NH_4^+, the cyanide ion, CN^-, and the hydroxide ion, OH^-.

Formula	Name
(NH₄)₂Se	ammonium selenide
Al(CN)₃	aluminum cyanide
Fe(CN)₃	iron(III) cyanide

Formula	Name
Mg(OH)₂	magnesium hydroxide
Cu(OH)₂	copper(II) hydroxide
NH₄I	ammonium iodide

Binary molecular compounds involve two nonmetals. Many nonmetals exhibit different oxidation states, but the oxidation numbers are **not** given by Roman numerals. Instead, prefixes are used to indicate elemental proportions of both elements. The first ten prefixes are: mono, di, tri, tetra, penta, hexa, hepta, octa, nona, and deca. The prefix mono is usually omitted, with carbon monoxide, CO, being a notable exception.

Formula	Name
N₂O	dinitrogen oxide
IF₃	iodine trifluoride
SF₆	sulfur hexafluoride

Formula	Name
N₂O₅	dinitrogen pentoxide
B₆Si	hexaboron silicide
SiC	silicon carbide

Binary acids are composed of hydrogen and a group VIA or group VIIA element. The compounds act as acids when dissolved to make aqueous solutions. The pure compounds are named as typical binary compounds except that no prefixes are used even when more than one hydrogen atom is present in the formula. Their aqueous solutions are named by modifying the characteristic stem of the nonmetal with the prefix **hydro** and the suffix **ic** followed by the word **acid**. The stem for sulfur is "sulfur" rather than "sulf" in this case.

Formula	Name of Compound	Name of Aqueous Solution
HBr	hydrogen bromide	hydrobromic acid, HBr(aq)
H₂Se	hydrogen selenide	hydroselenic acid, H₂Se(aq)
HCN	hydrogen cyanide	hydrocyanic acid, HCN(aq)

Ternary acids (oxoacids) are composed of hydrogen, oxygen, and a nonmetal. Nonmetals that exhibit more than one oxidation state form more than one ternary acid. The common ternary **ic** acids are tabulated in Table 4-14 of the text. The acid with one more oxygen than the **ic acid** is the **per...ic acid**. The acid with one less oxygen than the **ic acid** is the **ous acid**, while the acid with two less oxygens is the **hypo...ous acid**.

Ternary salts are formed by replacement of one or more of the acidic hydrogen atoms by a metal cation or an ammonium ion. An **...ic acid** makes an **...ate** anion, while an **...ous acid** makes an **...ite** anion. For sulfur, the "ur" is dropped and for phosphorus, the "or" is dropped in the ion name. H₂SO₃, sulfurous acid gives rise to SO₃²⁻, sulfite, ion. H₃PO₄, phosporic acid, is associated with PO₄³⁻, phospate, ion. The charge of the anion formed is numerically equal to the number of hydrogen atoms removed.

If not all the hydrogen atoms are replaced by metal ions, the salt is an **acidic salt**. The remaining hydrogen atoms are named, with a prefix if appropriate, when naming the anion. For example, arsenic acid, H_3AsO_4, is the source of three oxoanions: $H_2AsO_4^-$ (dihydrogen arsenate), $HAsO_4^{2-}$ (hydrogen arsenate), and AsO_4^{3-}, arsenate. Additional common oxoanions are found in Table 4-14 in the text. (Note that bromine and iodine form ions that are analogous to the ClO_4^-, ClO_3^-, ClO_2^-, and ClO^- ions listed in the table.)

When naming ternary salts, the name of the metal or ammonium is written first, followed by the anion name. Roman numerals, to indicate oxidation state, must be shown for those ions that exhibit more than one common charge.

Formula	Name	Formula	Name
$Mg(NO_3)_2$	magnesium nitrate	$FeCO_3$	iron(II) carbonate
$Ca(ClO)_2$	calcium hypochlorite	$Co(BrO_3)_3$	cobalt(III) bromate
$CrSO_4$	chromium(II) sulfate	$Pb(HC_2O_4)_4$	lead(IV) hydrogen oxalate
$NaNO_2$	sodium nitrite	$SnCr_2O_7$	tin(II) dichromate
$Al(HSO_4)_3$	aluminum hydrogen sulfate	$Ba_3(PO_3)_2$	barium phosphite

EXERCISES

The Periodic Table

1. Which of the following are alkali metals?

   (a) Li (b) Pb (c) Ca (d) Ag (e) Sn (f) K (g) Ti (h) Po

2. Which is more metallic?

   (a) Cs or K (b) Ca or Ni (c) Zn or Hg (d) P or Bi

3. In which group are the (a) halogens; (b) alkaline earth metals?

Aqueous Solutions

4. Which of the following are strong acids?

   (a) HI (b) H_2SeO_3 (c) HClO (d) HNO_3 (e) H_3AsO_3 (f) $HClO_4$

5. Which of the following are strong, soluble bases?

   (a) $Ba(OH)_2$ (b) $Cd(OH)_2$ (c) KOH (d) $Al(OH)_3$ (e) $Zr(OH)_4$

6. Based on the solubility rules given, show how each of the following substances would exist in aqueous solution.

   (a) AgCl (b) Na_3PO_4 (c) $(NH_4)_2SO_4$ (d) PbI_2 (e) $FeCO_3$ (f) CuO
   (g) $Zn(ClO_4)_2$ (h) $AlBr_3$ (i) RbF (j) $Ca_3(PO_4)_2$

Precipitation Reactions

7. Write balanced formula unit, total ionic, and net ionic equations that illustrate the reactions that occur when aqueous solutions of the following substances are mixed. If there is no insoluble product formed write "no reaction".

   (a) $AgNO_3$ and K_2SO_4
   (b) AlI_3 and RbOH
   (c) Li_2S and $CrCl_3$
   (d) $MgBr_2$ and NH_4ClO_4
   (e) Cs_3PO_4 and $CuBr_2$
   (f) Rb_2CO_3 and $SnCl_2$
   (g) NaCl and $FeSO_4$
   (h) $Hg(MnO_4)_2$ and K_2S

Acid-Base Reactions

8. Write balanced formula unit, total ionic and net ionic equations for the reactions between the following acids and bases. Be sure to consider if the acid or base is strong or weak and if the substances are soluble or insoluble.

   (a) $HClO_3$ and $Sr(OH)_2$
   (b) H_2SO_4 and $LiOH$
   (c) HNO_3 and $Al(OH)_3$
   (d) $HClO$ and $RbOH$
   (e) H_3AsO_4 and $Ba(OH)_2$
   (f) H_2CO_3 and KOH
   (g) HBr and $Ca(OH)_2$
   (h) HI and $CsOH$

Oxidation Numbers

9. Assign oxidation numbers to all elements in the substances below.

   (a) $KClO_4$
   (b) $NaHSeO_3$
   (c) $(NH_4)_3AsO_4$
   (d) $CaCrO_4$
   (e) Al_4C_3
   (f) $LiIO_3$
   (g) $Na_2Cr_2O_7$
   (h) P_4S_3
   (i) S_8
   (j) MgH_2
   (k) ClF_5
   (l) As_4O_6

10. Assign oxidation numbers to all elements in the ions below.

    (a) BrO_2^-
    (b) $S_2O_3^{2-}$
    (c) SbF_6^-
    (d) $Zn(H_2O)_6^{2+}$
    (e) $N_2H_5^+$
    (f) $HAsO_3^{2-}$
    (g) $HC_2O_4^-$
    (h) UO_2^{2+}
    (i) $Cu(NH_3)_4^{2+}$
    (j) N_3^-
    (k) HCO_2^-
    (l) $HSiO_5^-$

Oxidation-Reduction Reactions

11. Assign oxidation numbers to each element in the equations below and identify the oxidizing agent and the reducing agent in each.

    (a) $P_4(s) + 10F_2(g) \rightarrow 4PF_5(g)$
    (b) $Fe_2O_3(s) + 2Al(s) \rightarrow 2Fe(s) + Al_2O_3(s)$
    (c) $2H_2SO_4(aq) + S(s) \rightarrow 3SO_2(g) + 2H_2O(\ell)$
    (d) $As_4O_6(s) + 8HNO_3(aq) + 2H_2O(\ell) \rightarrow 4H_3AsO_4(aq) + 8NO_2(g)$
    (e) $F_2(g) + 2NaBr(aq) \rightarrow 2NaF(aq) + Br_2(\ell)$
    (f) $3I_2(s) + 5HClO_3(aq) + 3H_2O(\ell) \rightarrow 6HIO_3(aq) + 5 HCl(aq)$

Displacement Reactions

12. Which of the following metals will displace hydrogen from aqueous solutions of HCl. Write balanced net ionic equations for any reactions which occur.
    (a) Al (b) Mg (c) Ag (d) Cu (e) K (f) Pb (g) Sb

13. Which of the following metals will displace nickel from a solution of $Ni(NO_3)_2$. Write balanced net ionic equations for any reactions which occur.
    (a) Cd (b) Pb (c) Pt (d) Mg (e) Hg (f) Co (g) Al

Naming of Inorganic Compounds

14. Write formulas for the following compounds.
    (a) lithium hypoiodite
    (b) cesium selenide
    (c) ammonium thiocyanate
    (d) cobalt(II) arsenate
    (e) copper(I) dichromate
    (f) zirconium(IV) oxide
    (g) lead(II) acetate
    (h) calcium carbonate
    (i) gallium(III) perbromate
    (j) gold(III) sulfite
    (k) magnesium phosphide
    (l) nickel(II) oxalate
    (m) tin(IV) permanganate
    (n) manganese(VII) oxide
    (o) nitrogen trichloride
    (p) diphosphorus pentaselenide
    (q) xenon hexafluoride
    (r) dioxygen difluoride

15. Write names for the following compounds.
    (a) $LiHSeO_3$
    (b) $NaHSO_4$
    (c) $Sr(HCO_3)_2$
    (d) $Cd(SCN)_2$
    (e) $Co_3(PO_4)_2$
    (f) Li_2SO_3
    (g) Ag_3AsO_4
    (h) $W(ClO_4)_4$
    (i) $BiPO_4$
    (j) S_2F_6
    (k) $Sc(CH_3COO)_2$
    (l) $Sb_2(TeO_3)_3$
    (m) As_2Se_3
    (n) KCN
    (o) $Ti(OH)_4$
    (p) Rb_3P

16. Write formulas for the following compounds.
    (a) iron(III) fluoride
    (b) antimony(III) permanganate
    (c) rubidium oxalate
    (d) chromium(III) hypochlorite
    (e) cobalt(III) fluoride
    (f) disulfur decafluoride

(g) barium nitrate
(h) manganese(II) dichromate
(i) chromium(II) sulfate
(j) lithium silicate
(k) indium(III) telluride
(l) diiodine nonoxide
(m) dinitrogen trioxide
(n) tetraphosphorus decoxide
(o) tellurium hexafluoride
(p) titanium(IV) tellurate

17. Write names for the following compounds.
    (a) $Mn_2(Cr_2O_7)_3$ (b) $Ba(CH_3COO)_2$ (c) $Au(MnO_4)_3$ (d) $Sn(H_2PO_4)_2$
    (e) $ScBO_3$ (f) $Co(BrO_4)_2$ (g) $Ca(ClO)_2$ (h) $Hg(NO_2)_2$
    (i) WN_2 (j) TeS_2 (k) $KHCr_2O_7$ (l) Br_3O_8

18. Write formulas for the following acids.
    (a) hydrosulfuric acid
    (b) hydrocyanic acid
    (c) hydrofluoric acid
    (d) hydrobromic acid
    (e) selenic acid
    (f) periodic acid
    (g) hypoiodous acid
    (h) chloric acid
    (i) dichromic acid
    (j) permanganic acid
    (k) arsenic acid
    (l) tellurous acid

Miscellaneous Exercises

19. Write balanced net ionic equations for any reactions that occur when the following reactants are mixed. Consult activity and solubility tables as needed.
    (a) $H_3AsO_4(aq)$ and $LiOH(aq)$
    (b) $KBr(aq)$ and $Cl_2(g)$
    (c) $H_3PO_4(aq)$ and $Cr(OH)_3(s)$
    (d) $MnCl_2(aq)$ and $Fe(s)$
    (e) $Ba(OH)_2(aq)$ and $HNO_2(aq)$
    (f) $(NH_4)_2CO_3(aq)$ and $MnI_2(aq)$
    (g) $Cd(MnO_4)_2(aq)$ and $LiOH(aq)$
    (h) $CaCl_2(aq)$ and $Ni(s)$
    (i) $KI(aq)$ and $Au(s)$
    (j) $Hg_2(ClO_4)_2(aq)$ and $FeCl_3(aq)$
    (k) $Cr(OH)_3(s)$ and $HNO_3(aq)$
    (l) $HClO_4(aq)$ and $Mg(s)$

20. Name the compounds. (a) ClF_5 (b) Na_2S (c) CoP (d) As_4O_6 (e) $MgSO_4$ (f) XeF_4

21. Assign oxidation numbers to each element in the formulas below.
    (a) OsO_6^{4-} (b) $C_{12}H_{22}O_{11}$ (c) SiF_6^{2-} (d) $H_2W_{12}O_{42}^{10-}$

ANSWERS TO EXERCISES

1. Li, K   2. (a) Cs (b) Ca (c) Hg (d) Bi   3. (a) VIIA (b) IIA

4. HI, HNO_3, $HClO_4$ 5. $Ba(OH)_2$, KOH

6. (a) AgCl(s)
   (b) $3Na^+(aq)$, $PO_4^{3-}(aq)$
   (c) $2NH_4^+(aq)$, $SO_4^{2-}(aq)$
   (d) $PbI_2(s)$
   (e) $FeCO_3(s)$
   (f) CuO(s)
   (g) $Zn^{2+}(aq)$, $2ClO_4^-(aq)$
   (h) $Al^{3+}(aq)$, $3Br^-(aq)$
   (i) $Rb^+(aq)$, $F^-(aq)$
   (j) $Ca_3(PO_4)_2(s)$

7. (d) and (g) no reaction–no insoluble substances are formed

   (a) $2AgNO_3(aq) + K_2SO_4 \rightarrow Ag_2SO_4(s) + 2KNO_3(aq)$ formula unit

   $2[Ag^+(aq) + NO_3^-(aq)] + [2K^+(aq) + SO_4^{2-}(aq)] \rightarrow$
   $Ag_2SO_4(s) + 2[K^+(aq) + NO_3^-(aq)]$ total ionic

   $2Ag^+(aq) + SO_4^{2-}(aq) \rightarrow Ag_2SO_4(s)$ net ionic

   (b) $AlI_3(aq) + 3RbOH(aq) \rightarrow Al(OH)_3(s) + 3RbI(aq)$ formula unit

   $[Al^{3+}(aq) + 3I^-(aq)] + 3[Rb^+(aq) + OH^-(aq)] \rightarrow$
   $Al(OH)_3(s) + 3[Rb^+(aq) + I^-(aq)]$ total ionic

   $Al^{3+}(aq) + 3OH^-(aq) \rightarrow Al(OH)_3(s)$ net ionic

   (c) $3Li_2S(aq) + 2CrCl_3(aq) \rightarrow Cr_2S_3(s) + 6LiCl(aq)$ formula unit

   $3[2Li^+(aq) + S^{2-}(aq)] + 2[Cr^{3+}(aq) + 3Cl^-(aq)] \rightarrow$
   $Cr_2S_3(s) + 6[Li^+(aq) + Cl^-(aq)]$ total ionic

   $3S^{2-}(aq) + 2Cr^{3+}(aq) \rightarrow Cr_2S_3(s)$ net ionic

   (e) $2Cs_3PO_4(aq) + 3CuBr_2(aq) \rightarrow Cu_3(PO_4)_2(s) + 6CsBr(aq)$ formula unit

   $2[3Cs^+(aq) + PO_4^{3-}(aq)] + 3[Cu^{2+}(aq) + 2Br^-(aq)] \rightarrow$
   $Cu_3(PO_4)_2(s) + 6[Cs^+(aq) + Br^-(aq)]$ total ionic

   $2PO_4^{3-}(aq) + 3Cu^{2+}(aq) \rightarrow Cu_3(PO_4)_2(s)$ net ionic

Chapter 4 Some Types of Chemical Reactions 81

(f) $Rb_2CO_3(aq) + SnCl_2(aq) \rightarrow SnCO_3(s) + 2RbCl(aq)$ formula unit

$[2Rb^+(aq) + CO_3^{2-}(aq)] + [Sn^{2+}(aq) + 2Cl^-(aq)] \rightarrow$
$SnCO_3(s) + 2[Rb^+(aq) + Cl^-(aq)]$ total ionic

$CO_3^{2-}(aq) + Sn^{2+}(aq) \rightarrow SnCO_3(s)$ net ionic

(h) $Hg(MnO_4)_2(aq) + K_2S(aq) \rightarrow HgS(s) + 2KMnO_4(aq)$ formula unit

$[Hg^{2+}(aq) + 2MnO_4^-(aq)] + [2K^+(aq) + S^{2-}(aq)] \rightarrow$
$HgS(s) + 2[K^+(aq) + MnO_4^-(aq)]$ total ionic

$Hg^{2+}(aq) + S^{2-}(aq) \rightarrow HgS(s)$ net ionic

8.

(a) $2HClO_3(aq) + Sr(OH)_2(aq) \rightarrow Sr(ClO_3)_2(aq) + 2H_2O(\ell)$ formula unit

$2[H^+(aq) + ClO_3^-(aq)] + [Sr^{2+}(aq) + 2OH^-(aq)] \rightarrow$
$[Sr^{2+}(aq) + 2ClO_3^-(aq)] + 2H_2O(\ell)$ total ionic

$H^+(aq) + OH^-(aq) \rightarrow H_2O(\ell)$ net ionic

(b) $H_2SO_4(aq) + 2LiOH(aq) \rightarrow Li_2SO_4(aq) + 2H_2O(\ell)$ formula unit

$[2H^+(aq) + SO_4^{2-}(aq)] + 2[Li^+(aq) + OH^-(aq)] \rightarrow$
$[2Li^+(aq) + SO_4^{2-}(aq)] + 2H_2O(\ell)$ total ionic

$H^+(aq) + OH^-(aq) \rightarrow H_2O(\ell)$ net ionic

(c) $3HNO_3(aq) + Al(OH)_3(s) \rightarrow Al(NO_3)_3(aq) + 3H_2O(\ell)$ formula unit

$3[H^+(aq) + NO_3^-(aq)] + Al(OH)_3(s) \rightarrow$
$[Al^{3+}(aq) + 3NO_3^-(aq)] + 3H_2O(\ell)$ total ionic

$3H^+(aq) + Al(OH)_3(s) \rightarrow Al^{3+}(aq) + 3H_2O(\ell)$ net ionic

(d) $HClO(aq) + RbOH(aq) \rightarrow RbClO(aq) + H_2O(\ell)$ formula unit

$HClO(aq) + [Rb^+(aq) + OH^-(aq)] \rightarrow [Rb^+(aq) + ClO^-(aq)] + H_2O(\ell)$ total ionic

$HClO(aq) + OH^-(aq) \rightarrow ClO^-(aq) + H_2O(\ell)$ net ionic

(e) $2H_3AsO_4(aq) + 3Ba(OH)_2(aq) \rightarrow Ba_3(AsO_4)_2(s) + 6H_2O(\ell)$ formula unit

$2H_3AsO_4(aq) + 3[Ba^{2+}(aq) + 2OH^-(aq)] \rightarrow Ba_3(AsO_4)_2(s) + 6H_2O(\ell)$ total ionic

$2H_3AsO_4(aq) + 3Ba^{2+}(aq) + 6OH^-(aq) \rightarrow Ba_3(AsO_4)_2(s) + 6H_2O(\ell)$ net ionic

(f) $H_2CO_3(aq) + 2KOH(aq) \rightarrow K_2CO_3(aq) + 2H_2O(\ell)$ formula unit

$H_2CO_3(aq) + 2[K^+(aq) + OH^-(aq)] \rightarrow$
$[2K^+(aq) + CO_3^{2-}(aq)] + 2H_2O(\ell)$ total ionic

$H_2CO_3(aq) + 2OH^-(aq) \rightarrow CO_3^{2-}(aq) + 2H_2O(\ell)$ net ionic

(g) $2HBr(aq) + Ca(OH)_2(aq) \rightarrow CaBr_2(aq) + 2H_2O(\ell)$ formula unit

$2[H^+(aq) + Br^-(aq)] + [Ca^{2+}(aq) + 2OH^-(aq)] \rightarrow$
$[Ca^{2+}(aq) + 2Br^-(aq)] + 2H_2O(\ell)$ total ionic

$H^+(aq) + OH^-(aq) \rightarrow H_2O(\ell)$ net ionic

(h) $HI(aq) + CsOH(aq) \rightarrow CsI(aq) + H_2O(\ell)$ formula unit

$[H^+(aq) + I^-(aq)] + [Cs^+(aq) + OH^-] \rightarrow [Cs^+(aq) + I^-(aq)] + H_2O(\ell)$ total ionic

$H^+(aq) + OH^-(aq) \rightarrow H_2O(\ell)$ net ionic

9. (a) $\overset{+1\ +7\ -2}{KClO_4}$ (b) $\overset{+1\ +1\ +4\ -2}{NaHSeO_3}$ (c) $\overset{-3\ +1\ +5\ -2}{(NH_4)_3PO_4}$ (d) $\overset{+2\ +6\ -2}{CaCrO_4}$

(e) $\overset{+3\ -4}{Al_4C_3}$ (f) $\overset{+1\ +5\ -2}{LiIO_3}$ (g) $\overset{+1\ +6\ -2}{Na_2Cr_2O_7}$ (h) $\overset{+1.5\ -2}{P_4S_3}$

(i) $\overset{0}{S_8}$ (j) $\overset{+2\ -1}{MgH_2}$ (k) $\overset{+5\ -1}{ClF_5}$ (l) $\overset{+3\ -2}{As_4O_6}$

10. (a) $\overset{+3\ -2}{BrO_2^-}$ (b) $\overset{+2\ -2}{S_2O_3^{2-}}$ (c) $\overset{+5\ -1}{SbF_6^-}$ (d) $\overset{+2\ +1\ -2}{Zn(H_2O)_6^{2+}}$

(e) $\overset{-2\ +1}{N_2H_5^+}$ (f) $\overset{+1\ +3\ -2}{HAsO_3^{2-}}$ (g) $\overset{+1\ +3\ -2}{HC_2O_4^-}$ (h) $\overset{+6\ -2}{UO_2^{2+}}$

(i) $\overset{+2\ -3\ +1}{Cu(NH_3)_4^{2+}}$ (j) $\overset{-1/3}{I_3^-}$ (k) $\overset{+1\ +2\ -2}{HCO_2^-}$ (l) $\overset{+1\ +8\ -2}{HSiO_5^-}$

Chapter 4 Some Types of Chemical Reactions 85

11.

(a) $\overset{0}{P_4}(s) + \overset{0}{10F_2}(g) \rightarrow 4\overset{+5\ -1}{PF_5}(g)$
P$_4$ is the reducing agent; Cl$_2$ is the oxidizing agent.

(b) $\overset{+3\ -2}{Fe_2O_3}(s) + \overset{0}{2Al}(s) \rightarrow \overset{0}{2Fe}(s) + \overset{+3\ -2}{Al_2O_3}(s)$
Fe is the oxidizing agent; Al is the reducing agent.

(c) $\overset{+1\ +6\ -2}{2H_2SO_4}(aq) + \overset{0}{S}(s) \rightarrow 3\overset{+4\ -2}{SO_2}(g) + 2\overset{+1\ -2}{H_2O}(\ell)$
S is the reducing agent; H$_2$SO$_4$ is the oxiding agent.

(d) $\overset{+3\ -2}{As_4O_6}(s) + 8\overset{+1\ +5\ -2}{HNO_3}(aq) + 2\overset{+1\ -2}{H_2O}(\ell) \rightarrow 4\overset{+1\ +5\ -2}{H_3AsO_4}(aq) + 8\overset{+4\ -2}{NO_2}(g)$
As$_4$O$_6$ is the reducing agent; HNO$_3$ is the oxidizing agent.

(e) $\overset{0}{F_2}(g) + 2\overset{+1\ -1}{NaBr}(aq) \rightarrow 2\overset{+1\ -1}{NaF}(aq) + \overset{0}{Br_2}(\ell)$
NaBr is the reducing agent; F$_2$ is the oxidizing agent.

(f) $3\overset{0}{I_2}(s) + 5\overset{+1\ +5\ -2}{HClO_3}(aq) + 3\overset{+1\ -2}{H_2O}(\ell) \rightarrow 6\overset{+1\ +5\ -2}{HIO_3}(aq) + 5\overset{+1\ -1}{HCl}(aq)$
I$_2$ is the reducing agent; HClO$_3$ is the oxidizing agent.

12. (a) $2Al(s) + 6H^+(aq) \rightarrow 2Al^{3+}(aq) + 3H_2(g)$

(b) $Mg(s) + 2H^+(aq) \rightarrow Mg^{2+}(aq) + H_2(g)$

(e) $2K(s) + 2H^+(aq) \rightarrow 2K^+(aq) + H_2(g)$

(f) $Pb(s) + 2H^+(aq) \rightarrow Pb^{2+}(aq) + H_2(g)$

13. (a) $Cd(s) + Ni^{2+}(aq) \rightarrow Cd^{2+}(aq) + Ni(s)$

(d) $Mg(s) + Ni^{2+}(aq) \rightarrow Mg^{2+}(aq) + Ni(s)$

(f) $Co(s) + Ni^{2+}(aq) \rightarrow Co^{2+}(aq) + Ni(s)$

(g) $2Al(s) + 3Ni^{2+}(s) \rightarrow 2Al^{3+}(aq) + 3Ni(s)$

14. (a) LiIO   (b) Cs$_2$Se   (c) NH$_4$SCN   (d) Co$_3$(AsO$_4$)$_2$
    (e) Cu$_2$Cr$_2$O$_7$ (f) ZrO$_2$ (g) Pb(CH$_3$COO)$_2$ (h) CaCO$_3$
    (i) Ga(BrO$_4$)$_3$ (j) Au$_2$(SO$_3$)$_3$ (k) Mg$_3$P$_2$ (l) NiC$_2$O$_4$
    (m) Sn(MnO$_4$)$_4$ (n) Mn$_2$O$_7$ (o) NCl$_3$ (p) P$_2$Se$_5$
    (q) XeF$_6$ (r) O$_2$F$_2$

15. (a) lithium hydrogen selenite
    (b) sodium hydrogen sulfate
    (c) strontium hydrogen carbonate
    (d) cadmium thiocyanate
    (e) cobalt(II) phosphate
    (f) lithium sulfite
    (g) silver arsenate
    (h) tungsten(IV) perchlorate
    (i) bismuth(III) phosphate
    (j) disulfur hexafluoride
    (k) scandium(II) acetate
    (l) antimony(III) tellurite
    (m) diarsenic triselenide
    (n) potassium cyanide
    (o) titanium(IV) hydroxide
    (p) rubidium phosphide

16. (a) FeF_3 (b) $Sb(MnO_4)_3$ (c) $Rb_2C_2O_4$ (d) $Cr(ClO)_3$
    (e) CoF_3 (f) S_2F_{10} (g) $Ba(NO_3)_2$ (h) $MnCr_2O_7$
    (i) $CrSO_4$ (j) Li_4SiO_4 (k) In_2Te_3 (l) I_2O_9
    (m) N_2O_3 (n) P_4O_{10} (o) TeF_6 (p) $Ti(TeO_4)_2$

17. (a) manganese(III) dichromate
    (b) barium acetate
    (c) gold(III) permanganate
    (d) tin(II) dihydrogen phosphate
    (e) scandium(III) borate
    (f) cobalt(II) perbromate
    (g) calcium hypochlorite
    (h) mercury(II) nitrite
    (i) tungsten(VI) nitride
    (j) tellurium disulfide
    (k) potassium hydrogen dichromate
    (l) tribromine octoxide

18. (a) H_2S (b) HCN (c) HF (d) HBr
    (e) H_2SeO_4 (f) HIO_4 (g) HIO (h) $HClO_3$
    (i) $H_2Cr_2O_7$ (j) $HMnO_4$ (k) H_3AsO_4 (l) H_2TeO_3

19. (d), (h) and (i) have no reaction
    (a) $H_3PO_4(aq) + 3OH^-(aq) \rightarrow PO_4^{3-}(aq) + 3H_2O(\ell)$
    (b) $2Br^-(aq) + Cl_2(g) \rightarrow Br_2(\ell) + 2Cl^-(aq)$
    (c) $H_3PO_4(aq) + Cr(OH)_3(s) \rightarrow CrPO_4(s) + 3H_2O(\ell)$
    (e) $HNO_2(aq) + OH^-(aq) \rightarrow H_2O(\ell) + NO_2^-$
    (f) $CO_3^{2-}(aq) + Mn^{2+}(aq) \rightarrow MnCO_3(s)$
    (g) $Cd^{2+}(aq) + 2OH^-(aq) \rightarrow Cd(OH)_2(s)$
    (j) $Hg_2^{2+}(aq) + 2Cl^-(aq) \rightarrow Hg_2Cl_2(s)$
    (k) $Cr(OH)_3(s) + 3H^+(aq) \rightarrow Cr^{3+}(aq) + 3H_2O(\ell)$
    (l) $2H^+(aq) + Mg(s) \rightarrow H_2(g) + Mg^{2+}(aq)$

20. (a) chlorine pentafluoride
    (b) sodium sulfide
    (c) cobalt(III) phosphide
    (d) tetraarsenic hexoxide
    (e) magnesium sulfate
    (f) xenon tetrafluoride

21. (a) $\overset{+8\ -2}{OsO_6^{4-}}$ (b) $\overset{0\ +1\ -2}{C_{12}H_{22}O_{11}}$ (c) $\overset{+4\ -1}{SiF_6^{2-}}$ (d) $\overset{+1\ +6\ -2}{H_2W_{12}O_{42}^{10-}}$

Chapter 5: The Structure of Atoms

5-1 Fundamental Particles

The presently accepted picture of the atom is a nuclear model in which the protons and neutrons are located in a very small, very dense, positively charged nucleus at the center of the atom. Very lightweight electrons are diffusely distributed over relatively large distances from the nucleus. The radii of nuclei and atoms are approximately 10^{-4} nm and 10 nm, respectively. The masses and charges of the three fundamental particles are listed in Table 5-1.

Table 5-1 Fundamental Particles of Matter		
Particle	Mass (amu)	Charge
electron (e^-)	0.00055	-1
proton (p^+)	1.0073	+1
neutron ($n^°$)	1.0087	0

5-2 Atomic Number

The **atomic number** of an element is the number of protons in the nucleus of an atom of the element. All atoms of the same element have the same atomic number. For example, all atoms with 42 protons are molybdenum and all molybdenum atoms contain 42 protons. The number of protons and, thus, the atomic number is equal to the number of electrons in a neutral atom. Ions are formed by gaining or losing electrons, leaving the atomic number unchanged. A negative ion (anion) has gained the number of electrons equal to its charge, while a positive ion (cation) has lost the number of electrons equal to its charge. For example, a Co^{3+} ion has 27 protons and 24 electrons, while a P^{3-} ion has 15 protons and 18 electrons.

5-3 Mass Number and Isotopes

The **mass number** of an atom is an integer that is the sum of the number of protons and neutrons in its nucleus. The nuclide symbol of an atom shows both its mass number superscripted and its atomic number subscripted. For example, $^{21}_{10}Ne$ has an atomic number of 10 and a mass number of 21. It has 10 protons and 21-10 = 11 neutrons. Often the atomic number is omitted, because it must agree with the symbol of the element. Thus, the previous nuclide symbol could be written simply as ^{21}Ne.

Example 5-1 Subatomic Particles

How many protons, electrons and neutrons are in each of the following atoms or ions?
(a) $^{40}Ca^{2+}$ (b) ^{75}As (c) ^{104}Pd (d) $^{37}Cl^-$

Plan

We will use the information given by the nuclide designation and remember that we can use the periodic table to determine the atomic number which equals the number of protons of the element. The mass number minus the atomic number gives the number of neutrons. A negative charge means there are additional electrons, while a positive charge means that electrons have been removed.

Solution

(a) atomic number = 20, so: 20 protons, 40-20 = 20 neutrons, 20-2 = 18 electrons
(b) atomic number = 33, so: 33 protons, 75-33 = 42 neutrons, 33 electrons
(c) atomic number = 46, so: 46 protons, 104-46 = 58 neutrons, 46 electrons
(d) atomic number = 17, so: 17 protons, 37-17 = 20 neutrons, 17+1 = 18 electrons

---

Most elements occur naturally as mixtures of isotopes, that is, as atoms containing the same number of protons, but different numbers of neutrons. For example, naturally occuring lead is a mixture of $^{204}_{82}Pb$, $^{206}_{82}Pb$, $^{207}_{82}Pb$ and $^{208}_{82}Pb$. These isotopes have respectively, 204-82 = 122 neutrons, 206-82 = 124 neutrons, 207-82 = 125 neutrons, and 208-82 = 126 neutrons. The name of the isotope is reported as the element name followed by the mass number, therefore, ^{204}Pb is lead-24, ^{206}Pb is lead-206, etc.

Remember from Chapter 2 that one atomic mass unit (amu) is defined as exactly 1/12 the mass of one atom of carbon-12. A ^{204}Pb atom is 203.973/12 times as massive as a ^{12}C atom, while a ^{206}Pb atom is 205.9745/12 times as massive as a ^{12}C atom, a ^{207}Pb atom is 206.9759/12 times as massive as a ^{12}C atom, and a ^{208}Pb atom is 207.9766 times as massive as a ^{12}C atom. Therefore, the respective masses of the isotopes are reported as 203.973 amu, 205.9745 amu, 206.9759 amu, and 207.9766 amu.

5-4 The Atomic Weight Scale and Atomic Weights

Accurate and precise values of both the percent natural abundances and the masses of most known isotopes have been determined. The atomic weight of an element is the weighted average of the masses of its constituent isotopes. Therefore, the atomic weight may be thought of as the mass of an "average" atom of that element, even though for most elements no individual "average" atoms exist.

Example 5-2 Calculation of Atomic Weight

Naturally occurring lead discussed above is found to have natural percentage abundances of 1.48% ^{204}Pb, 23.6% ^{206}Pb, 22.6% ^{207}Pb, and 52.3% ^{208}Pb. Use this information to calculate the atomic weight of lead.

Plan
We multiple the fraction of each isotope by its mass and add these numbers to obtain the atomic weight of lead.

Solution

Isotope	Fractional Abundance	Mass (amu)	Contribution (amu)
^{204}Pb	0.0148	203.973	3.02
^{206}Pb	0.236	205.9745	48.6
^{207}Pb	0.226	206.9759	46.8
^{208}Pb	0.523	207.9766	109
			207

An "average" atom of Pb would have a mass of 207 amu. Note that mass numbers are integers, while atomic weights are not necessarily integers.

Example 5-3 Calculation of Isotopic Abundances

The atomic weight of carbon is 12.01115. Carbon occurs as ^{12}C with a mass defined as exactly 12 amu and ^{13}C with a mass of 13.00335 amu. What is the percent abundance of each isotope?

Plan

When any quantity is expressed in terms of fractions, the sum of the fractions is unity (1). We know that the fractions of each isotope times the mass must sum to the atomic weight. Let x be the fraction of the ^{12}C isotope; therefore, 1-x is the fraction of ^{13}C and:

$$(\text{fraction}_1)(\text{mass}_1) + (\text{fraction}_2)(\text{mass}_2) = \text{atomic weight}$$

Solution

(x)(12.00000 amu) + (1-x)(13.00335 amu) = 12.01115 amu

12.00000x + 13.00335 - 13.00335x = 12.01115

1.00335x = 0.99220

x = 0.98889

1-x = 1 - 0.98889 = 0.01111

Therefore, carbon is $\boxed{98.889\% \ ^{12}C}$ and $\boxed{1.111\% \ ^{13}C}$.

5-5 Electromagnetic Radiation and Energy

The most effective probe of electronic structure is electromagnetic radiation. Electrons are classified as matter since they have mass and occupy space, but they are also wave-like and interact with electromagnetic radiation. Each wave of electromagnetic radiation is characterized by a definite wavelength (λ) and frequency (ν). The wavelengths of electromagnetic radiation cover a wide range, from 10^{-10} m for x-rays up to 10^3 m for radiowaves. Frequency and wavelength are inversely proportional to each other. High frequency electromagnetic radiation has short wavelength and high energy. The product of the two is equal to the velocity of light, c.

$$\lambda \nu = c = 2.998 \times 10^8 \text{ m/s}$$

The energy of a photon of electromagnetic radiation of a given frequency or wavelength is given by the following equation in which h is Planck's constant with a value of 6.626×10^{-34} J·s.

$$E = h\nu = \frac{hc}{\lambda}$$

Example 5-4 Wavelength of Light

Calculate the wavelength and energy of blue-green light of frequency $6.167 \times 10^{14} \text{ s}^{-1}$.

Plan

We will use the equations above to calculate both quantities.

Solution

$$? \lambda = \frac{c}{\nu} = \frac{2.998 \times 10^8 \text{ ms}^{-1}}{6.167 \times 10^{-14} \text{ s}^{-1}} = \boxed{4.861 \times 10^{-7} \text{ m}}$$

$$? E = h\nu = (6.626 \times 10^{-34} \text{ J} \cdot \text{s})(6.167 \times 10^{14} \text{ s}^{-1}) = \boxed{4.086 \times 10^{-19} \text{ J}}$$

Radiation is absorbed by an atom only in discrete quanta or "packets" containing precisely the amount of energy required to promote an electron to a higher energy level. The same amount of energy is emitted by an atom as a photon when an electron returns to the lower energy level. These absorptions and emissions of energy provide a basis for studying the characteristics and distributions of electrons in atoms by means of absorption and emission spectroscopy.

Example 5-5 Energy of Light

One of the wavelengths of light emitted by excited barium atoms is 4.55×10^{-7} m. What is the energy separating the two electronic energy levels between which the electron moves in this emission? Calculate the energy difference in joules per atom and in kJ/mol of barium atoms.

Plan

We will use the relationship between energy and wavelength to determine the energy. We then use Avogadro's number to convert this result per atom to kilojoules per mole.

Solution

$$? \text{J} = E = \frac{hc}{\lambda} = \frac{(6.63 \times 10^{-34} \text{ J} \cdot \text{s})(2.998 \times 10^8 \text{ ms}^{-1})}{4.55 \times 10^{-7} \text{ m}} = \boxed{4.37 \times 10^{-19} \text{ J}}$$

This is the energy released by **one** atom.

$$? \text{ kJ/mol} = \frac{4.37 \times 10^{-19} \text{ J}}{\text{atom}} \times \frac{6.022 \times 10^{23} \text{ atoms}}{\text{mol}} \times \frac{1 \text{ kJ}}{1000 \text{ J}} = \boxed{263 \text{ kJ/mol}}$$

5-6 Quantum Numbers and Atomic Orbitals

The electronic structure of an element determines its physical properties and chemical characteristics and reactivity. Chemical reactions and bonding involve shifting or rearranging the outermost electrons in atoms. The best picture of the electronic structure of atoms is described by quantum mechanics using differential equations. The experimentally supported quantum mechanical treatment takes into consideration the dual particle-like, wave-like characteristics of electrons. The results derived from experimental data and solutions of quantum mechanical equations yield **atomic orbitals**. These are regions in space about the nucleus in which the probability of finding electrons is the greatest.

The orbital residence of an electron may be specified by a set of four quantum numbers, much as a person's address may be specified by four components of the address: state, city, street and street number. The four quantum numbers are:

1. **Principal quantum number, n.** This quantum number designates the major energy level in which an electron is located. It may take on positive integral values beginning with 1. There can be a maximum of $2n^2$ electrons in an energy level.

   $n = 1, 2, 3, 4 \ldots \ldots$

   Principle Quantum Number   **Maximum Number of Electrons in Level**

$n = 4$	(N shell)	32e-
$n = 3$	(M shell)	18e-
$n = 2$	(L shell)	8e-
$n = 1$	(K shell)	2e-

   ⇧ Energy Increases

2. **Subsidiary quantum number, ℓ.** This quantum number specifies a particular sublevel and "shape" within a major energy level. In their ground states, electrons in presently known atoms occupy one of four kinds of sublevels. The possible values of ℓ depend upon the value of n and can range from 0 to a maximum of **n-1**.

   $\ell = 0, 1, 2, 3, \ldots\ldots n-1$

There can be $2(2\ell + 1)$ electrons in a given sublevel at a given energy level. The sublevels are designated by letters that correspond to each ℓ as shown below.

$\ell = 0$ refers to an **s** sublevel, which can hold $2(2\cdot 0 + 1) = 2$ e-
$\ell = 1$ refers to a **p** sublevel, which can hold $2(2\cdot 1 + 1) = 6$ e-
$\ell = 2$ refers to a **d** sublevel, which can hold $2(2\cdot 2 + 1) = 10$ e-
$\ell = 3$ refers to an **f** sublevel, which can hold $2(2\cdot 3 + 1) = 14$ e-

Example 5-6 Quantum Numbers

In which energy level and sublevel is an electron found if its first two quantum numbers are $n = 3$ and $\ell = 1$?

Plan
We will consult the guidelines for quantum numbers.

Solution
Because $n = 3$, the electron is in the third energy level and with $\ell = 1$, it is in the p sublevel (specifically called the 3p sublevel).

Example 5-7 Quantum Numbers

Can an electron occupy a 2d sublevel?

Plan
We consult the guidelines for quantum numbers.

Solution
No, this would require that $n = 2$ and that $\ell = 2$. This would contradict the rule that the maximum value of ℓ is $n - 1$. On a physical basis, this indicates that there is no d sublevel until the third ($n = 3$) energy level.

3. **Magnetic quantum number, m_ℓ**. This quantum number designates the orientation of a specific orbital within a sublevel. The shapes and spatial orientations are described in your text. The values of m_ℓ for any given ℓ may be integers in the range of $-\ell...+\ell$.

$$m_\ell = -\ell, -(\ell - 1), -(\ell - 2), \cdots, 0, \cdots +(\ell - 2), +(\ell - 1), +\ell$$

This indicates that there can be $(2\ell + 1)$ orbitals for any given ℓ and that each orbital can hold up to 2 electrons. This is shown in the following table. The order of filling the orbitals of each type is also from $-\ell \cdots +\ell$. Table summarizes the sublevel information for the first four sublevels.

Table 5-2 Sublevel and Orbital Distribution in Atoms

ℓ	sublevel	orbitals	m_ℓ	#e⁻
$\ell = 0$	s	s	0	2
$\ell = 1$	p	p_x	-1	2
		p_y	0	2
		p_z	+1	2
$\ell = 2$	d	d_{z^2}	-2	2
		$d_{x^2-y^2}$	-1	2
		d_{xy}	0	2
		d_{xz}	+1	2
		d_{yz}	+2	2
$\ell = 3$	f	$f_{z^3-3/5zr^2}$	-3	2
		$f_{x^3-3/5xr^2}$	-2	2
		$f_{y^3-3/5yr^2}$	-1	2
		f_{xyz}	0	2
		$f_{y(x^2-z^2)}$	+1	2
		$f_{x(z^2-y^2)}$	+2	2
		$f_{z(x^2-y^2)}$	+3	2

4. **Spin quantum number, m_s**. This quantum number designates the spin of an electron. Each orbital can hold a maximum of two electrons, **if** they have opposed (paired) spins, that is, their magnetic fields must interact attractively. One electron occupying an orbital is assigned m_s = +1/2 and the other electron in the same orbital is assigned m_s = -1/2. The **Pauli Exclusion Principal** states that no two electrons in the same atom may have the same four quantum numbers. Such a situation would occur only if two electrons occupied the same orbital and had the same spin. **Paramagnetic** species have **at least one** electron unpaired (solely occupying an orbital) and are weakly attracted to a magnetic field. **Diamagnetic** species have **all** electrons paired and are weakly repelled by magnetic fields.

5-7 Electron Configurations

The **electron configuration** of an element shows the total number of electrons and the energy levels and sublevels that they occupy. By the Aufbau Principle, electrons fill available orbitals of lowest energy first. By Hund's rule, they occupy singly the orbitals of an energetically equivalent (degenerate) set, such as those of a particular p sublevel, before pairing in any one orbital of the sublevel. The relative energies of the orbitals are:

< 2s < 2p < 3s < 3p < 4s < 3d < 4p < 5s < 4d < 5p < 6s < 4f < 5d < 6p < 7s < 5f < 6d < 7p

A useful mnemonic device for predicting the order of filling is shown in Figure 5-29 in the text. What is more important is that the energy levels and sublevels can be correlated with positions on the periodic table and these positions can be used to determine electron configurations. Table 5-3 below is similar to Table 5-5 in the text. The annotations in bold face on the left side of each sublevel type designate the energy level being filled. At the top of each group is a designation of the general ending electron configuration for that group.

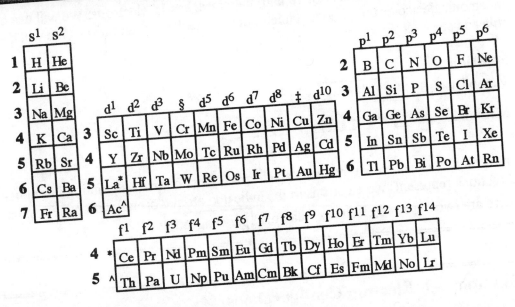

Table 5-3 Energy Levels and Electron configurations

§ These elements have an s^1d^5 configuration.
‡ These elements have an s^1d^{10} configuration.
Note that helium is placed by its configuration to stress the periodicity, rather than with the rest of the noble gases as is more typically shown.

Expected electron configurations will be slightly different for configurations ending in the "f" sublevel if the periodic table is used as opposed to the mnemonic device. With the chart, it is apparent that one 5d electron fills before the 4f block and one 6d electron before the 5f block. Note that Cr and other configurations, such as Cu, W, and Mo, involve half-filled and filled sets of d and s orbitals. It is observed that there is an added stability associated with those half-filled and filled sets. If you consult a complete listing of electron configurations, such as that in Appendix B of the text book, you will see that additional discrepancies occur among the elements beyond the fourth energy level whose configurations end in d or f electrons. The closeness in energy of the sublevels allows easy transition between levels for the electrons. The orbitals are said to undergo perturbation (their energy changes slightly) and the Aufbau Principle is not perfectly followed when this occurs. Electron configurations can be shown by either of two methods as illustrated in the next example. Remember that the sum of the superscripts **must** a to the total number of electrons in the species.

Example 5-8 Electron Configurations

Write out the electron configuration of silicon. Is silicon diamagnetic or paramagnetic?

Plan

Silicon has an atomic number of 14 and, therefore, has 14 electrons. We will use either the mnemonic device or the periodic table to determine the configuration. We will use Hund's rule and see if all the electrons are paired.

Solution

Method 1: $_{14}$Si: $\boxed{1s^2 2s^2 2p^6 3s^2 3p^2}$

Method 2: $_{14}$Si: $\underset{1s}{\uparrow\downarrow}\ \underset{2s}{\uparrow\downarrow}\ \underset{2p}{\uparrow\downarrow\ \uparrow\downarrow\ \uparrow\downarrow}\ \underset{3s}{\uparrow\downarrow}\ \underset{3p}{\uparrow\ \uparrow\ \_}$

In method 2, each arrow represents an electron. Pairs of arrows pointing in opposite directions represent two electrons in the indicated atomic orbital with paired spins. Because there are two unpaired electrons, silicon is $\boxed{\text{paramagnetic}}$.

Example 5-9 Electron Configurations

What is the electron configuration for vanadium? Is vanadium paramagnetic or diamagnetic?

Plan

Vanadium has an atomic number of 23 and, therefore, has 23 electrons.

Solution

$_{23}$V = $\boxed{1s^2 2s^2 2p^6 3s^2 3p^6 4s^2 3d^3}$ or $\boxed{[Ar] 4s^2 3d^3}$ or

[Ar]$\underset{4s}{\uparrow\downarrow}\ \underset{3d}{\uparrow\ \uparrow\ \uparrow\ \_\ \_}$

Vanadium is $\boxed{\text{paragmagnetic}}$ because it has 3 unpaired electrons.

Attention is usually focused on the outermost electrons of an atom since those electrons are involved in chemical bonding. To emphasis the type and number of outer electrons, the core electrons are often abbreviated as **[noble gas]**, where [noble gas] stands for the electron configuration of that noble gas. Noble gases are used as the shorthand because they share the common characteristic that all except helium have completely filled outer s and p sublevels. (Helium has only a filled s sublevel because it has no p electrons.) These configurations lend exceptional stability to the noble gases and account for their relative chemical inertness.

Example 5-10 Shorthand Electron Configurations

Use the noble gas shorthand to write out a method 1 electron configuration for the elements with atomic numbers 19, 25, 27, 34, 40, 50, 62, 86, 89 and 97.

Plan

We will follow the same techniques as in previous examples, letting the noble gas stand for as many electrons as possible.

Solution

$_{19}$K: $[Ar]4s^1$

$_{25}$Mn: $[Ar]4s^23d^5$

$_{27}$Co: $[Ar]4s^23d^7$

$_{34}$Se: $[Ar]4s^23d^{10}4p^4$

$_{40}$Zr: $[Kr]5s^24d^2$

$_{50}$Sn: $[Kr]5s^24d^{10}5p^2$

$_{62}$Sm: $[Xe]6s^25d^14f^5$

$_{86}$Rn: $[Xe]6s^25d^{10}4f^{14}6p^6$ or $[Rn]$

$_{89}$Ac: $[Rn]7s^26d^1$

$_{97}$Bk: $[Rn]7s^26d^15f^8$

Example 5-11 Electron Configurations and Quantum Numbers

Write an acceptable set of four quantum numbers to describe each electron in a boron atom.

Plan

Boron has atomic number 5 and, therefore, has 5 electrons. We will follow the filling order as predicted by one of the methods above. We will also correlate this with our knowledge of what the quantum numbers represent.

Solution

Electron	n	ℓ	m_ℓ	m_s	Configuration
1	1	0	0	+1/2	$1s^2$
2	1	0	0	-1/2	

Electron	n	ℓ	m_ℓ	m_s	Configuration
3	2	0	0	+1/2	
4	2	0	0	-1/2	$2s^2$
5	2	1	-1	+1/2	$2p^1$

Example 5-12 Electron Configurations and Quantum Numbers

Write an acceptable set of four quantum numbers that describe each electron in a nickel atom.

Plan

Ni has atomic number 28 and, thus, has 28 electrons. We will follow the filling order as predicted by one of the methods above. We will also correlate this with our knowledge of what the quantum numbers represent.

Solution

Electron	n	ℓ	m_ℓ	m_s	Configuration
1,2	1	0	0	±1/2	$1s^2$
3,4	2	0	0	±1/2	$2s^2$
5,8	2	1	-1	±1/2	
6,9	2	1	0	±1/2	
7,10	2	1	+1	±1/2	$2p^6$
11,12	3	0	0	±1/2	$3s^2$
13,16	3	1	-1	±1/2	
14,17	3	1	0	±1/2	
15,18	3	1	+1	±1/2	$3p^6$
19,20	4	0	0	±1/2	$4s^2$
21,26	3	2	-2	±1/2	
22,27	3	2	-1	±1/2	
23,28	3	2	0	±1/2	
24	3	2	1	+1/2	
25	3	2	+2	+1/2	$3d^8$

EXERCISES

Atomic Number, Mass Number, and Isotopes

1. Write the complete nuclide symbol for the atoms or ions containing the following numbers of subatomic particles.

	protons	electrons	neutrons
(a)	5	5	6
(b)	9	10	10
(c)	14	14	16
(d)	16	18	20
(e)	25	22	30
(f)	33	36	42
(g)	38	36	46
(h)	47	46	68
(i)	58	58	82
(j)	65	63	90
(k)	72	72	97
(l)	85	86	124
(m)	88	88	134
(n)	92	86	143
(o)	100	100	158

2. How many electrons, neutrons and protons are contained in each of the following atoms or ions?

   (a) $^{191}Au^{3+}$ (b) $^{224}Fr^+$ (c) ^{143}La (d) $^{74}Ga^{3+}$
   (e) ^{234}Th (f) $^{18}O^{2-}$ (g) $^{95}Sr^{2+}$ (h) $^{74}Se^{2-}$

Atomic Weights

3. Naturally occuring germanium consists of the following isotopes: ^{70}Ge(69.9243 amu, 20.52%), ^{72}Ge(71.9217 amu, 27.43%), ^{73}Ge (72.9234 amu, 7.76%), ^{74}Ge (73.9212 amu, 36.54%), and ^{76}Ge (75.9214 amu, 7.76%). What is the atomic weight of germanium?

4. Naturally occuring titanium consists of the following isotopes: ^{46}Ti (45.95623 amu; 7.93%), ^{47}Ti (46.9518 amu; 7.28%), ^{48}Ti (47.94795 amu; 73.94%), ^{49}Ti (48.94787 amu; 5.51%) and ^{50}Ti (49.9448 amu; 5.34%). What is its atomic weight?

5. The atomic weight of europium is 151.965 amu. Naturally occuring europium consists of two isotopes, ^{153}Eu (152.9209 amu) and ^{151}Eu (150.9196 amu). What is the percent abundance of each of the two isotopes?

6. The atomic weight of silver is 107.87 amu. Naturally occuring silver consists of two isotopes, ^{107}Ag (106.9041 amu) and ^{109}Ag (108.9047). What is the percent abundance of each of the two isotopes?

Electromagnetic Radiation and Energy

7. Convert the following wavelengths to frequencies.
   (a) 4.22 pm
   (b) 5.26 nm
   (c) 100 Å

8. Convert the following frequencies to wavelengths.
   (a) $3.79 \times 10^{15} s^{-1}$
   (b) $2.41 \times 10^{18} s^{-1}$
   (c) $5.01 \times 10^{13} s^{-1}$

9. Calculate the frequency and wavelength of light having the following energies.
   (a) 4.38×10^{-19} J/photon
   (b) 383.6 kJ/mo
   (c) 8.72 eV/photon

10. One of the wavelengths absorbed by neon atoms is 4.33×10^{-7} m. Convert this wavelength to angstroms and to nanometers? How large is the energy separation (in J/atom and in kJ/mol of atoms) of the electronic energy levels between which the electron moves during this absorption?

11. The emission spectrum of gold shows a line at 267.6 nanometers. What is the wavelength in centimeters? How much energy is emitted by the excited electron as it falls to a lower energy level (in J/atom and kJ/mol of atoms)?

Electron configurations and Quantum Numbers

12. Show complete electron configurations for the following atoms. Use both methods described in this book. Tell if each is paramagnetic or diamagnetic.
    (a) Be (b) O (c) S (d) Mn (e) Sn (f) Hg (g) Re (h) Bi

13. Show the outermost electron configuration in condensed form for each of the atoms below. Example: As = [Ar] $4s^2 3d^{10} 4p^3$.
    (a) B (b) Mg (c) Cl (d) K (e) As (f) Br (g) Y
    (h) Tc (i) Cd (j) Te (k) Ir (l) Es (m) Ra (n) At

14. Write an acceptable set of the four quantum numbers (n, ℓ, m_ℓ, m_s) for the electrons in the following atoms.

   (a) Ar   (b) Sc   (c) As

Miscellaneous Exercises

15. The following sets of quantum numbers describe electrons of some neutral elements. Some sets are impossible. Which ones? Why?

	n	ℓ	m_ℓ	m_s
(a)	1	0	0	+1/2
(b)	3	2	-3	-1/2
(c)	5	3	-2	-1/2
(d)	4	4	0	+1/2
(e)	6	3	2	-1/2
(f)	5	3	+1	0

16. Which **elements** have the ground states that is described by the following electron configurations? Some of the following configurations are invalid for an element in the ground state or are otherwise not acceptable combinations. Which ones are incorrect and why?

    (a) $1s^2 2s^2$ (b) $1s^2 2s^2 2p^4$ (c) $1s^2 2s^3 2p^4$

    (d) $1s^2 2s^1 3p^5$ (e) $1s^2 2s^2 2p^6 3s^2 3p^1 3d^2$ (f) $1s^2 2s^2 2p^6 3s^2 3p^5$

    (g) $1s^2 2s^2 2p^6 3s^2 3p^6 3d^8$ (h) $1s^2 2s^2 2p^6 3s^2 3p^6 4s^2 3d^{10} 4p^6 5s^2 4d^{10} 5p^6 6s^1$

    (i) $1s^2 2s^2 2p^6 3s^2 3p^6 4s^2 3d^{10} 4p^6 5s^2 4d^{10} 5p^6 6s^2 5d^1 4f^{14}$

17. An isotope of uranium has 140 neutrons.

    (a) Write the nuclide designation of the ion of this isotope that has only 89 electrons.

    (b) What is the ground state electron configuration of uranium?

18. Naturally occurring lithium exists as two isotopes ^{6}Li (mass 6.01513 amu) and ^{7}Li (mass 7.01601 amu).

    (a) What is the atomic weight of lithium?

    (b) How many neutrons are in each isotope of lithium?

    (c) Calculate the natural percent abundance of each isotope.

19. The following ground state electron configurations are exceptions to the normal Aufbau filling order. What element does each configuration represent?

    (a) [Xe]$6s^1 4f^{14} 5d^{10}$ (b) [Kr]$4d^{10}$ (c) [Ar]$4s^1 3d^{10}$

ANSWERS TO EXERCISES

1. (a) $^{11}_{5}B$ (b) $^{19}_{9}F^-$ (c) $^{30}_{14}Si$ (d) $^{36}_{16}S^{2-}$ (e) $^{55}_{25}Mn^{3+}$

   (f) $^{75}_{33}As^{3-}$ (g) $^{84}_{38}Sr^{2+}$ (h) $^{115}_{47}Ag^+$ (i) $^{140}_{58}Ce$ (j) $^{155}_{65}Tb^{2+}$

   (k) $^{169}_{72}Hf$ (l) $^{209}_{85}At^-$ (m) $^{222}_{88}Rn$ (n) $^{235}_{92}U^{6+}$ (o) $^{258}_{100}Fm$

2.
	protons	electrons	neutrons
(a)	79	76	112
(b)	87	86	137
(c)	57	57	86
(d)	31	28	43
(e)	90	90	144
(f)	8	10	10
(g)	38	36	57
(h)	34	36	40

3. 72.64 amu   4. 47.88 amu   5. 52.22% ^{153}Eu, 47.78% ^{151}Eu

6. 51.73% ^{107}Ag, 48.27% ^{109}Ag

7. (a) 7.10×10^{27} s^{-1} (b) 5.70×10^{16} s^{-1} (c) 3.00×10^{16} s^{-1}

8. (a) 7.91×10^{-8} m (b) 1.24×10^{-10} m (c) 5.98×10^{-6} m

9. (a) $\nu = 6.61 \times 10^{14}$ s^{-1}, $\lambda = 4.54 \times 10^{-7}$ m (b) $\nu = 9.612 \times 10^{14}$ s^{-1}, $\lambda = 3.119 \times 10^{-7}$ m
   (c) $\nu = 2.10 \times 10^{15}$ s^{-1}, $\lambda = 1.43 \times 10^{-7}$ m

10. 4.33×10^3 Å, 433 nm, $\Delta E = 4.59 \times 10^{-19}$ J, $\Delta E = 276$ kJ/mol

11. 2.676×10^{-5} cm, $E = 7.428 \times 10^{-19}$ J/atom, 447.3 kJ/mol

12. (a) $_4$Be: $1s^2 2s^2$ or ↑↓ ↑↓   (diamagnetic)
                                1s   2s

    (b) $_8$O: $1s^2 2s^2 2p^4$ or ↑↓ ↑↓ ↑↓ ↑ ↑   (paramagnetic)
                                   1s 2s   2p

    (c) $_{16}$S: $1s^2 2s^2 2p^6 3s^2 3p^4$ or [Ne] ↑↓ ↑↓ ↑ ↑   (paramagnetic)
                                                    3s   3p

(d) $_{25}$Mn: $1s^22s^22p^63s^23p^64s^23d^5$ or [Ar]$\uparrow\downarrow$ $\underset{3d}{\uparrow\ \uparrow\ \uparrow\ \uparrow\ \uparrow}$ (paramagnetic)
$\phantom{(d)\ _{25}Mn:\ 1s^22s^2}\underset{4s}{}$

(e) $_{50}$Sn: $1s^22s^22p^63s^23p^64s^23d^{10}4p^65s^24d^{10}5p^2$ or   (paramagnetic)

[Kr]$\underset{5s}{\uparrow\downarrow}$ $\underset{4d}{\uparrow\downarrow\ \uparrow\downarrow\ \uparrow\downarrow\ \uparrow\downarrow\ \uparrow\downarrow}$ $\underset{5p}{\uparrow\ \uparrow\ \_}$

(f) $_{80}$Hg: $1s^22s^22p^63s^23p^64s^23d^{10}4p^65s^24d^{10}5p^66s^24f^{14}5d^{10}$ or

[Xe]$\underset{6s}{\uparrow\downarrow}$ $\underset{4f}{\uparrow\downarrow\ \uparrow\downarrow\ \uparrow\downarrow\ \uparrow\downarrow\ \uparrow\downarrow\ \uparrow\downarrow\ \uparrow\downarrow}$ $\underset{5d}{\uparrow\downarrow\ \uparrow\downarrow\ \uparrow\downarrow\ \uparrow\downarrow\ \uparrow\downarrow}$ (diamagnetic)

(g) $_{75}$Re: $1s^22s^22p^63s^23p^64s^23d^{10}4p^65s^24d^{10}5p^66s^24f^{14}5d^5$ (paramagnetic)

[Xe]$\underset{6s}{\uparrow\downarrow}$ $\underset{4f}{\uparrow\downarrow\ \uparrow\downarrow\ \uparrow\downarrow\ \uparrow\downarrow\ \uparrow\downarrow\ \uparrow\downarrow\ \uparrow\downarrow}$ $\underset{5d}{\uparrow\ \uparrow\ \uparrow\ \uparrow\ \uparrow}$

(h) $_{83}$Bi: $1s^22s^22p^63s^23p^64s^23d^{10}4p^65s^24d^{10}5p^66s^24f^{14}5d^{10}6p^3$ or

[Xe]$\underset{6s}{\uparrow\downarrow}$ $\underset{4f}{\uparrow\downarrow\ \uparrow\downarrow\ \uparrow\downarrow\ \uparrow\downarrow\ \uparrow\downarrow\ \uparrow\downarrow\ \uparrow\downarrow}$ $\underset{5d}{\uparrow\downarrow\ \uparrow\downarrow\ \uparrow\downarrow\ \uparrow\downarrow\ \uparrow\downarrow}$ $\underset{6p}{\uparrow\ \uparrow\ \uparrow}$ (paramagnetic)

13. (a) $_5$B: [He]$2s^22p^1$ (b) $_{13}$Al: [Ne]$3s^23p^1$ (c) $_{17}$Cl: [Ne]$3s^23p^5$
    (d) $_{20}$Ca: [Ar]$4s^2$ (e) $_{31}$Ga: [Ar]$4s^23d^{10}4p^1$ (f) $_{35}$Br: [Ar]$4s^23d^{10}4p^5$
    (g) $_{39}$Y: [Kr]$5s^24d^1$ (h) $_{43}$Tc: [Kr]$5s^24d^5$ (i) $_{48}$Cd: [Kr]$5s^24d^{10}$
    (j) $_{52}$Te: [Kr]$5s^24d^{10}5p^4$ (k) $_{77}$Ir: [Xe]$6s^24f^{14}5d^7$ (l) $_{99}$Es: [Rn]$7s^26d^15f^{10}$
    (m) $_{88}$Ra: [Rn]$7s^2$ (n) $_{85}$At: [Xe]$6s^24f^{14}5d^{10}6p^5$

14. (a) $_{18}$Ar

# of electron	n	ℓ	m_ℓ	m_s
1,2	1	0	0	±1/2
3,4	2	0	0	±1/2
5,8	2	1	-1	±1/2
6,9	2	1	0	±1/2
7,10	2	1	+1	±1/2
11,12	3	0	0	±1/2
13,16	3	1	-1	±1/2
14,17	3	1	0	±1/2
15,18	3	1	+1	±1/2

(b) $_{21}$Sc, the first 18 electrons are the same as for Ar.

# of electron	n	ℓ	m_ℓ	m_s
19,20	4	0	0	±1/2
21	3	2	-2	+1/2

(c) $_{32}$Ge, the first 18 electrons are the same as for Ar.

# of electron	n	ℓ	m_ℓ	m_s
19,20	4	0	0	±1/2
21,26	3	2	-2	±1/2
22,27	3	2	-1	±1/2
23,28	3	2	0	±1/2
24,29	3	2	+1	±1/2
25,30	3	2	+2	±1/2
31	4	1	-1	+1/2
32	4	1	0	+1/2

15. (b) impossible because $m_\ell < -1$.
    (d) impossible because $\ell > n - 1$.
    (f) impossible because m_s for an electron must be either +1/2 or -1/2.

16. (a) Be (b) O (c) impossible, s only holds 2 e-
    (d) not a ground state configuration, because the 2s sublevel should be filled before 3p begins filling
    (e) not a ground state configuration, because the 3p and 4s sublevels should be filled before 3d begins filling
    (f) Cl
    (g) not a ground state configuration, because the 4s sublevel should be filled before 3d begins filling
    (h) Cs (i) Lu

17. (a) ^{232}U^{3+} (b) [Rn] 7s^{2}6d^{1}5f^3

18. (a) The atomic weight of lithium is 6.941 amu.
    (b) Lithium has 3 protons. ^{6}Li has 6 - 3 = 3 neutrons; ^{7}Li has 7 - 3 = 4 neutrons
    (c) 7.49% ^{6}Li, 92.51% ^{7}Li

19. (a) $_{79}$Au (b) $_{46}$Pd (c) $_{29}$Cu

Chapter 6: Chemical Periodicity

6-1 The Periodic Table

Review the discussion of the periodic table in Chapter 4. Groups can now be discussed in terms of their electron configurations. The "A" Groups, also known as **representative elements** have outermost electrons in the **s** and **p** sublevels. The **noble gases** have filled **s** and **p** sublevels in their highest occupied energy level ($1s^2$ for helium and ns^2np^6 for the other noble gases). The "B" Groups, or **d-transition elements**, have outermost electrons in **d** sublevels. The lanthanides and actinides, or **f-transition elements**, have outermost electrons filling **f** sublevels.

6-2 Atomic Radii

The radii of neutral atoms decrease from left to right across a period. This happens because the increasing positive nuclear charge attracts the outer shell electrons more strongly, pulling them in closer to the nucleus. The radii increase from top to bottom within groups as electrons occupy energy levels further from the nucleus. See Figure 6-1 in the text.

Example 6-1 Trends in Atomic Radii

Arrange the following elements in order of increasing atomic radii: Li, C, Be, O, N.

Plan
These are all elements of the second period and their radii decrease from left to right in the periodic table.

Solution
Increasing Radius: $\boxed{O < N < C < Be < Li}$

Example 6-2 Trends in Atomic Radii

Arrange the following elements in order of increasing atomic radii: Br, F, I, Cl.

Plan

These are all in the halogen family. The radius increases down the group as higher energy levels, further from the nucleus, are being filled.

Solution

Increasing Radius: $F < Cl < Br < I$

Example 6-3 Atomic Volume

Which of the following elements has the smallest atomic volume and which has the largest?
Na, Cl, Xe, Rb, Te

Plan

The volume of a sphere is $4/3\pi r^3$, where r is the radius. We compare within rows and periods, considering the trends discussed above.

Solution

Cl has the smallest radius and smallest volume.
Rb has the largest radius and volume.

6-3 Ionization Energy

The amount of energy required to remove the most loosely held electron from an isolated, neutral gaseous atom is called the first ionization energy of the element. The first ionization energy can be represented by the following equation:

$$\text{Atom}°\ (g) + \text{Ionization Energy} \rightarrow \text{Ion}^+\ (g) + e^-$$

Higher ionization energy means that it is more difficult to remove an electron. In general, first ionization energies increase from left to right across a period and from bottom to top within a group. This parallels the trend for decreasing radii for "A" group elements. In a smaller atom, the electrons are held more tightly by the nucleus and, thus, are harder to remove. The very stable electron configurations of the noble gases are difficult to disrupt, and, therefore, the noble gas will

Chapter 6 Chemical Periodicity

have the highest first ionization energy of any element in the period. Group IA elements have the most loosely held electrons and, consequently, the lowest first ionization energies in the period.

Exceptions to these trends occur at the IIIA and VIA elements of the first four periods. The IIIA and VIA elements have slightly lower ionization energies than the preceding elements from IIA and VA. This occurs because the IIIA elements have only one electron in their outermost p sublevel, which is more easily removed than one of the electrons in the filled s sublevel in the same shell. The VIA elements have two paired and two unpaired electrons in their outermost p sublevel. Removal of one of the paired electrons to produce a relatively stable half-filled p sublevel requires less energy than removal of an unpaired electron of a VA element, which disrupts a half-filled p sublevel.

Example 6-4 Trends in First Ionization Energies

Arrange the Group VA elements in order of increasing first ionization energy.

Plan
Ionization energy generally decreases as we go down a group.

Solution
Increasing First Ionization Energy: $\boxed{Bi < Sb < As < P < N}$

Example 6-5 Trends in First Ionization Energies

Arrange the following elements in order of increasing first ionization energy:
K, Ca, Ga, Ge, As, Br, Se, Kr.

Plan
We find the elements on the periodic table and apply the tendencies for trends. Note that Ga from Group IIIA has a lower first ionization energy than Ca from Group IIA, while Se from Group VIA has approximately the same value as As from Group VA for reasons of exceptions to the general trends as cited above. (See Table 6-1 in the text.)

Solution
Increasing first ionization energy: $\boxed{K < Ga < Ca < Ge < As < Se < Kr}$

Subsequent removal of additional electrons requires more and more energy to be expended. Second ionization energies are higher than first ionization energies because the remaining electrons are more tightly held by the higher effective nuclear charge. Third ionization energies are higher than second for the same reason.

Example 6-6 Ionization Energies

Match these experimental ionization energies of 2.62×10^3 kJ/mol, 717 kJ/mol, 5.02×10^3 kJ/mol and 1558 kJ/mol with the species Mn^{2+}, Mn, Mn^{3+}, and Mn^+.

Plan

A species which has lost more electrons (more positive) will hold on to its remaining electrons more tightly and, thus, have a higher ionization energy.

Solution

| Mn = 717 kJ/mol | Mn^+ = 1558 kJ/mol | Mn^{2+} = 2.62×10^3 kJ/mol |

| Mn^{3+} = 5.02×10^3 kJ/mol |

6-4 Electron Affinity

Electron affinity is defined as the amount of energy involved when an electron is added to a neutral gaseous atom. Electron affinity is assigned a negative value when energy is released.

$$Atom°(g) + e^- \rightarrow Ion^-(g) + \text{Electron Affinity}$$

Highly negative electron affinities are associated with atoms having the greatest tendencies to form simple anions. There are several irregular variations in trends, but, in general, electron affinities become more negative from left to right across a period (with the major exceptions of noble gases and alkaline earth metals, which have high positive electron affinities) and from bottom to top within a group. See Table 6-2 in the text.

6-5 Ionic Radii

Simple cations always have smaller radii than the neutral atoms from which they are derived and, for a given element, the radius decreases with increasing positive charge. For example, the radii of Pb, Pb^{2+}, and Pb^{4+} are 1.75 Å, 1.21 Å, and 0.84 Å, respectively. This decrease is due to the reduction in screening of a constant positive nuclear charge by decreasing numbers of electrons. See Figure 6-1 in the text.

simple anions always have larger radii than the neutral atoms from which they are derived. For example, the radii of oxygen atoms and oxide, O^{2-}, ions are 0.66 Å and 1.40 Å, respectively. This increase is due to the greater number of electrons, which repel each other, but are attracted by a constant positive nuclear charge. See Figure 6-1 in the text.

For isoelectronic ions, such as N^{3-}, O^{2-}, F^-, Na^+, Al^{3+}, and Mg^{2+}, all of which have ten electrons, the species with the most negative charge has the largest radius. The N^{3-} has the fewest number of protons and, thus, has the smallest positive charge to attract the electrons. It has the largest radius. Al^{3+} has the smallest radius in this isoelectronic series. The most positive ion has the most protons per electron and has the smallest radius. Al^{3+} has the smallest radius in this isoelectronic series.

Example 6-7 Ionic Radii

Rank each of the following in terms of increasing radius.

(a) Fe^{2+}, Fe, Fe^{3+} (b) Cl^- or Cl (c) S^{2-} or Cl^- (d) Mg^{2+} or Mg

Plan
We will apply the rules for ion sizes as noted above.

Solution
(a) $\boxed{Fe^{3+} < Fe^{2+} < Fe}$ (ions made from the same atom, the most positive is the smallest)
(b) $\boxed{Cl < Cl^-}$ (the anion is larger than the atom from which it was made)
(c) $\boxed{Cl^- < S^{2-}}$ (isoelectronic, more negative is larger)
(d) $\boxed{Mg^{2+} < Na^+}$ (isoelectronic, more positive is smaller)

6-6 Electronegativity

The electronegativity of an element is a measure of its ability to attract electrons to itself when bonding with another element. It can be thought of as a rough average of ionization energy and electron affinity. Elements with the greatest attraction for electrons in chemical bonds have the highest electronegativities, which in general increase from left to right across periods (again, with the major exception of the noble gases which have very little tendency to enter into chemical bonds) and from bottom to top within groups. See Table 6-3 in the text.

Example 6-8 Trends in Electronegativity

Arrange the Group IIA elements in order of increasing electronegativity.

Plan
Electronegativity decreases down a group.

Solution
Increasing Electronegativity: $\boxed{Ra < Ba < Sr < Ca < Mg < Be}$

Example 6-9 Trends in Electronegativity

Arrange the following elements in order of increasing electronegativity: Po, Pb, Tl, Cs, Ba.

Plan
These are elements of the sixth period. Electronegativity increases left to right across a period.

Solution
Increasing Electronegativity: $\boxed{Cs < Ba < Tl < Pb < Po}$

Example 6-10 Trends in Electronegativity

Arrange the following elements in order of increasing electronegativity: F, C, Be, Rb, Li.

Plan
These elements are from different parts of the periodic table, but we can still determine the correct order by comparing their relative positions.

Solution
Increasing Electronegativity: $\boxed{Rb < Li < Be < C < F}$

6-7 Hydrides and Oxides of Elements

Many substances combine with hydrogen to produce hydrides. In general, a more electronegative, less metallic element will yield a hydride with more molecular character and more acidic properties. A less electronegative, more metallic element will yield an hydride with more ionic character and more basic properties.

Most substances combine with oxygen to produce oxides. In general, a more electronegative, less metallic element will yield an oxide with more molecular character and more acidic properties. A less electronegative, more metallic element will yield an oxide with more ionic character and more basic properties.

Example 6-11 Acidity of Oxides

Rank these compounds in order of increasing acidic properties.
SrO, SeO_3, Ga_2O_3, Br_2O_7

Plan
All of these elements combined with oxygen are in the fourth period. They become more electronegative and, thus, less metallic and the oxides more acidic from left to right.

Solution
Increasing acidity: $\boxed{SrO < Ga_2O_3 < SeO_3 < Br_2O_7}$

Inspection of the periodic chart reveals that the following general tendencies are true:

1. Non-metals tend to have negative electron affinities, high ionization potentials and high electronegativities. Because of this they have a tendency to become anions. They tend to make molecular, acidic oxides.

2. Metals tend to have less negative or positive electron affinities, low ionization potentials and low electronegativities. Because of this, they have a tendency to lose electrons and become cations. They tend to make ionic, basic oxides and ionic, basic hydrides.

3. Noble gases have positive electron affinities and high ionization potentials and, thus, have no tendency to create ions. They are less likely to become involved in chemical bonding because their electron configurations are extremely stable without any transfer or sharing of electrons.

EXERCISES

Trends on the Periodic Chart

1. Arrange the members of the following groups in order of increasing atomic radius.
   (a) Rb, Y, I, Sn, Sr
   (b) N, As, Bi, Sb, P
   (c) H, O, Cl, F, I
   (d) Sr, Zr, Mo, Rb, Ru

2. Arrange the members of the following groups in order of increasing ionic radii.
   (a) O^{2-}, Se^{2-}, Te^{2-}, S^{2-}
   (b) Tl^{3+}, Ga^{3+}, Al^{3+}, In^{3+}
   (c) Cr^{3+}, Cr^{+}, Cr^{4+}, Cr^{2+}
   (d) Ca^{2+}, S^{2-}, P^{3-}, K^{+}

3. Arrange the following species in order of increasing atomic or ionic volume.
   (a) Mg, Mg^{2+}, Na
   (b) Cu, Cu^{+}, Cu^{2+}
   (c) I^{-}, Ba^{2+}, La^{3+}, Te^{2-}

4. Arrange the members of the following groups of elements in order of increasing first ionization energy.
   (a) C, Ge, Si, Pb, Sn
   (b) Ar, P, Na, Al, Cl
   (c) Li, Be, B, C, N, O, F
   (d) Ba, Cs, Pb, Po, Tl

5. Explain any deviations in the general trends observed in Exercise 4.

6. Arrange the members of the following groups of elements in order of increasing electronegativity.
   (a) Li, Be, B, C, N, O, F
   (b) Ge, Ga, Ca, K, Br
   (c) As, P, S, Ge, F
   (d) I, Sn, Te, Sb, Sr

7. Arrange the members of each group in terms of increasing acidic character.
   (a) SO_3, TeO_3, SeO_3
   (b) MgO, SO_3, P_4O_{10}
   (c) N_2O_5, Sb_2O_5, As_2O_5

8. Which of the following hydrides are likely to be basic?
   (a) KH
   (b) CH_4
   (c) MgH_2

ANSWERS TO EXERCISES

1. (a) I < Sn < Y < Sr < Rb
   (b) N < P < As < Sb < Bi
   (c) H < F < O < Cl < I
   (d) Ru < Mo < Zr < Sr < Rb

2. (a) $O^{2-} < S^{2-} < Se^{2-} < Te^{2-}$
   (b) $Al^{3+} < Ga^{3+} < In^{3+} < Tl^{3+}$
   (c) $Cr^{4+} < Cr^{3+} < Cr^{2+} < Cr^{+}$
   (d) $Ca^{2+} < K^{+} < S^{2-} < P^{3-}$

3. (a) $Mg^{2+} < Mg < Na$
   (b) $Cu^{2+} < Cu^{+} < Cu$
   (c) $La^{3+} < Ba^{2+} < I^{-} < Te^{2-}$

4. (a) Pb < Sn < Ge < Si < C
   (b) Na < Al < P < Cl < Ar
   (c) Li < B < Be < C < O < N < F
   (d) Cs < Ba < Tl < Pb < Po

5. Generally, first ionization energies increase from left to right across a period and from bottom to top within a group. In 4(c), the ionization energy of B is less than that of Be. This is because the energy required to remove the most loosely held electron, the first one in a set of **p** orbitals, in the Group IIIA elements, is less than that for the IIA elements, which must lose an electron from a filled s orbital. Completely filled sets of orbitals are relatively stable configurations. Additionally, the s electron is closer to the nucleus and more tightly held, making it more difficult to remove.

6. (a) Li < Be < B < C < N < O < F
   (b) K < Ca < Ga < Ge < Br
   (c) Ge < As < P < S < F
   (d) Sr < Sn < Sb < Te < I

7. (a) $TeO_3 < SeO_3 < SO_3$
   (b) $MgO < P_4O_{10} < SO_3$
   (c) $Sb_2O_5 < As_2O_5 < N_2O_5$

8. KH and MgH_2, the metal hydrides, are most likely to be basic.

Chapter 7: Chemical Bonding

7-1 Lewis Dot Representation of Atoms

Variations of properties are much more regular among the representative elements than among the d- or f- transition elements, so attention will be focused on the "A" group elements. It is often convenient to describe the electron configurations of these elements with Lewis formulas that show only the outermost **s** and **p** electrons. These electrons are called **valence electrons** and are responsible for chemical bonding. The Lewis formulas of the different representative elements are shown in Table 7-1 in the text. Note that the number of valence electrons is the same as the "A" group number. For bonding purposes, the noble gases are treated as though they are in group VIIIA. (Helium, with only two electrons, has a Lewis formula with just two dots.)

7-2 Chemical Bonding

The attractive forces between atoms or ions in compounds are called chemical bonds. There are two major classes of bonds, ionic and covalent. Ionic bonds are formed by the transfer of one or more electrons from one atom or group of atoms to another atom or group of atoms. Covalent bonds are when two atoms share one or more electron pairs. Compounds that contain predominantly ionic bonds are called ionic compounds, whereas those with primarily covalent bonds are called covalent compounds.

7-3 Formation of Ionic Compounds

Ionic bonds form when there is a sufficient electronegativity difference between elements. Metals, at the extreme left of the periodic table, lose electrons easily to form positive ions, while nonmetals, located at the right of the periodic table, gain electrons easily to form negative ions. In each case, the ions formed by representative elements tend to be isoelectronic with a noble gas. The ions that are formed then combine in proper ratios to form a neutral compound.

Potassium (Group IA) and fluorine (Group VIIA) react explosively to form the ionic solid, potassium fluoride.

$$2K(s) + F_2(g) \rightarrow 2KF(s) + heat$$

This reaction involves molecular fluorine rather than atomic fluorine; however, it is useful to consider the reaction in terms of the Lewis formulas of the atoms.

$$K\cdot + :\ddot{F}\cdot \rightarrow K^+ \left[:\ddot{F}:\right]^-$$

Both potassium and fluorine achieve noble gas configurations in this transfer as follows:

K: [Ar]↑   →   K$^+$:   [Ar]
     4s

F: [He]↑↓ ↑↓ ↑↓ ↑   →   F$^-$: [He]↑↓ ↑↓ ↑↓ ↑↓ = [Ne]
        2s   2p   2s   2p

The potassium ions are isoelectronic with Ar atoms (18e$^-$), while the fluoride ions are isoelectronic with Ne atoms (10e$^-$). Ions and compounds formed by elements in other groups can be determined in the same manner.

Example 7-1 Formulas of Ionic Compounds

Write the balanced equation for the reaction of calcium with bromine. Show the electron configurations and the Lewis formulas of the atoms and ions.

Plan
Both elements will form ions that are isoelectronic to a noble gas; calcium by losing two electrons and each bromine atom by gaining one electron. The compound must then consist of two bromide ions for every one calcium ion. The net reaction is $Ca + Br_2 \rightarrow CaBr_2$.

Solution
Ca: [Ar]↑↓   → Ca^{2+}: [Ar]
         4s

2Br: [Ar]↑↓ 3d^{10} ↑↓ ↑↓ ↑   → Br$^-$: [Ar]↑↓ 3d^{10} ↑↓ ↑↓ ↑↓ = [Kr]
         4s   4p   4s   4p

$$Ca: + 2 :\ddot{Br}\cdot \rightarrow Ca^{2+}, 2\left[:\ddot{Br}:\right]^-$$

Example 7-2 Formulas of Ionic Compounds

Write the balanced equation for the reaction of magnesium and sulfur. Write the electron configurations and the Lewis formulas of the atoms and ions.

Plan

We will approach this in the same manner as in Example 7-1. The net reaction is:
$Mg + S \rightarrow MgS$.

Solution

Mg: [Ne]↑↓   → Mg^{2+}: [Ne]
      3s

S: [Ne]↑↓ ↑↓ ↑ ↑   → S^{2-}: [Ne]↑↓ ↑↓ ↑↓ ↑↓ = [Ar]
     3s   3p   3s   3p

Mg: + :S:• → Mg^{2+}, [:S:]$^{2-}$

Example 7-3 Formulas of Ionic Compounds

Write the balanced equation for the reaction of aluminum with oxygen. Show the electron configurations and the Lewis formulas of the atoms and ions.

Plan

We will solve this as in the previous examples, remembering that oxygen is diatomic. The net reaction will be: $4Al + 3O_2 \rightarrow 2Al_2O_3$.

Solution

4Al: [Ne]↑↓ ↑ _ _   → $4Al^{3+}$: [Ne]
      3s   3p

6O: [He]↑↓ ↑↓ ↑ ↑   → $3O^{2-}$: [He]↑↓ ↑↓ ↑↓ ↑↓ = [Ne]
     2s   2p   2s   2p

4 Al: + 6 :Ö• → $4Al^{3+}$, 6[:Ö:]$^{2-}$

Table 7-2 in the text summarizes the general formulas of the simple binary ionic compounds resulting from combinations of elements from different groups in the periodic table.

7-4 Lewis Formulas for Molecules and Polyatomic Ions

Covalent bonds are formed by the reaction of two or more nonmetals. Because the electronegativities are similar, transfer cannot take place. Instead, the electrons must be shared to achieve a stable configuration. If two electrons (one pair) are shared, the bond is called a **single covalent bond** or simply a **single bond**. Four electrons (two pairs) shared constitutes a **double bond** and six electrons (three pairs) shared constitutes a **triple bond**.

The **octet rule** states that most representative elements tend to achieve the s^2p^6 or **stable octet** configuration characteristic of the noble gases (except for helium) in their compounds. Elements in covalent compounds achieve the stable octet by sharing electrons. The relationship below can be used for compounds in which the octet rule is followed. The total number of electrons **shared** (S) equals the sum of the number of outer shell electrons **needed** (N) by all the atoms to attain a noble gas configuration minus the total number of electrons **available** (A) in valence shells.

$$S = N - A$$

The "A" group number gives the number of electrons **available** in the valence shell of a representative element. In polyatomic ions, one electron is subtracted from the number available for each positive charge and one electron is added to the number available for each negative charge. Eight electrons (the stable octet) are **needed** to attain a noble gas configuration (except for hydrogen, which only needs two electrons to achieve the helium $1s^2$ configuration). Therefore:

$$N = 8(\text{number of atoms except H}) + 2(\text{number of H atoms})$$

General Rules for Writing Lewis Formulas:

1. Select a reasonable "skeleton" for the molecule.
   (a) Generally, the least electronegative atom is the central atom and the structure is as symmetrical as possible.
   (b) Oxygen atoms rarely bond to each other except in O_2, O_3, and O_2^{2-} (peroxides).
   (c) Hydrogen usually bonds to oxygen in ternary acids, which are composed of oxygen, hydrogen, and some other element.
   (d) In ions or molecules with more than one central atom, the most symmetrical structure possible is used.
2. Calculate N, the number of electrons needed by all atoms to complete octets.
3. Calculate A, the number of valence electrons available.
4. Calculate S, the number of electrons shared from S = N - A.
5. Place S electrons into the skeleton as shared pairs, using multiple bonds only when necessary.
6. Distribute the remaining available electrons to complete the octet of every A group element except hydrogen which can only share two electrons.

Formal charges (FC) are a useful tool for determining a correct Lewis structure when several possibilities seem to exist for a molecule. The rules for assigning formal charges are as follows:

1. the sum of the formal charges in a molecule is equal to 0.
2. the sum of the formal charges in a polyatomic ion is equal to the charge on the ion.
3. the formal charge of an atom is equal to:

    FC = Group number - (number of bonds + number of unshared e⁻)

    (For noble gases, the group number is considered to be 8.)
4. Generally, the most favorable structure for a molecule is one in which the formal charge on each atom is zero.
5. When charges must be assigned, the more electronegative atom should be given the negative formal charge.
6. Adjacent atoms should not have the same formal charge.

In writing Lewis formulas, dashes are often used to indicate shared pairs of electrons and occasionally unshared or "lone" pairs of electrons are omitted.

Example 7-4 Lewis Formulas

Write the Lewis formula for carbon dioxide, CO_2.

Plan

We will follow the stepwise procedure given above. Formal charges will be used to determine which of several possible structures is more likely.

Solution

$N = 1 \times 8$(C atom) $+ 2 \times 8$ (O atoms) $= 24$ e⁻
$A = 1 \times 4$(C atom) $+ 2 \times 6$(O atoms) $= 16$ e⁻
$S = N - A = 24 - 16 = 8$ e⁻

$\ddot{\text{O}}::\text{C}::\ddot{\text{O}}$ or $\ddot{\text{O}}=\text{C}=\ddot{\text{O}}$ $\ddot{\text{O}}:\text{C}:::\text{O}:$ or $:\ddot{\text{O}}-\text{C}\equiv\text{O}:$

FC = Group # - (# bonds + # unshared electrons)
C: FC = 4 - 4 = 0.
Right O: FC = 6 - (2 + 4) = 0
Left O: FC = 6 - (2 + 4) = 0

C: FC = 4 - 4 = 0
Right O: FC = 6 - (1 + 6) = -1
Left O: FC = 6 - (3 + 2) = +1

The left hand structure is both more symmetrical and preferred on the basis of formal charges because all atoms have a formal charge of zero.

Example 7-5 Lewis Formulas

Write the Lewis formula for phosphine, PH_3, a poisonous, foul-smelling gas.

Plan
We will follow the stepwise procedure given above.

Solution
$N = 8 \times 1(\text{P atom}) + 2 \times 3(\text{H atoms}) = 14 \text{ e}^-$
$A = 5 \times 1(\text{P atom}) + 1 \times 3(\text{H atoms}) = 8 \text{ e}^-$
$S = N - A = 14 - 8 = 6 \text{ e}^-$

$$H:\overset{..}{\underset{H}{P}}:H \quad \text{or} \quad H-\overset{..}{\underset{H}{P}}-H$$

P: FC = 5 - (3 + 2) = 0   Each H: FC = 1 - 1 = 0

Example 7-6 Lewis Formulas

Write the Lewis formula for 1,1-dichloroethene, $C_2H_2Cl_2$, a gas used as a fumigant. The carbon atoms are bonded together, the two hydrogen atoms are bonded to one carbon, and the two chlorine atoms are bonded to the other carbon atoms.

Plan
We will follow the stepwise procedure given above. Formal charges will be used to determine which of several possible structures is more likely.

Solution
$N = 8 \times 2(\text{C atoms}) + 8 \times 2(\text{Cl atoms}) + 2 \times 2(\text{H atoms}) = 36 \text{ e}^-$
$A = 4 \times 2(\text{C atoms}) + 7 \times 2(\text{Cl atoms}) + 1 \times 2(\text{H atoms}) = 24 \text{ e}^-$
$S = N - A = 36 - 24 = 12 \text{ e}^-$

$$H-C=C-\overset{..}{\underset{..}{Cl}}: \atop \underset{H}{|} \underset{:\overset{..}{Cl}:}{|}$$

Each H: FC = 1 - 1 = 0

$$H-\overset{..}{\underset{..}{C}}-C=\overset{..}{\underset{..}{Cl}} \atop \underset{H}{|} \underset{:\overset{..}{Cl}:}{|}$$

Each H: FC = 1 - 1 = 0

Each C: FC = 4 - 4 = 0
Each Cl: FC = 7 - (1 + 6) = 0

Left C: FC = 4 - (3+2) = -1
Right C: FC = 4 - 4 = 0
Top Cl: FC = 7 - (1 + 6) = 0
Right Cl: FC = 7 - (2 + 4) = +1

Formal charges preclude the structure on the right because the formal charges are not all zero. Additionally, in the right hand structure, the more electronegative chlorine would have a positive formal charge while the less electronegative carbon would have a negative formal charge. This occurrence is not likely.

Example 7-7 Lewis Formulas

Draw the Lewis formula for the cyanate ion, OCN^-.

Plan
We will follow the stepwise procedure given above, adding one electron to the number available because of the ion charge. Formal charges will be used to determine which of several possible structures is more likely.

Solution

$N = 8 \times 1(O\ atom) + 8 \times 1(C\ atom) + 8 \times 1(N\ atom)$ = 24 e^-
$A = 6 \times 1(O\ atom) + 4 \times 1(C\ atom) + 5 \times 1(N\ atom) + 1\ (charge)$ = 16 e^-
$S = N - A = 24 - 16$ = 8 e^-

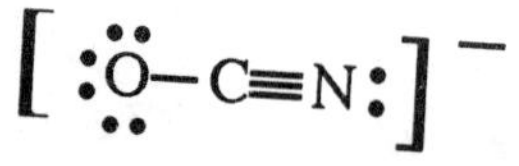

$[:\ddot{O}-C\equiv N:]^-$ $[:\ddot{O}=C=\ddot{N}:]^-$ $[:O\equiv C-\ddot{\ddot{N}}:]^-$

O: FC = 6 - (1 + 6) = -1
C: FC = 4 - 4 = 0
N: FC = 5 - (3 + 2) = 0

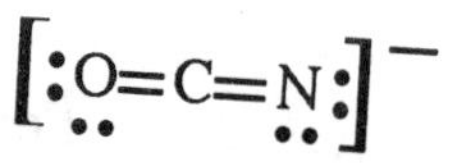

O: FC = 6 - (2 + 4) = 0
C: FC = 4 - 4 = 0
N: FC = 5 - (2 + 4) = -1

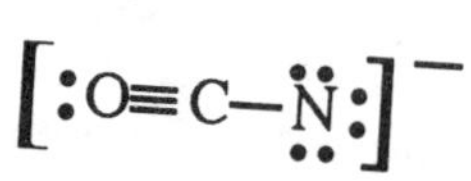

O: FC = 6 - (3+2) = +1
C: FC = 4 - 4 = 0
N: FC = 5 - (1+ 6) = -2

The middle structure is eliminated because the most electronegative atom (oxygen) does not have the negative formal charge. The right hand structure is eliminated for the same reason and, additionally, has fewer atoms with zero formal charge.

7-5 Resonance

Some molecules and complex ions cannot be represented by a single Lewis formula that accurately reflects their properties. Such species are said to exhibit **resonance**. Several structures called

onance structures can be drawn with the understanding that the real structure of the
lecule or ion is an average of the various resonance structures.

Example 7-8 Lewis Formulas, Resonance

Draw the three equivalent Lewis formulas for sulfur trioxide, SO_3. Sulfur is the central atom.

Plan
We will follow the stepwise procedure given above.

Solution

N = 8 × 1(S atom) + 8 × 3(O atoms)   = 32 e⁻
A = 6 × 1(S atom) + 6 × 3(O atoms)   = 24 e⁻
S = N - A   = 8 e⁻

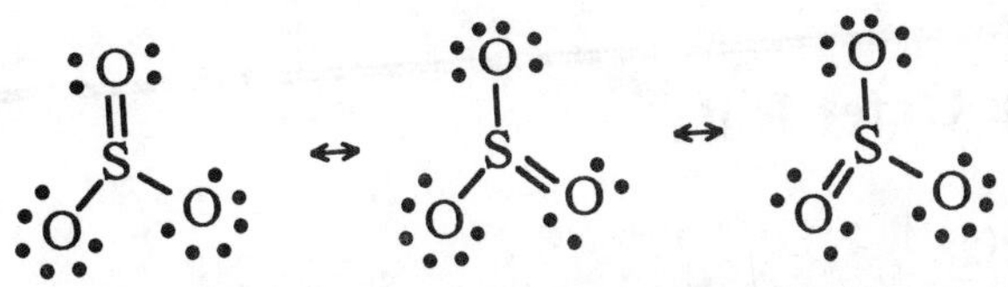

Experimental data indicate that there are no formal double or single bonds in SO_3 molecules. All bonds are identical and have one-third double bond character and two-third single bond character as if each were a 1 1/3 total bond.

Example 7-9 Lewis Formulas, Resonance

Draw the resonance structures for the nitrite ion, NO_2^-. Nitrogen is the central element.

Plan
We will follow the stepwise procedure given above, adding 1 electron to the number available because of the ion charge.

Solution

N = 8 × 1(N atom) + 8 × 3(O atoms)   = 32 e⁻
A = 5 × 1(N atom) + 6 × 3(O atoms) + 1(charge) = 24 e⁻
S = N - A = 32 - 24   = 8 e⁻

$$\left[\begin{array}{c}:\ddot{O}:\\ \cdot\cdot\diagup\diagdown\\ \cdot N\\ \diagdown\ddot{O}:\\ \cdot\cdot\end{array}\right]^{-} \leftrightarrow \left[\begin{array}{c}:\ddot{O}:\\ \cdot\cdot\diagup\diagdown\\ \cdot N\\ \diagdown\ddot{O}\cdot\\ \cdot\cdot\end{array}\right]^{-}$$

7-6 Limitations of the Octet Rule

The octet rule and, therefore, $S = N - A$ does not apply to the following special cases:
1. most covalent compounds of beryllium.
2. most covalent compounds of Group IIIA elements
3. compounds in which the central atom must have a share of more that eight electrons to accommodate all substituents
4. compounds containing **d**- or **f**- transition elements
5. species containing an odd number of electrons

Example 7-10 Limitations of the Octet Rule

Write the covalent Lewis formula for $BeBr_2$.

Plan

We will follow the stepwise procedure given above.

Solution

$N = 8 \times 1(\text{Be atom}) + 8 \times 2(\text{Br atoms}) = 24$ e$^-$
$\underline{A = 2 \times 1(\text{Be atom}) + 7 \times 2(\text{Cl atoms}) = 16}$ e$^-$
$S = N - A = 24 - 16 = 8$ e$^-$

The structure appears to be: $\ddot{B}r=Be=\ddot{B}r$

This is similar to the CO_2 structure in Example 7-4; however, inspection of the formal charges yields: Be: FC = 2 - 4 = -2   Each Br: FC = 7 - (2 + 4) = +1
Br is more electronegative than the Be and should have the negative formal charge, if either does. The accepted Lewis formula of $BeBr_2$ is:

$:\ddot{B}r-Be-\ddot{B}r:$

In this Lewis formula: Be: FC = 2 - 2 = 0   Each Br: FC = 7 - (1 + 6) = 0
Be generally forms just two single bonds in its covalent compounds.

Example 7-11 Limitations of the Octet Rule

Draw the Lewis formula of AsBr$_5$.

Plan
Inspection reveals that the central As must have five pairs of electrons around it to bond each of the bromine atoms. We will find the number of available electrons as usual, then use five pairs of electrons to bond the bromine atoms to the central atom. We place octets around the attached atoms. Finally, we place any additional electron pairs around the central atom.

Solution
A = 5 x 1(As atom) + 7 x 5(Br atoms) = 40 e$^-$

$$\begin{array}{c}
\ddot{\text{Br}} \\
| \\
\ddot{\text{Br}} - \text{As} - \ddot{\text{Br}} \\
| \\
\ddot{\text{Br}} \quad \ddot{\text{Br}}
\end{array}$$

As: FC = 5 - 5 = 0   Each Br: FC = 7 - (1 + 6) = 0

7-7 Polar and Nonpolar Covalent Bonds

Covalent bonds can be either nonpolar or polar. Nonpolar covalent bonds result from the equal sharing of electrons by two atoms of equal electronegativity. These bonds occur primarily in homonuclear diatomic molecules such as H$_2$, N$_2$, and I$_2$. Nonpolar bonds are also present in more complex molecules when like atoms are bonded such as in P$_4$ and in many organic compounds that have carbon-carbon bonds. Polar covalent bonds result when two atoms with different electronegativities unequally share electrons. They are intermediate in character between ionic and nonpolar covalent bonds. The greater the electronegativity difference, the greater the charge separation and the more polar the bond. Electronegativities are tabulated in Table 6-3 in the text.

Two designations are used to indicate the dipolar nature of the bond. A partial charge transfer or unequal sharing is indicated by δ+ and δ-. Crossed arrows are also used to indicate that sharing is unequal. In this notation the head of the arrow is directed toward the more electronegative atom. When different species are compared, a longer arrow means that there is a larger electronegativity difference.

```
     +----->      +---->              δ+  δ-    δ+  δ-
      H - Cl       H - Br     or      H - Cl    H - Br
EN   2.1  3.0     2.1  2.8
ΔEN    0.9           0.7
```

In this example, the electronegativity difference is greater for HCl than for HBr, so the HCl bond is slightly more polar than the HBr bond.

Example 7-12 Polarity of Bonds

Consider the following covalent compounds, CH_4, CF_4, CCl_4, CBr_4, and CI_4. Carbon is the central atom in each case and all bonds are single bonds. Determine the direction and the order of the bond polarities in these compounds.

Plan

We need to consult the electronegativity table and determine electronegativity differences between the bonded atoms. A greater electronegativity difference means that the bond is more polar.

Solution

The electronegativities are C (2.5), H (2.1), I (2.5), F (4.0), Cl (3.0) and Br (2.8). The differences in electronegativities, ΔEN are:

	<----+	+-------->	+---->	+--->	
	C - H	C - F	C - Cl	C - Br	C - I
ΔEN	0.4	1.5	0.5	0.3	0

Note that C is more electronegative than H but less electronegative than the halogen atoms. This fact accounts for the directional difference in polarity. To two significant figures, the electronegativities of C and I are the same, and the C-I bond is essentially nonpolar. The magnitude of bond polarity increases in the order:

$$\boxed{C\text{-}I < C\text{-}Br < C\text{-}H < C\text{-}Cl < C\text{-}F}$$

The fact that molecules such as CF_4 have very polar bonds does not mean that they are polar molecules. Instead, the CF_4 molecule and all of the other molecules in Example 7-12 are nonpolar because the individual bond dipoles are symmetrically located and therefore they cancel. Molecules such as HF and HCl contain asymmetrical dipoles and, therefore, are polar molecules. In order for a molecule to be polar, it must have: (1) at least one polar bond; and, (2) the polar bonds must be asymmetrically arranged. The polarity of an entire molecule, as opposed to an individual bond, is described in terms of dipole moments. These measure the tendency of a molecule to align itself with an applied external field. Dipole moments (μ) must be experimentally measured and are usually expressed in Debyes (D). Dipole moments give valuable information about a molecule's three dimensional geometry (discussed in Chapter Eight).

EXERCISES

Ionic Bonding

1. Show the transfer of electrons with ↑↓ notation in the reactions between the following pairs of elements to create an ionic compound. Show the elements as atoms, even though they may exist as diatomic or polyatomic species.

   (a) Ba and F   (b) K and I   (c) Cs and S   (d) Ca and As
   (e) Li and N   (f) Rb and Se   (g) Sr and P   (h) Al and Br

2. Write Lewis dot structures for atoms and ions of each of the reactants and products in Exercise 1.

Lewis Formulas for Covalent Compounds and Polyatomic Ions

3. Draw a Lewis formula for each of the following covalent compounds. The central atom(s) is (are) underlined.

   (a) $\underline{Cl}F_3$ (b) $\underline{C}Se_2$ (c) $\underline{N}Cl_3$ (d) $\underline{As}Br_5$ (e) $\underline{C}OCl_2$
   (f) $H_2\underline{Te}$ (g) $\underline{Ge}Cl_4$ (h) $\underline{C_2}Br_4$ (i) $\underline{C_2}F_2$ (j) $\underline{N_2}O_3$
   (k) $H\underline{C}O_2Cl$ (l) $\underline{Se}F_6$ (m) $CH_3\underline{S}H$ (n) $\underline{Te}F_4$ (o) $\underline{Br}F_5$
   (p) $\underline{Kr}F_2$

   In **j**, the N's are bonded together with 2 O's on one N and one O on the other N. In **k**, H is bonded to the C and the Cl to one of the O's. In **m**, one H is bonded to the S.

4. Draw a Lewis formula for each of the ions below. The central atoms are underlined.

   (a) $\underline{Cl}O_2^-$ (b) $\underline{N_3}^-$ (c) $\underline{S}O_3^{2-}$ (d) $\underline{Al}(OH)_4^-$ (e) $\underline{O_2}^{2-}$
   (f) $\underline{C}N^-$ (g) $\underline{Se}O_4^{2-}$ (h) $\underline{Si}O_3^{2-}$ (i) $\underline{I}O_4^-$ (j) $\underline{I}F_4^+$

5. Draw resonance formulas for the following species. The central element is underlined.

   (a) $CH_3\underline{N}O_2$ (N bonded to C, both O's to N, and all H's to C.
   (b) $H\underline{C}O_3^-$ (H bonded to O)
   (c) $\underline{N_2}O_4$ (N's bonded together; two O's bonded to each N)

6. Why are the following structures unlikely on the basis of formal charges? Draw a more appropriate structure for each.

(a) :Cl≡C–C̈–H   (b) H–C̈=N̈–H with :Ö: and H on C   (c) [:S̈–C≡N:]⁻

Polar and Nonpolar Covalent Bonds

7. Interhalogen compounds exist with formulas of ClF_3, BrF_3, IF_3 and ICl_3. In which of these compounds are the halogen-halogen bonds most polar and in which are they least polar? Refer to the electronegativity table if necessary.

8. Hydrogen forms compounds with the group VA elements of NH_3, SbH_3, PH_3, and AsH_3. In which of these compounds are the bonds most polar and in which are they least polar?

9. Predict, using only the periodic table, whether bonds between the following atoms would be primarily ionic or primarily covalent.

(a) N and C   (b) B and Cl   (c) Mg and Cl   (d) S and Li   (e) I and Te

Miscellaneous Exercises

10. Draw a Lewis dot structure for sulfuryl chloride, SO_2Cl_2, a common organic reagent. All atoms are bonded to the central S atom. In chemistry texts, this species is usually drawn with two double bonds. Redraw the compound in this manner and use formal charges to explain why a structure that does not obey the octet rule is preferred.

11. Repeat the process of Exercise 10 to minimize formal charges for the recently synthesized compound Br_2O_3, in which the three O atoms and the other Br atom are bonded to a central Br atom.

12. What are the formulas of the ionic compounds formed from the elements listed?

(a) Rb and N   (b) Ca and S   (c) Sr and N   (d) Ra and P

13. Which of the compounds below are likely to be ionic?

(a) ClF_3 (b) Na_2O (c) P_4O_{10} (d) Fe_2S_3 (e) BaF_2 (f) S_4N_2

14. "Fill in the dots" in the skeleton structures below to create Lewis structures that have minimized formal charges.

(a) N N N N O   (b) Si C C   (c) O Xe O   (d) H C C C N with O above second C and H H below

ANSWERS TO EXERCISES

1. (a) BaF$_2$

    Ba: [Xe]↑↓   → Ba^{2+}: [Xe]
         6s

    2F: [He]↑↓ ↑↓ ↑↓ ↑   → 2F$^-$: [He]↑↓ ↑↓ ↑↓ ↑↓ = [Ne]
         2s   2p   2s   2p

    (b) KI

    K: [Ar]↑   → K$^+$: [Ar]
         4s

    I: [Kr]↑↓ 4f^{14} ↑↓ ↑↓ ↑   → I$^-$: [Kr]↑↓ 4f^{14} ↑↓ ↑↓ ↑↓ = [Xe]
         5s   5p   5s   5p

    (c) Cs$_2$S

    2Cs: [Xe]↑   → 2Cs$^+$: [Xe]
          6s

    S: [Ne]↑↓ ↑↓ ↑ ↑   → S^{2-}: [Ne]↑↓ ↑↓ ↑↓ ↑↓ = [Ar]
         3s   3p   3s   3p

    (d) Ca$_3$As$_2$

    3Ca: [Ar]↑↓   → 3Ca^{2+}: [Ar]
          4s

    2As: [Ar]↑↓ 3d^{10} ↑ ↑ ↑   → 2As^{3-}: [Ar]↑↓ 3d^{10} ↑↓ ↑↓ ↑↓ = [Kr]
          4s   4p   4s   4p

    (e) Li$_3$N

    3Li: [He]↑   → 3Li$^+$: [He]
          2s

    N: [He]↑↓ ↑ ↑ ↑   → N^{3-}: [He]↑↓ ↑↓ ↑↓ ↑↓ = [Ne]
         2s   2p   2s   2p

    (f) Rb$_2$Se

    2Rb: [Kr]↑   → Rb$^+$: [Kr]
          5s

    Se: [Ar]↑↓ 3d^{10} ↑↓ ↑ ↑   → Se^{2-}: [Ar]↑↓ 3d^{10} ↑↓ ↑↓ ↑↓ = [Kr]
         4s   4p   4s   4p

(g) Sr_3P_2

3Sr: [Kr]↑↓
       5s
→ 3Sr^{2+}: [Kr]

2P: [Ne]↑↓ ↑ ↑ ↑
      3s   3p
→ 2P^{3-}: [Ne]↑↓ ↑↓ ↑↓ ↑↓ = [Ar]
                3s   3p

(h) $AlBr_3$

Al: [Ne]↑↓ ↑ _ _
      3s   3p
→ Al^{3+}: [Ne]

3Br: [Ar]↑↓ 3d^{10} ↑↓ ↑↓ ↑
       4s   4p
→ 3Br$^-$: [Ar]↑↓ 3d^{10} ↑↓ ↑↓ ↑↓ = [Kr]
              4s   4p

2. (a) Ba: + 2 :F̈· → Ba^{2+}, 2 $[:\ddot{\underset{..}{F}}:]^-$

(b) K· + :Ï· → K$^+$, $[:\ddot{\underset{..}{I}}:]^-$

(c) 2Cs· + :S̈· → 2Cs$^+$, $[:\ddot{\underset{..}{S}}:]^{2-}$

(d) 3 Ca: + 2 ·Äs· → 3Ca^{2+}, 2$[:\ddot{\underset{..}{As}}:]^{3-}$

(e) 3Li· + ·N̈· → 3Li$^+$, $[:\underset{..}{N}:]^{3-}$

(f) 2Rb· + :S̈e· → 2Rb$^+$, $[:\underset{..}{Se}:]^{2-}$

(g) 3 Sr: + 2 ·P̈· → 3Sr^{2+}, 2$[:\underset{..}{P}:]^{3-}$

(h) Al: + 3 :Ï· → Al^{3+}, 3$[:\ddot{\underset{..}{I}}:]^-$

3. (a) N - A = S does not apply
         A = 7 + 3(7)   = 28 e⁻

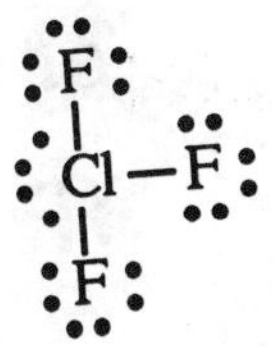

   (b) N = 3(8)   = 24 e⁻
        A = 4 + 2(6)   = 16 e⁻
        S   = 8 e⁻

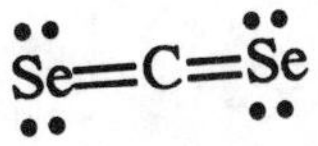

$$\ddot{\text{Se}}=C=\ddot{\text{Se}}$$

   (c) N = 4(8)   = 32 e⁻
        A = 5 + 3(7)   = 26 e⁻
        S   = 6 e⁻

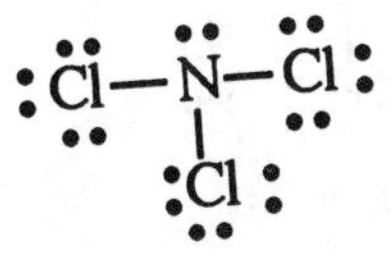

   (d) N - A = S does not apply;
        As needs 5 atoms attached
        A = 5 + 5(7)   = 40 e⁻

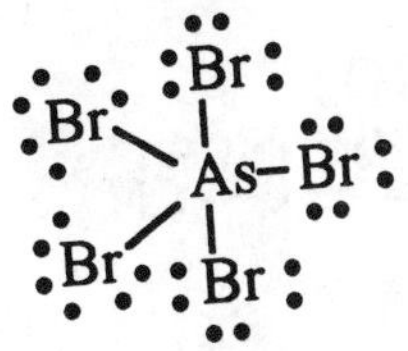

   (e) N = 4(8) +   = 32 e⁻
        A = 4 + 6 + 2(7) = 24 e⁻
        S   = 6 e⁻

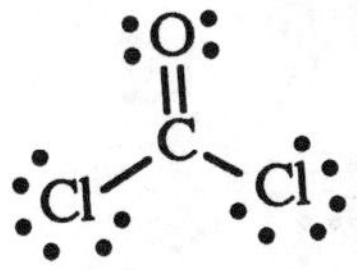

   (f) N = 2(2) + 8   = 12 e⁻
        A = 2(1) + 6   = 8 e⁻
        S   = 4 e⁻

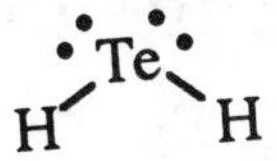

   (g) N = 5(8)   = 40 e⁻
        A = 4 + 4(7)   = 32 e⁻
        S   = 8 e⁻

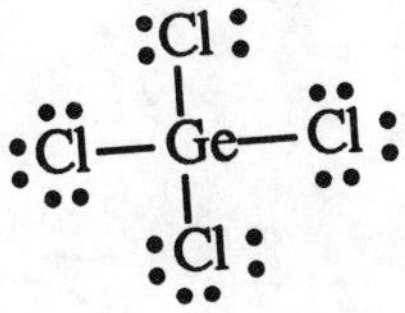

   (h) N = 6(8)   = 48 e⁻
        A = 2(4) + 4(7) = 36 e⁻
        S   = 12 e⁻

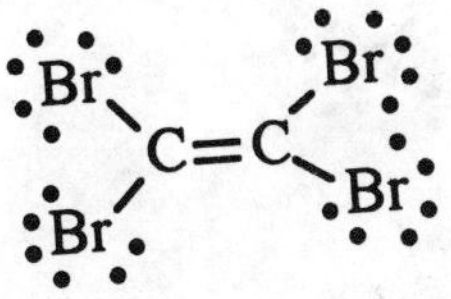

(i) $N = 4(8)$ = 32 e⁻
$\underline{A = 2(4) + 2(7)}$ = 22 e⁻
S = 10 e⁻

$:\!\ddot{F}\!-\!C\!\equiv\!C\!-\!\ddot{F}\!:$

(j) $N = 5(8)$ = 40 e⁻
$\underline{A = 2(5) + 3(6)}$ = 28 e⁻
S = 12 e⁻

$\ddot{O}\!=\!\ddot{N}\!-\!N\!\underset{\ddot{O}:}{\overset{\ddot{O}:}{\diagup\!\!\!\diagdown}}$

(k) $N = 1(2) + 4(8)$ = 34 e⁻
$\underline{A = 1(1) + 4 + 2(6) + 1(7)}$ = 24 e⁻
S = 10 e⁻

$H\!-\!\underset{}{\overset{\ddot{O}:}{C}}\!-\!\ddot{O}\!-\!\ddot{C}\!l\!:$

(l) $N - A = S$ does not apply; Se needs 6 atoms attached
$A = 1(6) + 6(7)$ = 48 e⁻

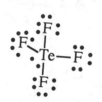

(m) $N = 2(8) + 4(2)$ = 24 e⁻
$\underline{A = 4 + 6 + 4(1)}$ = 14 e⁻
S = 10 e⁻

$H\!-\!\underset{H}{\overset{H}{C}}\!-\!\ddot{S}\!-\!H$

(n) $N - A = S$ does not apply
$A = 1(6) + 4(7)$ = 34 e⁻

$:\!\ddot{F}\!\underset{:\ddot{F}:}{\overset{:\ddot{F}:}{\diagdown\!Te\!-\!\ddot{F}:}}$

(o) $N - A = S$ does not apply
$A = 1(7) + 5(7)$ = 42 e⁻

$:\!\ddot{F}\!\underset{:\ddot{F}:\;:\ddot{F}:}{\overset{:\ddot{F}:}{\diagdown\!Br\!\diagdown}}$

(p) $N - A = S$ does not apply
$A = 1(8) + 2(7)$ = 22 e⁻

$:\!\ddot{F}\!:$
$\;\;|$
$:\!Kr\!:$
$\;\;|$
$:\!\ddot{F}\!:$

4. (a) $N = 3(8)$ $= 24\ e^-$
$\underline{A = 7 + 2(6) + 1} = 20\ e^-$
$S = 4\ e^-$

$$\left[\begin{array}{c} :\!\ddot{Cl}\!-\!\ddot{\underset{..}{O}}\!: \\ | \\ :\!\ddot{\underset{..}{O}}\!: \end{array} \right]^{-}$$

(b) $N = 3(8) = 24\ e^-$
$\underline{A = 3(5) + 1} = 16\ e^-$
$S = 8\ e^-$

$$\left[:\!\ddot{N}\!=\!N\!=\!\ddot{N}\!: \right]^{-}$$

(c) $N = 4(8) = 32\ e^-$
$\underline{A = 6 + 3(6) + 2} = 26\ e^-$
$S = 6\ e^-$

$$\left[:\!\ddot{\underset{..}{O}}\!-\!\underset{\underset{:\ddot{O}:}{|}}{S}\!-\!\ddot{\underset{..}{O}}\!: \right]^{2-}$$

(d) $N = 4(2) + 5(8) = 48\ e^-$
$\underline{A = 4(1) + 3 + 4(6) + 1} = 32\ e^-$
$S = 16\ e^-$

$$\left[\begin{array}{c} H \\ | \\ :\ddot{O}: \\ | \\ H\!-\!\ddot{\underset{..}{O}}\!-\!Al\!-\!\ddot{\underset{..}{O}}\!-\!H \\ | \\ :\ddot{O}: \\ | \\ H \end{array} \right]^{-}$$

(e) $N = 2(8) = 16\ e^-$
$\underline{A = 2(6) + 2} = 14\ e^-$
$S = 2\ e^-$

$$\left[:\!\ddot{\underset{..}{O}}\!-\!\ddot{\underset{..}{O}}\!: \right]^{2-}$$

(f) $N = 2(8) = 16\ e^-$
$\underline{A = 4 + 5 + 1} = 10\ e^-$
$S = 6\ e^-$

$$\left[:C\!\equiv\!N\!: \right]^{-}$$

(g) $N = 5(8) = 40\ e^-$
$\underline{A = 6 + 4(6) + 2} = 32\ e^-$
$S = 8\ e^-$

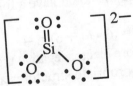

(h) $N = 4(8) = 32\ e^-$
$\underline{A = 4 + 3(6) + 2} = 24\ e^-$
$S = 8\ e^-$

$$\left[\begin{array}{c} :\!\ddot{O}\!: \\ \| \\ :\ddot{\underset{..}{O}}\!\!\diagdown\!\!\underset{Si}{}\!\!\diagup\!\!\ddot{\underset{..}{O}}\!: \end{array} \right]^{2-}$$

(i) $N = 5(8) = 40$ e$^-$
$A = 7 + 4(6) + 1 = 32$ e$^-$
$S = 8$ e$^-$

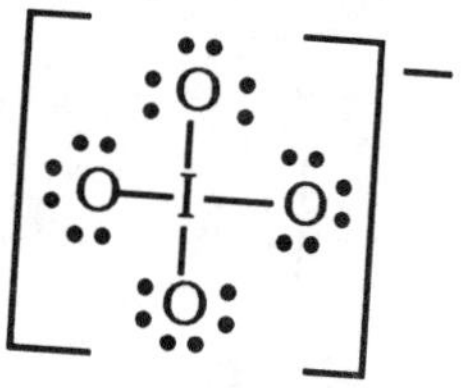

(j) N - A = S does not apply
$A = 7 + 4(7) - 1 = 34$ e$^-$

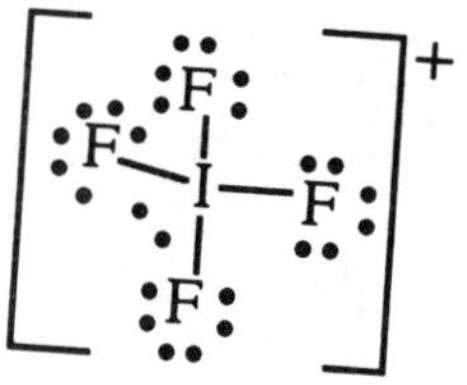

5. (a) [structure: H–C(H)(H)–N with two O, resonance between single/double bonded O's]

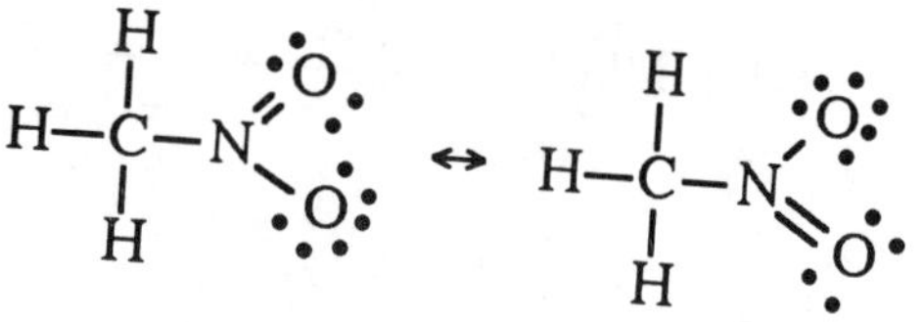

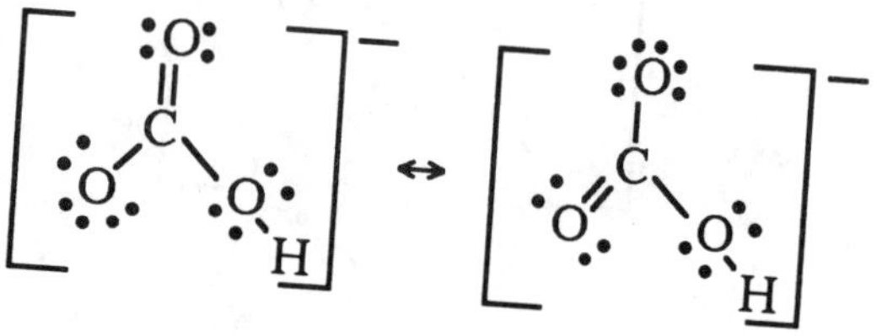

(b) [carbonate-like structure with OH, two resonance forms]

(c) [N–N with multiple O's, four resonance structures]

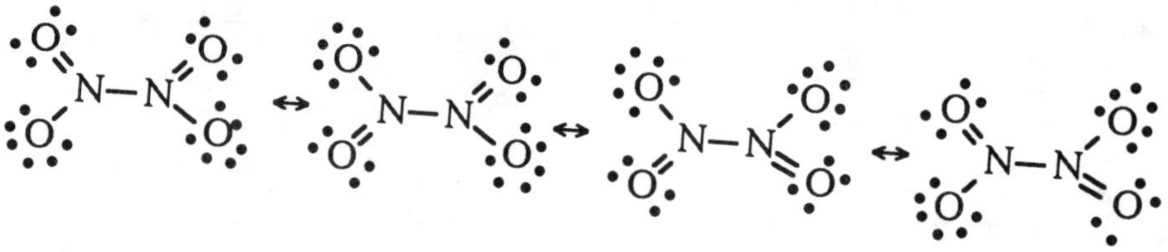

6. (a) Right hand C: FC = 4 - (2+4) = -2   FC of Cl = 7 - (3+2) = +2
Each atom should have a formal charge of 0 if possible.

(b) FC of O = 6 - (1 + 6) = -1   FC N = 5 - 4 = +1
Each atom should have a formal charge of 0 if possible.

(c) S: FC = 6 - (1 + 6) = -1   N: FC = 5 - (3 + 2) = 0
N is more electronegative than S and should have the negative FC.
The more appropriate structures are:

(a) 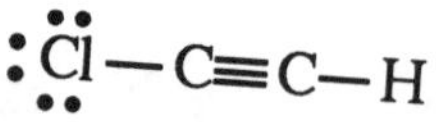 $:\ddot{Cl}-C\equiv C-H$ (b) H–C(=O)–N(H)–H with O above C   (c) $[:\ddot{S}=C=\ddot{N}:]^-$

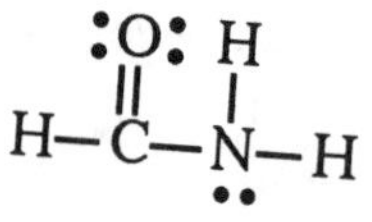

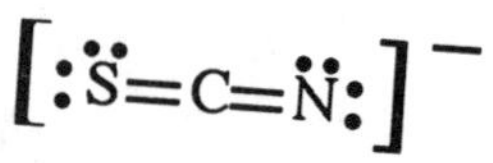

7. I-Cl is the least polar bond ($\Delta EN = 0.5$). I-F is the most polar bond ($\Delta EN = 1.5$).

8. NH₃ has the most polar bonds (ΔEN = 0.9), while AsH₃ has the least polar bonds (ΔEN = 0). (In SbH3, the partial negative charge is on the hydrogen (EN=2.1), instead of on the antimony (EN = 1.9).

9. (a) covalent (b) covalent (c) ionic (d) ionic (e) covalent

10. N = 5(8) = 40 e⁻
    A = 6 + 2(6) + 2(7) = 32 e⁻
    S = 8 e⁻

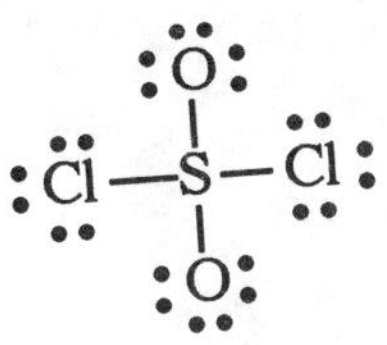

    [Left structure: :Ö: above S, S bonded to two Cl and another O below]

    S: FC = 6 - 4 = +2
    Each O: FC = 6 - (1 + 6) = -1

    [Right structure: double bonds from S to both O atoms]

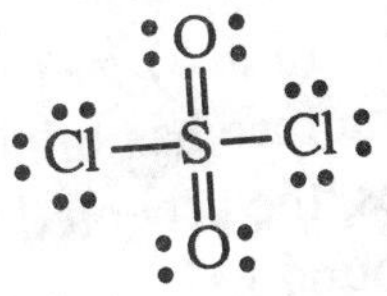

    S: FC = 6 - 6 = 0
    Each O: FC = 6 - (2 + 4) = 0

    The right hand structure is preferred because all atoms have a formal charge of zero. This also agrees with measured bond lengths.

11. N = 5(8) = 40 e⁻
    A = 2(7) + 3(6) = 32 e⁻
    S = 8 e⁻

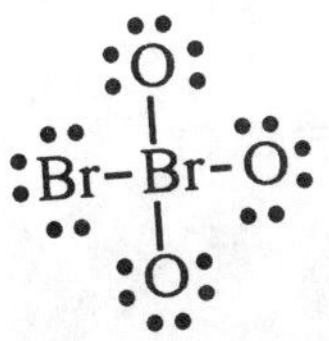

    [Left structure: Br-Br-O with O above and below, single bonds]

    Cental Br: FC = 7 - 4 = +3
    Each O: FC = 6 - (1 + 6) = -1

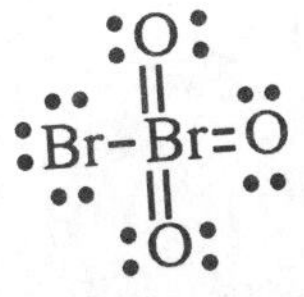

    [Right structure: Br-Br=O with double bonds to two O atoms]

    Central Br: FC = 7 - 7 = 0
    Each O: FC = 6 - (2 + 4) = 0

    The right hand structure is preferred because all atoms have a formal charge of zero.

12. (a) Rb⁺, N³⁻ = Rb₃N   (b) Ca²⁺, S²⁻ = CaS   (c) Sr²⁺, N³⁻ = Sr₃N₂
    (d) Ra²⁺, P³⁻ = Ra₃P₂

13. (b) Na₂O, (d) Fe₂S₃, and (e) BaF₂ are likely to be ionic because they are the metal/nonmetal compounds.

14. (a) N = 5(8)   = 40 e⁻
    A = 4(5) + 1(6) = 26 e⁻
    S   = 14 e⁻

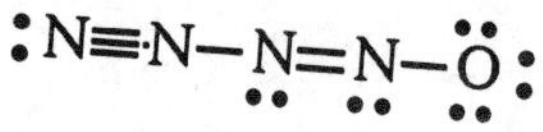

    :N≡N–N=N–Ö:

    FC N's left to right:
    FC = 5 - (3 + 2) = 0
    FC = 5 - 4 = +1
    FC = 5 - (3 + 2) = 0
    FC = 5 - (3 + 2) = 0
    For O FC = 6 - (1 + 6) = -1
    While this structure has minimized formal charges, the accepted structure for the compound is:

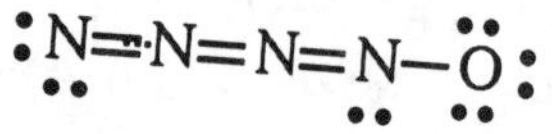

    :N=·N=N=N–Ö:

(b) Exception to octet rule.
    A = 1(4) + 2(4) = 12 e⁻

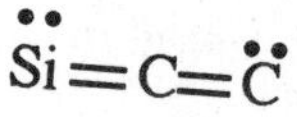

    Si=C=C̈

    Si: FC = 4 - (2 + 2) = 0
    Left C: FC = 4 - 4 = 0
    Right C: FC = 4 - (2 + 2) = 0

(c) A = 8 + 3(6) = 32 e⁻

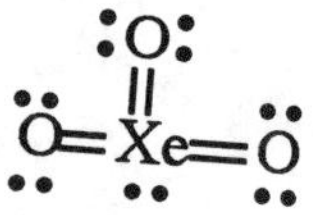

    :Ö:
    ‖
    Ö=Xe=Ö

    FC each O = 6 - (2 + 4) = 0
    FC Xe = 8 - (6 + 2) = 0

(d) N = 3(2) + 4(8)   = 38 e⁻
    A = 3(1) + 3(4) + 5   = 20 e⁻
    S =   = 18 e⁻

    H–C=C–C≡N:
      | |
      H H

    FC each C = 4 - 4 = 0
    FC each H = 1 - 1 = 0
    FC N = 5 - (3 + 2) = 0

Chapter 8: Molecular Structure and Covalent Bonding Theories

8-1 Geometry of Molecules and Polyatomic Ions

The geometry of a molecule is indicated by its polarity or lack of polarity. As discussed in Chapter Seven, nonpolar molecules either have only nonpolar bonds or their polar bonds are symmetrically arranged such that they do not have dipole moments. Polar molecules have asymmetric polar bonds and, thus, have dipole moments. The most important factor in determining the geometry of molecules and complex ions is the number of sets of electrons (regions of high electron density) around the central element. A set of electrons may consist of: a single bond, a double bond, a triple bond, or a lone pair (unshared pair) of electrons in the valence shell of the central atom.

8-2 Valence Shell Electron Pair Repulsion Theory

Electronic geometry describes the arrangement of electron groups around the central atom. These electronic geometries are predicted (and experimentally observed) from the **Valence Shell Electron Pair Repulsion Theory (VSEPR)**. VSEPR theory states: groups of electrons arrange themselves around the central atom in the way that minimizes repulsions among sets of electrons. Table 8-1 summarizes the electronic shapes.

Molecular Geometry describes the arrangement of the atoms surrounding the central element. The electronic geometry and the molecular geometry are the same if, and only if, there are no lone pairs of electrons around the central element. Table 8-2 in this book summarizes the molecular shapes associated with different electronic geometries. The convention used is that **A** stands for the central atom, **B** for bonded atoms and **U** for unshared pairs. Different substituents distort the bond angles associated with the repulsion of the electron groups. The amount of distortion depends on the magnitude of the repulsions. Repulsions between two lone pairs (lp) are greater than the repulsions between a lone pair and a bonded pair (bp) which, in turn, are greater than those between bonded pairs: **lp-lp > lp-bp > bp-bp**

8-3 Valence Bond Theory

Valence bond theory describes covalent bonding in terms of the overlap of orbitals on separate atoms. Bonds are formed as electrons in these orbitals pair up with each other. To provide the necessary atomic orbitals for observed shape, the concept of hybrid orbitals about the central atom is used. The number of high electron density regions suggests the kind of hybrid orbitals to be used. The designation given the hybrid reflects the number and type of hybrid orbitals. The hybrid orbitals have some of the characteristics of the component atomic orbitals, yet they are not the same as atomic orbitals. Sets of hybrid orbitals have the same total electronic capacity as their component atomic orbitals, even though the number of lobes may be different. Table 8-1 indicates the electronic geometries and bond angles corresponding to different numbers of electron sets and the common hybridization associated with the geometries described. The text provides pictures of the contribution of the atomic orbitals to the hybrid orbitals of each type.

Table 8-1 Electron Sets About the Central Atom and Electronic Geometry

Electron Sets	Electronic Geometry		Bond Angles	Hybridization
2	linear		180°	sp
3	trigonal planar		120°	sp^2
4	tetrahedral		109.5°	sp^3
5	trigonal bipyrimidal		90°, 120°, 180°	sp^3d
6	octahedral		90°, 180°	sp^3d^2

Chapter 8 Molecular Structure and Covalent Bonding Theories 135

In Table 8-1, the angles listed are made by drawing lines through the nucleus and the centers of high electron density.

Table 8-2 Molecular or Ionic Geometry

Number of Electron Sets	Representation	Molecular or Ionic Shape
2	AB_2	linear
3	AB_3	trigonal planar
4	AB_4	tetrahedral
4	AB_3U	trigonal pyramid
4	AB_2U_2	angular
5	AB_5	trigonal bipyramid
5	AB_4U	seesaw
5	AB_3U_2	T-shaped
5	AB_2U_3	linear
6	AB_6	octahedral
6	AB_5U	square pyramid
6	AB_4U_2	square planar

In the representation in Table 8.2 **A** stands for the central atom, **B** represents a bonded atom, and **U** represents an unshared or "lone" pair of electrons.

Example 8-1 Geometry, Hybridization, and Polarity

Describe the electronic and molecular geometries, the polarity, and the hybridization in a BeI_2 molecule.

Plan
We will first draw the dot structure, then analyze it using the relationships above.

Solution

:Ï–Be—Ï:

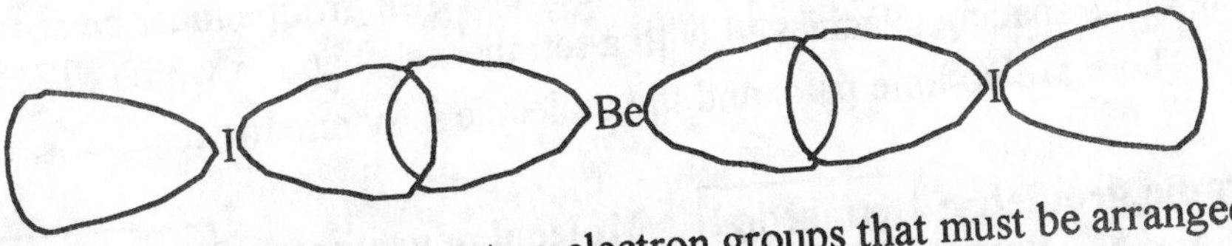

VSEPR: This is an AB_2 molecule. There are two electron groups that must be arranged 180° apart to minimize repulsion. The molecule is symmetrical about the central atom.

Electronic geometry: linear   **Molecular geometry:** linear
The molecule is nonpolar.

Valence Bond: Be has an electronic configuration of [He]2s². It uses a 2p orbital to hybridize with the 2s to form two equivalent sp hybrid orbitals.

Be: [He]↑↓ → [He]↑ ↑
      2s   sp sp

The sp hybrid orbitals can hold a maximum of four electrons, the same number that could be contained by the original s and p from which the hybrids were formed. The hybrid orbitals are aligned 180° apart and contain one electron in each orbital. These orbitals can be overlapped and paired with an electron in the 5p orbital of iodine.

Example 8-2 Geometry, Hybridization, and Polarity

Describe the electronic and molecular geometries, the polarity, and the hybridization of an SiH_4 molecule.

Plan
We will first draw the dot structure, then analyze it using the relationships above.

Solution

VSEPR: This is an AB_4 molecule. The four electron groups must be spaced as far apart as possible. This spacing is achieved with a tetrahedral shape in which the H-Si-H angles are all 109.5°. There are no lone pairs and the molecule is symmetrical.

Electronic geometry: tetrahedral   **Molecular geometry:** tetrahedral
The molecule is nonpolar.

Chapter 8 Molecular Structure and Covalent Bonding Theories

Valence Bond: Si has an electronic configuration of $[Ne]3s^23p^2$. It uses an additional p orbital to hybridize to form four equivalent $\boxed{sp^3}$ hybrid orbitals.

Si: $[Ne]\uparrow\downarrow\ \uparrow\ \uparrow\ \_\ \rightarrow\ [Ne]\uparrow\ \uparrow\ \uparrow\ \uparrow$
    3s   3p sp^3

Each of the four sp^3 hybrid orbitals contains one electron that can pair with a 1s electron of a hydrogen to fill the sp^3 orbitals to capacity.

Example 8-3 Geometry, Hybridization, and Polarity

Describe the electronic and molecular geometries, the polarity, and the hybridization of an H_2S molecule.

Plan

We will first draw the dot structure, then analyze it using the relationships above.

Solution

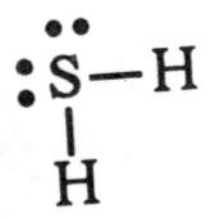

:S—H
 |
 H

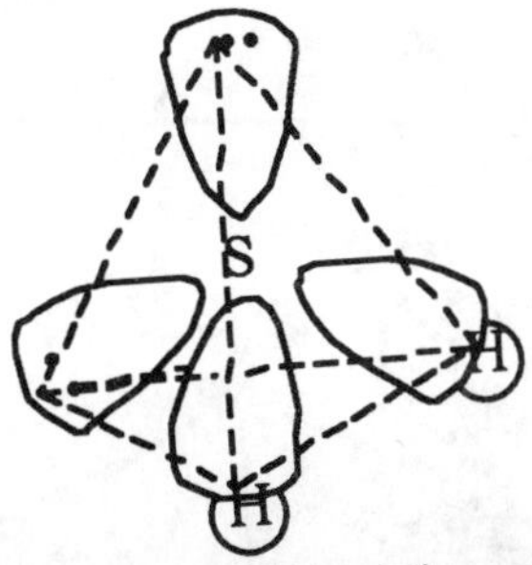

VSEPR: This is an AB_2U_2 molecule. There are four regions of electron density around the central atom, so the electrons are in a distorted tetrahedral arrangement. The lone pairs show slightly greater repulsion than the bonding pairs, so their bond angle is greater than 109.5°, while the angle between the bonding pairs is much less than 109.5°. The arrangement of atoms and, therefore, the molecular shape is simply angular. There are dipoles directed toward the more electronegative S atom from the H atoms. There are also dipoles directed from the S atom toward the lone pairs. The net dipole (vector sum of the individual dipoles) which gives rise to a polar molecule with a dipole moment is shown above.

Electronic geometry: $\boxed{\text{tetrahedral}}$ **Molecular geometry:** $\boxed{\text{angular}}$

The molecule is $\boxed{\text{polar}}$.

Valence Bond: S has an electronic configuration of [Ne]$3s^2 3p^4$. It hybridizes to form four equivalent $\boxed{sp^3}$ hybrid orbitals.

S: [Ne]↑↓ ↑↓ ↑ ↑ → [Ne]↑↓ ↑↓ ↑ ↑
    3s   3p   sp³

The sp^3 hybridization provides the proper equivalent orbitals and shape for maximum separation of electron pairs. The electrons that will be the unshared pairs are already paired in two of the orbitals while the other two sp^3 orbitals are open to allow pairing and overlap with the 1s orbital of the hydrogen atom.

Example 8-4 Geometry, Hybridization, and Polarity

Describe the electronic and molecular geometries, the polarity, and the hybridization of a XeF_4 molecule.

Plan

We will first draw the dot structure, then analyze it using the relationships above.

Solution

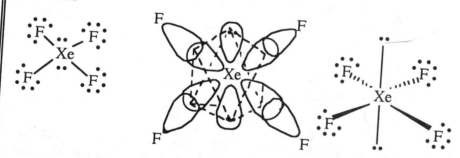

VSEPR: This is an AB_4U_2 molecule. There are six regions of electron density around the central atom. The regions are aligned at the corners of an octahedral with 90° angles between the bonding pairs and between the bonding pairs and the lone pairs. The lone pairs are 180° apart to minimize repulsions. The molecular shape is square planar and the molecule is nonpolar because the dipoles due to the lone pairs cancel each other as do the dipoles directed toward each of the highly electronegative fluorine atoms.

Electronic geometry: $\boxed{\text{octahedral}}$ Molecular geometry: $\boxed{\text{square planar}}$

The molecule is $\boxed{\text{nonpolar}}$.

Chapter 8 Molecular Structure and Covalent Bonding Theories 139

Valence Bond: Xe has an electronic configuration of $[Kr]5s^2 4d^{10} 5p^6$. It hybridizes to form six equivalent sp^3d^2 hybrid orbitals.

$$Xe: [Kr]\underset{5s}{\uparrow\downarrow}\ 4d^{10}\ \underset{5p}{\uparrow\downarrow\ \uparrow\downarrow\ \uparrow\downarrow} \rightarrow [Kr]\underset{sp^3d^2}{\uparrow\downarrow\ \uparrow\downarrow\ \uparrow\ \uparrow\ \uparrow\ \uparrow}\ 4d^{10}$$

The sp^3d^2 hybridization provides the proper equivalent orbitals and shape for maximum separation of electron pairs. The electrons that will be the unshared pairs are already paired in two of the orbitals while the other four sp^3d^2 orbitals are open to allow pairing and overlap with the 2p orbital of the fluorine atom.

Example 8-5 Geometry, Hybridization, and Polarity

Describe the electronic and molecular geometries, the polarity, and hybridization of an SO_2 molecule.

Plan
We will first draw the dot structure, then analyze it using the relationships above.

Solution
The dot formula for SO_2 is found to be a combination of two resonance structures:

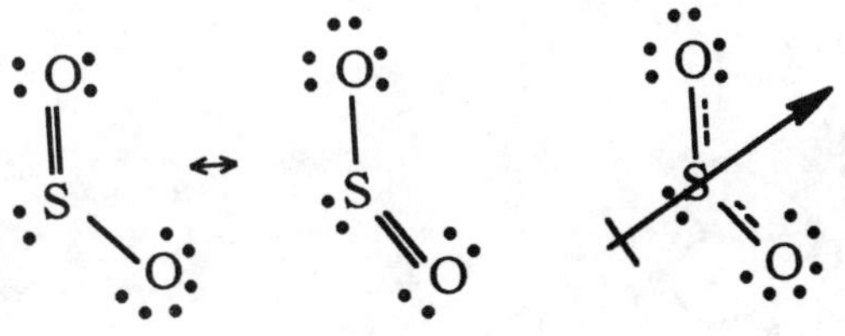

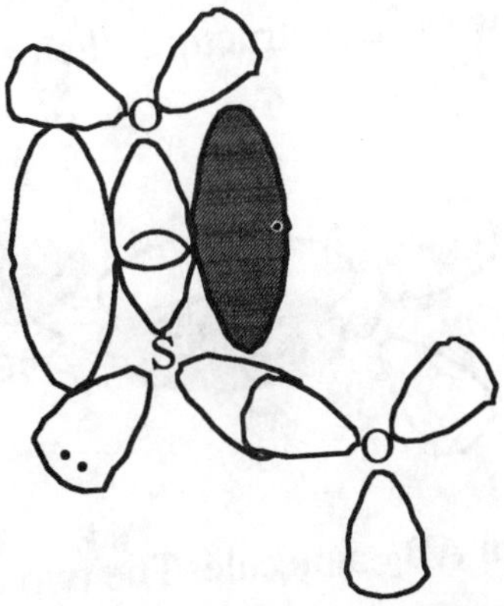

VSEPR: This is an AB_2U molecule. There are three regions of electron density around the central atom, so the electrons are in a distorted trigonal planar arrangement with slightly less than 120° angles between each of the bonding pairs. The angle between the lone pair and the bonding pairs is slightly greater than between the bonding pairs. The angular molecule is polar because the lone pair causes the molecule to be asymmetrical.

Electronic geometry: trigonal planar   **Molecular geometry:** angular

The molecule is |polar|.

Valence Bond: S has an electronic configuration of [Ne]$3s^2 3p^4$. The hybridization is |sp^2| to provide the proper geometry for the three regions of high electron density. One of the sp^2 orbitals forms a conventional bond with the doubly bonded oxygen. The second sp^2 hybrid holding two paired electrons forms a coordinate covalent bond with the singly bonded oxygen. The final sp^2 hybrid holds the lone pair. It may be considered that the remaining $3p$ orbital is reserved to form a pi (π) bond with a reserved $2p$ of one of the oxygen atoms. In reality, the bonds are not formally localized and each sulfur-oxygen bond is the same. By the valence bond approach, one resonance hybrid of SO_2 would appear to be as on the right. Note that the oxygen to which the double bond is shown also must be sp^2 hybridized.

S: [Ne]↑↓ ↑↓ ↑ ↑ → [Ne]↑↓ ↑↓ ↑   ↑
    3s   3p sp^2 "reserved" $3p$ orbital for π bond

Example 8-6 Geometry, Hybridization, and Polarity

Describe the electronic and molecular geometries, polarity, and hybridization of a CO_2 molecule.

Plan

We will first draw the dot structure, then analyze it using the relationships above.

Solution

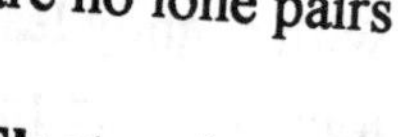

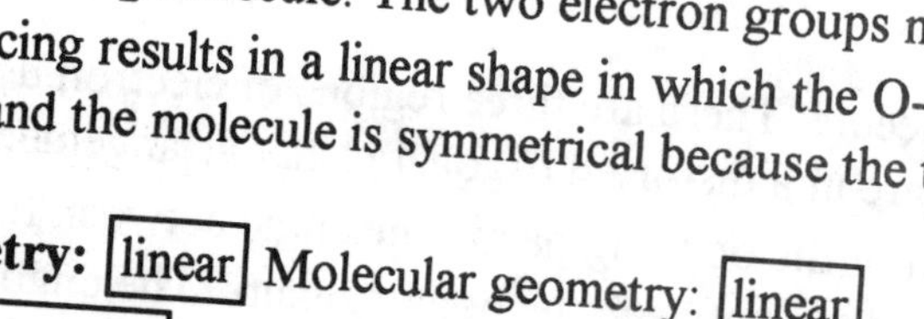

:Ö=C=Ö:

VSEPR: This is an AB_2 molecule. The two electron groups must be spaced as far apart as possible. This spacing results in a linear shape in which the O-C-O angles are all 180°. There are no lone pairs and the molecule is symmetrical because the two polar C-O bonds cancel.

Electronic geometry: |linear| **Molecular geometry:** |linear|
The molecule is |nonpolar|.

Valence Bond: Carbon has an electronic configuration of [He]$2s^22p^2$. It hybridizes to form two equivalent $\boxed{sp}$ hybrid orbitals. Carbon forms bonds by overlapping sp orbitals with each of the oxygen atoms, which are sp^2 hybridized. The remaining two p orbitals of the carbon form the side-to-side overlap for π bonds with p orbitals from each oxygen. Note that the p orbitals of the carbon and, thus, the two π bonds are 90° to each other. This could be accomplished with a p_y-p_y overlap with one oxygen atom and a p_z-p_z overlap with the other oxygen atom.

C: [He]↑↓ ↑ ↑ __ → [He]↑ ↑   ↑ ↑
   2s   2p   sp   "reserved" for 2 π bonds

Example 8-7 Geometry and Hybridization

Describe the electronic and ionic geometries, and the hybridization of a ClO_3^- ion.

Plan

We will first draw the dot structure, then analyze it using the relationships above.

Solution

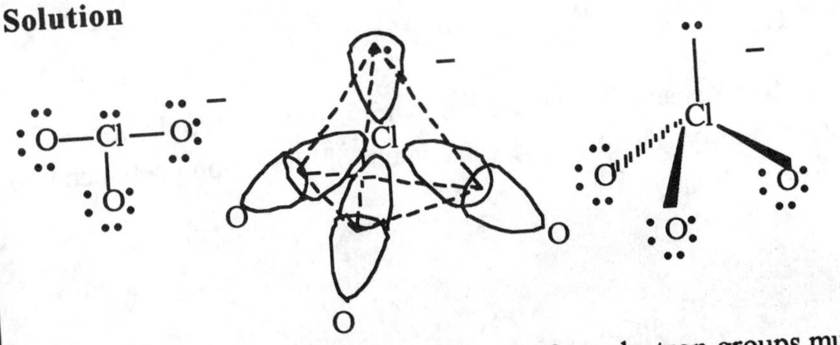

VSEPR: This is an AB_3U molecule. The four electron groups must be spaced as far apart as possible. This spacing is achieved with a slightly distorted tetrahedral shape in which the O-Cl-O angles are all slightly less than 109.5° because of the influence of the lone pair.

Electronic geometry: $\boxed{\text{tetrahedral}}$ **Ionic geometry:** $\boxed{\text{trigonal pyramid}}$

Valence Bond: Cl has an electronic configuration of [Ne]$3s^23p^5$. It hybridizes to form four equivalent $\boxed{sp^3}$ hybrid orbitals.

Cl: [Ne]↑↓ ↑↓ ↑↓ ↑ → [Ne]↑↓ ↑↓ ↑↓ ↑↓
       3s   3p   sp³

The chlorine hybridizes to provide four equivalent orbitals of the proper geometry. The "extra" electron from the negative charge may be placed for pairing in the fourth sp³ orbital. Chlorine forms coordinate covalent bonds overlapping its filled sp³ orbitals with an empty p orbital from each oxygen.

Example 8-8 Hybridization

The structure below is for aspartame, the artificial sweetener in Nutrasweet®. What is the hybridization about each carbon in the molecule? How many π bonds are in this molecule?

Plan

The number of regions of electron density determines the hybridization. Multiple bonds count as one region of electron density. There can be a maximum of one σ bond between two atoms; additional bonds must be π bonds.

Solution

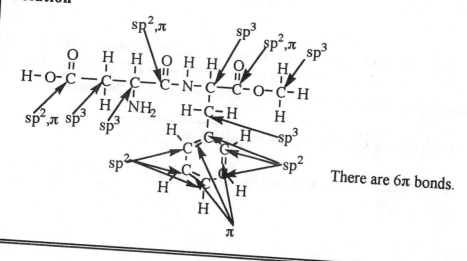

There are 6π bonds.

Chapter 8 Molecular Structure and Covalent Bonding Theories 143

EXERCISES

Valence Shell Electron Pair Repulsion Theory

1. Tabulate the electronic and molecular geometry relative to the central element(s) for the molecules listed below. The Lewis structures of these molecules were drawn in Chapter Seven, Exercise 3. Tell if each molecule is polar or non-polar.

   (a) $\underline{Cl}F_3$ (b) $C\underline{Se}_2$ (c) $\underline{N}Cl_3$ (d) $\underline{As}Br_5$ (e) $\underline{C}OCl_2$

   (f) $H_2\underline{Te}$ (g) $\underline{Ge}Cl_4$ (h) $\underline{C_2}Br_4$ (i) $\underline{C_2}F_2$ (j) $\underline{N_2}O_3$

   (k) $H\underline{C}O_2Cl$ (l) $\underline{Se}F_6$ (m) $\underline{C}H_3SH$ (n) $\underline{Te}F_4$ (o) $\underline{Br}F_5$

   (p) $\underline{Kr}F_2$

2. Repeat the process in Exercise 1 for the polyatomic ions below, omitting the polarity portion of the problem. The Lewis structures of these ions were drawn in Chapter Seven, Exercise 4.

   (a) $\underline{Cl}O_2^-$ (b) $\underline{N}_3^-$ (c) $\underline{S}O_3^{2-}$ (d) $\underline{Al}(OH)_4^-$ (e) $\underline{O}_2^{2-}$

   (f) $\underline{C}N^-$ (g) $\underline{Se}O_4^{2-}$ (h) $\underline{Si}O_3^{2-}$ (i) $\underline{I}O_4^-$ (j) $\underline{I}F_4^+$

Valence Bond Theory

3. What is the hybridization of the central atom of each molecule in Exercise 1? How many π bonds are present in each molecule, if any?

4. What is the hybridization of the central atom of each ion in Exercise 2? How many π bonds are present in each ion, if any?

Mixed Exercises

5. Draw the dot structures of the species below, emphasizing geometry around the central atom. Tabulate the hybridization, the electronic geometry, and the molecular or ionic geometry for each. The central atom is underlined.

   (a) $\underline{P}OBr_3$ (b) $\underline{I}F_4^-$ (c) $\underline{As}Cl_2^-$ (d) $\underline{Se}F_4$ (e) $\underline{Xe}F_2$

   (f) $\underline{Be}H_2$ (g) $\underline{I}F_3$ (h) $\underline{N}H_2Br$ (i) $\underline{As}H_4^+$ (j) $\underline{Br}Cl_2^-$

   (k) $\underline{Al}Br_3$ (l) $\underline{Xe}OF_4$ (m) $\underline{Te}F_6$ (n) $\underline{P}O_4^{3-}$ (o) $\underline{N}O_2^-$

6. Seconal, which has the structure shown below, is a prescription drug used as a tranquilizer. What is the hybridization about each carbon and nitrogen atom in the molecule? How many π bonds are in this molecule?

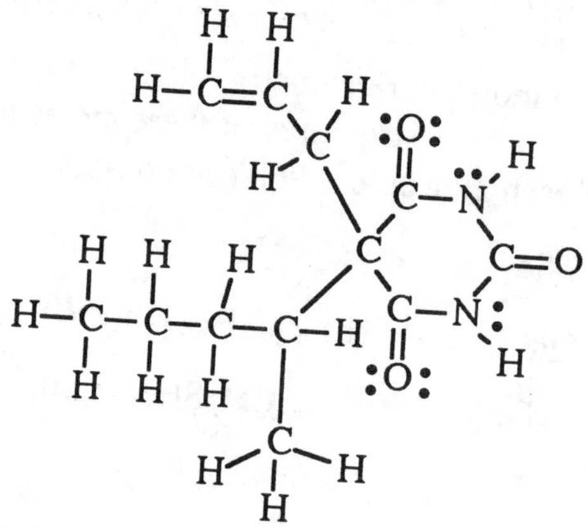

7. Propargyl alcohol, which is a skin irritant with a mild geranium odor, has the structure shown below. Label each carbon atom with its proper hybridization. How many π bonds are in the molecule?

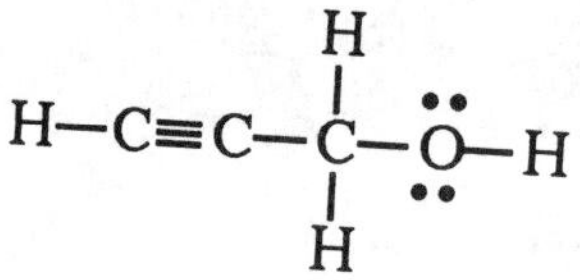

ANSWERS TO EXERCISES

& 3.

	Type	Electronic Geometry	Molecular Geometry	Polarity	Hybrid Orbitals
(a)	AB_3U_2	trigonal bipyramid	T-shaped	Polar	sp^3d
(b)	AB_2	linear (2 π bonds)	linear	Nonpolar	sp
(c)	AB_3U	tetrahedral	trigonal pyramid	Polar	sp^3
(d)	AB_5	trigonal bipyramid	trigonal bipyramid	Nonpolar	sp^3d
(e)	AB_2	trigonal planar; (1 π bond)	trigonal planar	Polar	sp^2
(f)	AB_2U_2	tetrahedral	angular	Polar	sp^3
(g)	AB_4	tetrahedral	tetrahedral	Nonpolar	sp^3
(h)	Both C AB_3	trigonal planar about each C (1 π bond)	trigonal planar about each C	Nonpolar	sp^2 about each C
(i)	Both C AB_2	linear about each C (2 π bonds)	linear about each C	Nonpolar	sp about each C
(j)	Both N AB_3	trigonal planar about both N (1 π bond on each; 2 total)	trigonal planar about both N	Polar	sp^2 about both N
(k)	AB_2	trigonal planar (1 π bond)	trigonal planar	Polar	sp^2
(l)	AB_6	octahedral	octahedral	Nonpolar	sp^3d^2
(m)	AB_4	tetrahedral	tetrahedral	Polar	sp^3
(n)	AB_4U	trigonal bipyramid	seesaw	Polar	sp^3d
(o)	AB_5U	octahedral	square pyramid	Polar	sp^3d^2
(p)	AB_2U_3	trigonal bipyramid	linear	Nonpolar	sp^3d

2. & 4.

	Type	Electronic Geometry	Ionic Geometry	Hybrid Orbitals
(a)	AB_4	tetrahedral		sp^3
(b)	AB_2	linear (2 π bonds)	angular	sp
			linear	
(c)	AB_3U	tetrahedral	trigonal pyramid	sp^3
(d)	AB_4	tetrahedral	tetrahedral	sp^3
(e)	AB	linear	linear	p overlap
(f)	AB	linear (2 π bonds)	linear	sp
(g)	AB_4	tetrahedral	tetrahedral	sp^3
(h)	AB_3	trigonal planar (1 π bond)	trigonal planar	sp^2
(i)	AB_4	tetrahedral	tetrahedral	sp^3
(j)	AB_4U	trigonal bipyramid	seesaw	sp^3d

5.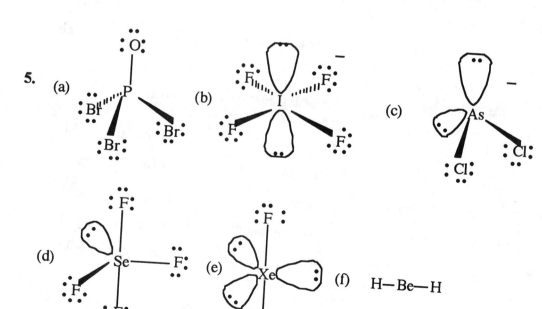

Chapter 8 Molecular Structure and Covalent Bonding Theories 147

(g) [structure of IF$_3$ with lone pairs on I, bonded to three F atoms]

(h) [structure of NBrH$_2$ with lone pair on N, bonded to Br and two H]

(i) [structure of AsH$_3$ with H$^+$, three H atoms bonded to As]

(j) [structure of BrCl$_2^-$ with two lone pairs on Br, bonded to two Cl]

(k) [structure of AlBr$_3$, Al bonded to three Br]

(l) [structure of XeOF$_3$ with lone pair on Xe, bonded to three F and one O]

(m) [structure of TeF$_6$, Te bonded to six F]

(n) [structure of PO$_4^{3-}$, P bonded to four O]

(o) [structure of NO$_2^-$ with lone pair on N, bonded to two O]

	Hybridization (Central Atom)	Electronic Geometry	Molecular or Ionic Geometry
(a)	sp^3	tetrahedral	tetrahedral
(b)	sp^{3}d^2	octahedral	square planar
(c)	sp^3	tetrahedral	angular
(d)	sp^{3}d	trigonal bipyramid	seesaw
(e)	sp^{3}d	trigonal bipyramid	linear
(f)	sp	linear	linear
(g)	sp^{3}d	trigonal bipyramid	T-shaped
(h)	sp^3	tetrahedral	trigonal pyramid
(i)	sp^3	tetrahedral	tetrahedral
(j)	sp^{3}d	trigonal bipyramid	linear
(k)	sp^2	trigonal planar	trigonal planar
(l)	sp^{3}d^2	octahedral	square pyramid

	Hybridization (Central Atom)	Electronic Geometry	Molecular or Ionic Geometry
(m)	sp^3d^2	octahedral	octahedral
(n)	sp^3	tetrahedral	tetrahedral
(o)	sp^2	trigonal planar	angular

6.

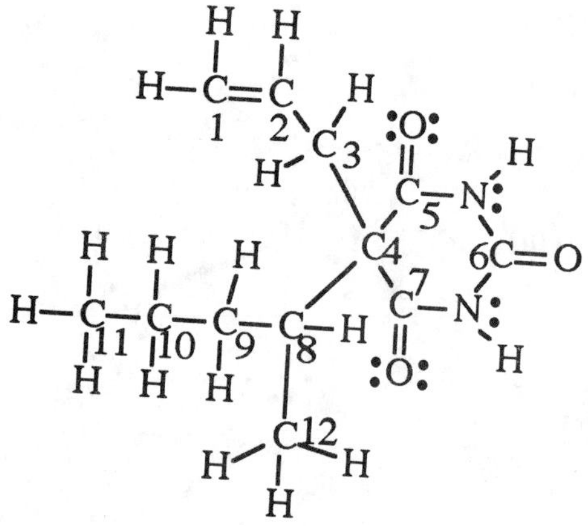

[Structure with carbons labeled C1–C12]

Both N = sp^3

$C_1 = sp^2$, with a π bond
$C_2 = sp^2$, with a π bond
$C_3 = sp^3$
$C_4 = sp^3$
$C_5 = sp^2$, with a π bond
$C_6 = sp^2$, with a π bond
$C_7 = sp^2$, with a π bond
$C_8 = sp^3$
$C_9 = sp^3$
$C_{10} = sp^3$
$C_{11} = sp^3$
$C_{12} = sp^3$
Both N = sp^3

7.

H−C≡C−C(H)(H)−Ö−H

sp, π ; sp^3

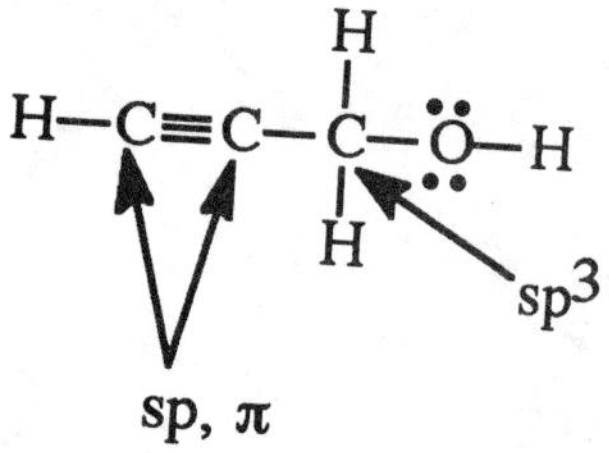

Chapter 9: Molecular Orbitals in Chemical Bonding

9-1 Molecular Orbitals

Molecular Orbital (MO) theory provides an alternative to valence bond theory with respect to covalent bonding. The basic idea of molecular orbital theory is that when atoms bond new molecular orbitals are created. These molecular orbitals belong to the molecule as a whole, or to a portion of the molecule, rather than to individual atoms. In each case, two types of molecular orbitals (MO) are formed. One, the **bonding MO**, is lower in energy than the individual atomic orbitals, while the other, the **antibonding MO**, is higher in energy than the atomic orbitals. The overlap of an **s** orbital on one atom with an **s** orbital on another atom results in a bonding σ_s orbital and an antibonding σ_s^*. (The asterisk superscript signifies an antibonding orbital). In the diagram below, the dots (•) represent nuclei.

Two **p** orbitals of the same spatial orientation can overlap "head-on" to form σ_p and σ_p^* molecular orbitals. For example, two p_x orbitals could overlap as shown below.

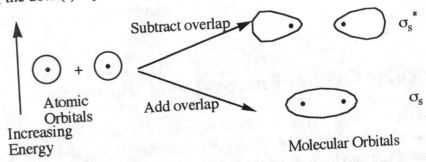

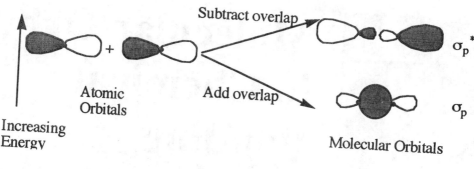

The remaining atomic orbitals of the same orientations (such as the two p_y or the two p_z) can only overlap side-on to form two pi molecular orbitals, π_p and π_p^*. The overlap of two p_z orbitals is shown below.

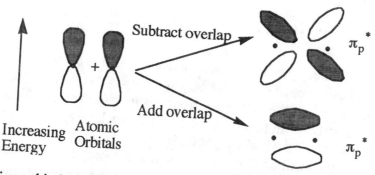

Because the bonding orbitals are lower in energy than the atomic orbitals from which they are formed, electrons occupying them are lower in energy than they would be in the pure atomic orbitals. Thus, molecules or complex ions are said to be stabilized by electrons in bonding orbitals. Conversely, the occupation of the higher energy antibonding orbitals by electrons tends to destabilize species.

9-2 Molecular Orbital Energy-Level Diagrams

The diagram on the next page shows the relative ordering of energies of molecular orbitals resulting from combinations of s and p orbitals of the first two energy levels for homonuclear diatomic molecules and ions from H_2 up through N_2. The order of energies of the σ_{2p} and π_{2p_y}, π_{2p_z} is reversed for O_2, F_2 and Ne_2. Electrons fill molecular orbitals in accordance with the Aufbau Principle and Hund's Rule. To draw molecular orbital energy level diagrams, the following rules must be applied:

1. Draw or select the appropriate molecular orbital energy level diagram,
2. Use the periodic table and the charge, if any, to determine the **total** number of electrons in the species. **All** electrons, not just the valence electrons are used.
3. Add the electrons to the diagram, placing each electron into the lowest energy level available. Remember that a maximum of two electrons can occupy any orbital, and

then only if they have opposite spin. Electrons must occupy all orbitals of the same energy singly before pairing begins. Unpaired electrons have parallel spins.

When molecular orbital electron configurations are written, these same rules apply, but the spins are not explicitly shown.

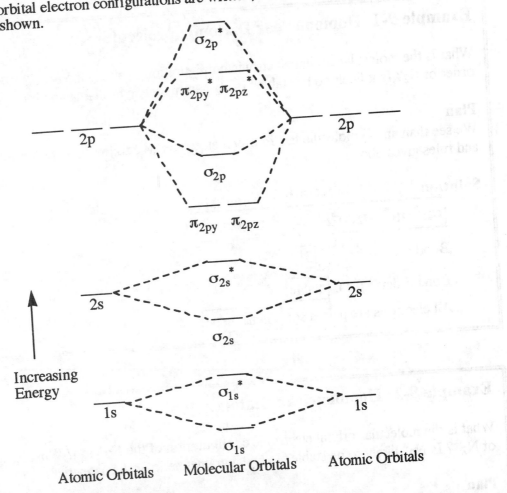

9-3 Bond Order and Bond Stability

The **bond order** corresponds to the number of covalent bonds determined by Valence Bond theory. It is defined as half the difference in the number of electrons in bonding orbitals and the number of electrons in antibonding orbitals.

$$\text{Bond Order} = \frac{\{\text{number of electrons in bonding orbitals}\} - \{\text{number of electrons in antibonding orbitals}\}}{2}$$

A greater bond order of a diatomic molecule or ion indicates a more stable species. A greater bond order also indicates a shorter bond length and a higher bond energy. (See Table 9.1 in the text for a correlation of bond orders with stability, bond length, etc., for some diatomic species.) The net charge of the species will also affect its stability.

Example 9-1 Homonuclear Diatomic Molecule

What is the molecular orbital electronic configuration of an N_2 molecule? What is the bond order of N_2? Is it likely to be stable? Is N_2 paramagnetic or diamagnetic?

Plan

We see than an N_2 molecule has fourteen electrons. We follow the appropriate filling order and rules given above.

Solution

$$\boxed{\sigma_{1s}^2 \; \sigma_{1s}^{*2} \; \sigma_{2s}^2 \; \sigma_{2s}^{*2} \; \pi_{2p_y}^2 \; \pi_{2p_z}^2 \; \sigma_{2p}^2}$$

Bond order $= \dfrac{10-4}{2} = \boxed{3}$

Bond order > 0 so: $\boxed{\text{stable}}$

All electrons are paired so: $\boxed{\text{diamagnetic}}$

Example 9-2 Homonuclear Diatomic Ion

What is the molecular orbital electronic configuration of the N_2^+ ion? What is the bond order of N_2^+? Is N_2^+ likely to be stable? Is N_2^+ paramagnetic or diamagnetic?

Plan

The N_2^+ has one less high energy electron than does the N_2 in Example 9-1. We follow the same procedure.

Solution

$$\boxed{\sigma_{1s}^2 \; \sigma_{1s}^{*2} \; \sigma_{2s}^2 \; \sigma_{2s}^{*2} \; \pi_{2p_y}^2 \; \pi_{2p_z}^2 \; \sigma_{2p}^1}$$ Bond order $= \dfrac{9-4}{2} = \boxed{2.5}$

Bond order > 0 so: $\boxed{\text{stable}}$

One electron is unpaired so: $\boxed{\text{paramagnetic}}$

4 MO Diagram for Heteronuclear Diatomic Molecules

The molecular orbital energy level diagrams of heteronuclear diatomic molecules are skewed because the atomic orbitals of the more electronegative element are lower in energy than those of the element with which it combines. The molecular orbital energy level diagram for NO is:

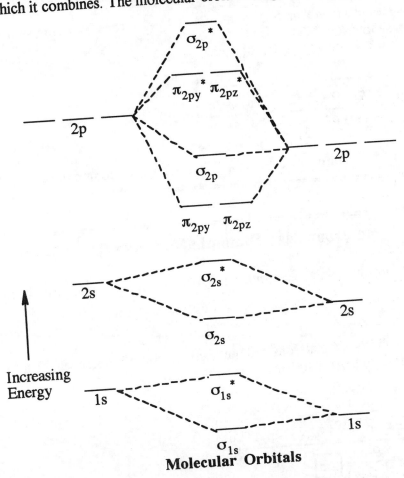

Occupancy within the levels shown is:

$$\sigma_{1s}^2 \; \sigma_{1s}^{*2} \; \sigma_{2s}^2 \; \sigma_{2s}^{*2} \; \pi_{2p_y}^2 \; \pi_{2p_z}^2 \; \sigma_{2p}^2 \; \pi_{2p_y}^{*1}$$

The bond order of NO is calculated to be: $\frac{10-5}{2} = 2.5$. NO should be stable because the bond order is greater than zero. NO is paramagnetic because it has an unpaired electron.

Example 9-3 Heteronuclear Diatomic Ion

What are the molecular orbital electronic configuration and the bond order of NO^+?

Plan

NO^+ has one less electron than NO for a total of 14 electrons. We can use the same energy level diagram as for NO.

Solution

$$\boxed{\sigma_{1s}^2 \; \sigma_{1s}^{*2} \; \sigma_{2s}^2 \; \sigma_{2s}^{*2} \; \pi_{2p_y}^2 \; \pi_{2p_z}^2 \; \sigma_{2p}^2}$$

Bond order = $\dfrac{10-4}{2}$ = $\boxed{3}$

The NO^+ ion is quite stable and exists in a number of compounds.

Example 9-4 Heteronuclear Diatomic Molecule

What are the molecular orbital electronic configuration and the bond order of the neutral molecule CN?

Plan

The carbon contributes six electrons and the nitrogen contributes seven for a total of thirteen. We can use an energy level diagram similar to that for NO.

Solution

$$\boxed{\sigma_{1s}^2 \; \sigma_{1s}^{*2} \; \sigma_{2s}^2 \; \sigma_{2s}^{*2} \; \pi_{2p_y}^2 \; \pi_{2p_z}^2 \; \sigma_{2p}^1}$$

Bond order = $\dfrac{9-4}{2}$ = $\boxed{2.5}$

This molecule is known, but has a tendency to dimerize to cyanogen, $(CN)_2$, which has no unpaired electrons and is one of the pseudohalogens.

9-5 Delocalization and the Shape of Molecular Orbitals

Structures exhibiting resonance are not treated adequately by Valence Bond theory, because valence bond resonance structures indicate shifting double bonds and changing hybridizations about atoms. Molecular orbital theory does not need resonance, but instead utilizes the concept

Chapter 9 Molecular Orbitals in Chemical Bonding

delocalized electrons. The shapes of the molecular orbitals in which electron delocalization occurs can be determined by averaging all of the contributing atomic orbitals.

Example 9-5 Delocalized Molecular Orbitals

Sketch the delocalized molecular orbital for the formate ion, HCO_2^-.

Plan

First, we must draw the two resonance structures according to the rules in Chapter 7. Neither structure has been shown to have an independent existence. Molecular orbital theory postulates the formation of a new molecular orbital that is created by simultaneously overlapping the $2p_z$ orbitals of the carbon and both oxygen atoms. It gives the same shape that would be achieved by averaging the contributing valence bond structures.

Solution

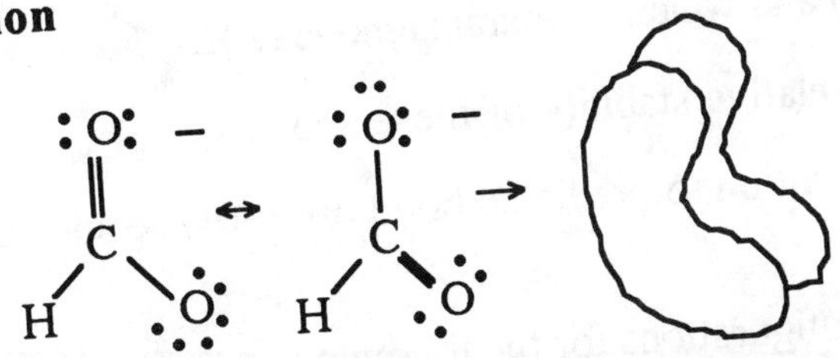

EXERCISES

Homonuclear Diatomic Molecules and Ions

1. From memory, draw the molecular orbital energy level diagram for homonuclear diatomic molecules of the first and second period, showing the π_{2p} molecular orbitals below that of the σ_{2p} molecular orbitals in energy.

2. Use the sets of molecular orbitals from Exercise 1 to write out the molecular orbital electronic configurations for the following molecules and ions in the form $\sigma_{1s}^2 \sigma_{1s}^{*2}$ etc.

   (a) He_2^+ (b) Li_2^{2+} (c) Li_2^{2-} (d) Be_2 (e) Be_2^{2+} (f) B_2 (g) B_2^{2+}
   (h) C_2 (i) C_2^- (j) C_2^{2+}

3. (a) What is the bond order for each species in Exercise 2?
   (b) Which species are diamagnetic (D) and which are paramagnetic (P)?
   (c) Apply MO theory to predict the relative stability of these species.
   (d) What else must be considered in addition to predicted bond order of the species to predict stability?

4. Write out molecular orbital electronic configurations for the following species, remembering that the σ_{2p} is lower in energy than π_{2p} for O_2, F_2 and Ne_2.

   (a) O_2 (b) O_2^{2+} (c) O_2^- (d) F_2 (e) F_2^{2+} (f) Ne_2^{2+} (g) Ne_2^-

5. (a) What is the bond order of each species in Exercise 4?
   (b) Which are diamagnetic (D) and which are paramagnetic (P)?
   (c) Predict the relative stabilities of the species.

Heteronuclear Diatomic Molecules and Ions

6. From memory, draw the molecular orbital energy level diagram for a heteronuclear diatomic molecule, XY, in which both elements are from the second period and Y is more electronegative than X. Show the σ_{2p} molecular orbitals lower in energy than the π_{2p} molecular orbitals.

7. Write out molecular orbital configurations for the following heteronuclear species, using the molecular orbital diagram from Exercise 6 for all but (e) and (f), which have π_{2p} lower in energy than σ_{2p}.

   (a) NO^{2-} (b) CN^{+2} (c) NF (d) NF^- (e) CN^- (f) BeB (g) OF^+

8. (a) What is the bond order of each species in problem 7?
   (b) Which are diamagnetic (D) and which are paramagnetic (P)?
   (c) Predict the relative stability of the species.

Delocalization and the Shapes of Molecular Orbitals

9. Draw Lewis dot structures indicating resonance from the Valence Bond point of view, then sketch molecular orbitals for the delocalized π systems.

   (a) CH_3NO_2 (b) NO_3^- (c) N_2O_4

ANSWERS TO EXERCISES

1. See page 151.

2. (a) $\sigma_{1s}^2 \sigma_{1s}^{*1}$

   (b) $\sigma_{1s}^2 \sigma_{1s}^{*2}$

   (c) $\sigma_{1s}^2 \sigma_{1s}^{*2} \sigma_{2s}^2 \sigma_{2s}^{*2}$

   (d) $\sigma_{1s}^2 \sigma_{1s}^{*2} \sigma_{2s}^2 \sigma_{2s}^{*2}$

   (e) $\sigma_{1s}^2 \sigma_{1s}^{*2} \sigma_{2s}^2$

   (f) $\sigma_{1s}^2 \sigma_{1s}^{*2} \sigma_{2s}^2 \sigma_{2s}^{*2} \pi_{2p_y}^1 \pi_{2p_z}^1$

   (g) $\sigma_{1s}^2 \sigma_{1s}^{*2} \sigma_{2s}^2 \sigma_{2s}^{*2}$

   (h) $\sigma_{1s}^2 \sigma_{1s}^{*2} \sigma_{2s}^2 \sigma_{2s}^{*2} \pi_{2p_y}^2 \pi_{2p_z}^2$

   (i) $\sigma_{1s}^2 \sigma_{1s}^{*2} \sigma_{2s}^2 \sigma_{2s}^{*2} \pi_{2p_y}^2 \pi_{2p_z}^2 \sigma_{2p}^1$

   (j) $\sigma_{1s}^2 \sigma_{1s}^{*2} \sigma_{2s}^2 \sigma_{2s}^{*2} \pi_{2p_y}^1 \pi_{2p_z}^1$

3.

	Species	Bond Order	D or P	Predicted Stability
(a)	He_2^+	0.5	P	marginally stable
(b)	Li_2^{2+}	0	D	unstable
(c)	Li_2^{2-}	0	D	unstable
(d)	Be_2	0	D	unstable
(e)	Be_2^{2+}	1	D	stable
(f)	B_2	1	P	stable
(g)	B_2^{2+}	0	D	unstable
(h)	C_2	2	D	stable
(i)	C_2^-	2.5	P	stable
(j)	C_2^{2+}	1	P	stable

   (d) Some of the conclusions in (c) are not reliable, because the charges **must** be considered.

4. (a) $\sigma_{1s}^2 \sigma_{1s}^{*2} \sigma_{2s}^2 \sigma_{2s}^{*2} \sigma_{2p}^2 \pi_{2p_y}^2 \pi_{2p_z}^2 \pi_{2p_y}^{*1} \pi_{2p_z}^{*1}$

   (b) $\sigma_{1s}^2 \sigma_{1s}^{*2} \sigma_{2s}^2 \sigma_{2s}^{*2} \sigma_{2p}^2 \pi_{2p_y}^2 \pi_{2p_z}^2$

   (c) $\sigma_{1s}^2 \sigma_{1s}^{*2} \sigma_{2s}^2 \sigma_{2s}^{*2} \sigma_{2p}^2 \pi_{2p_y}^2 \pi_{2p_z}^2 \pi_{2p_y}^{*2} \pi_{2p_z}^{*1}$

   (d) $\sigma_{1s}^2 \sigma_{1s}^{*2} \sigma_{2s}^2 \sigma_{2s}^{*2} \sigma_{2p}^2 \pi_{2p_y}^2 \pi_{2p_z}^2 \pi_{2p_y}^{*2} \pi_{2p_z}^{*2}$

(e) $\sigma_{1s}^2 \sigma_{1s}^{*2} \sigma_{2s}^2 \sigma_{2s}^{*2} \sigma_{2p}^2 \pi_{2py}^2 \pi_{2pz}^2 \pi_{2py}^{*1} \pi_{2pz}^{*1}$

(f) $\sigma_{1s}^2 \sigma_{1s}^{*2} \sigma_{2s}^2 \sigma_{2s}^{*2} \sigma_{2p}^2 \pi_{2py}^2 \pi_{2pz}^2 \pi_{2py}^{*2} \pi_{2pz}^{*2}$

(g) $\sigma_{1s}^2 \sigma_{1s}^{*2} \sigma_{2s}^2 \sigma_{2s}^{*2} \sigma_{2p}^2 \pi_{2py}^2 \pi_{2pz}^2 \pi_{2py}^{*2} \pi_{2pz}^{*2} \sigma_{2p}^{*2} \sigma_{3s}^1$

5.

	Species	Bond Order	D or P	Predicted Stability
(a)	O_2	2	P	very stable
(b)	O_2^{2+}	3	D	very stable
(c)	O_2^-	1.5	P	stable
(d)	F_2	1	D	stable
(e)	F_2^{2+}	2	P	very stable
(f)	Ne_2^{2+}	1	D	stable
(g)	Ne_2^-	0.5	P	marginally stable

Note that the stability will also be dependent on the charge.

6. See page 153, reversing the order of σ_{2p} and π_{2p}.

7. (a) $\sigma_{1s}^2 \sigma_{1s}^{*2} \sigma_{2s}^2 \sigma_{2s}^{*2} \sigma_{2p}^2 \pi_{2py}^2 \pi_{2pz}^2 \pi_{2py}^{*2} \pi_{2pz}^{*1}$

(b) $\sigma_{1s}^2 \sigma_{1s}^{*2} \sigma_{2s}^2 \sigma_{2s}^{*2} \sigma_{2p}^2 \pi_{2py}^1$

(c) $\sigma_{1s}^2 \sigma_{1s}^{*2} \sigma_{2s}^2 \sigma_{2s}^{*2} \sigma_{2p}^2 \pi_{2py}^2 \pi_{2pz}^2 \pi_{2py}^{*1} \pi_{2pz}^{*1}$

(d) $\sigma_{1s}^2 \sigma_{1s}^{*2} \sigma_{2s}^2 \sigma_{2s}^{*2} \sigma_{2p}^2 \pi_{2py}^2 \pi_{2pz}^2 \pi_{2py}^{*2} \pi_{2pz}^{*1}$

(e) $\sigma_{1s}^2 \sigma_{1s}^{*2} \sigma_{2s}^2 \sigma_{2s}^{*2} \pi_{2py}^2 \pi_{2pz}^2 \sigma_{2p}^2 \pi_{2py}^{*2} \pi_{2pz}^{*1}$

(f) $\sigma_{1s}^2 \sigma_{1s}^{*2} \sigma_{2s}^2 \sigma_{2s}^{*2} \pi_{2py}^1$

(g) $\sigma_{1s}^2 \sigma_{1s}^{*2} \sigma_{2s}^2 \sigma_{2s}^{*2} \sigma_{2p}^2 \pi_{2py}^2 \pi_{2pz}^2 \pi_{2py}^{*1} \pi_{2pz}^{*1}$

8.

	Species	Bond Order	D or P	Predicted Stability
(a)	NO^{2-}	1.5	P	stable
(b)	CN^{+2}	1.5	P	stable
(c)	NF	2	P	very stable
(d)	NF^-	1.5	P	stable
(e)	CN^-	3	D	very stable
(f)	BeB	0.5	P	marginally stable
(g)	OF^+	2	P	very stable

9. (a)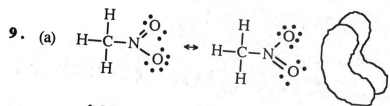

(b)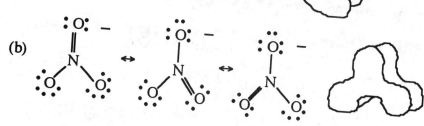

(c)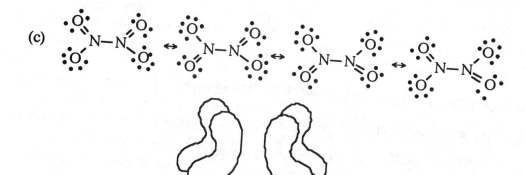

Chapter 10: Reactions in Aqueous Solutions I: Acids, Bases, and Salts

10-1 The Arrhenius Theory

In the **Arrhenius theory** of acids and bases, an **acid** is any substance containing hydrogen that ionizes in aqueous solutions to produce hydrogen ions, H^+. In water the hydrogen ions are hydrated, $H^+(H_2O)_n$. The hydrated hydrogen ion is often represented as H_3O^+, the hydronium ion. A **base** is any substance containing OH groups that dissociates in aqueous solution to produce hydroxide ions, OH^-. **Neutralization** is defined as the combination of H^+ and OH^- to produce water. Some examples are:

Arrhenius acid:
$$HNO_3(aq) \rightarrow H^+(aq) + NO_3^-(aq) \text{ or}$$
$$HNO_3(aq) + H_2O(\ell) \rightarrow H_3O^+(aq) + NO_3^-(aq)$$

Arrhenius base:
$$Ca(OH)_2(aq) \rightarrow Ca^{2+}(aq) + 2OH^-(aq)$$

Arrhenius neutralization:
$$2HNO_3(aq) + Ca(OH)_2(aq) \rightarrow Ca(NO_3)_2(aq) + 2H_2O(\ell)$$

10-2 The Brønsted-Lowry Theory

According to the **Brønsted-Lowry theory**, an **acid** is defined as a proton (H^+) donor. A **base** is defined as a proton (H^+) acceptor. **Neutralization** is defined as the transfer of a proton from an acid to a base. Therefore, acid-base reactions involve transfer from one species to another instead of simply being dissociation reactions. In the reaction below, HI is the acid because it is donating a proton to water. Water is the base because it is accepting that proton.

$$\underset{\text{acid}}{HI(aq)} + \underset{\text{base}}{H_2O(\ell)} \rightarrow H_3O^+(aq) + I^-(aq)$$

If a weak acid such as HF is involved, the reverse reaction proceeds to an appreciable extent and is itself an acid-base reaction.

$$\text{HF(aq)} + \text{H}_2\text{O}(\ell) \rightleftharpoons \text{H}_3\text{O}^+(\text{aq}) + \text{F}^-(\text{aq})$$
$$\text{acid}_1 \qquad \text{base}_2 \qquad\qquad \text{acid}_2 \qquad \text{base}_1$$

In this forward reaction, the HF is the acid because it transfers a proton to water. Water is the base because it accepts that proton. In the reverse reaction, the H_3O^+ is the acid because it donates a proton to F^-. F^- is the base because it accepts the proton. The subscripts indicate **conjugate pairs** of acids and bases. Conjugates differ by one H^+. Note that water is a base by this broader definition but not by the Arrhenius definition.

In another example, ammonia, NH_3, is dissolved in water.

$$\text{NH}_3(\text{aq}) + \text{H}_2\text{O}(\ell) \rightleftharpoons \text{NH}_4^+(\text{aq}) + \text{OH}^-(\text{aq})$$
$$\text{base}_1 \qquad \text{acid}_2 \qquad\qquad \text{acid}_1 \qquad \text{base}_2$$

In the forward reaction, H_2O, the acid, transfers a proton to NH_3, the base. In the reverse reaction, NH_4^+, the conjugate acid of NH_3, transfers a proton to OH^-, the conjugate base of H_2O.

These examples illustrate that the acidic or basic action of a substance depends upon both components in the system being considered. Substances that can accept protons from a stronger acid or donate protons to a stronger base are called **amphoteric** or **amphiprotic**. Pure water **autoionizes** to a certain extent, with one water molecule functioning as the acid and another as the base.

$$\text{H}_2\text{O}(\ell) + \text{H}_2\text{O}(\ell) \rightleftharpoons \text{H}_3\text{O}^+(\text{aq}) + \text{OH}^-(\text{aq})$$
$$\text{base}_1 \qquad \text{acid}_2 \qquad\qquad \text{acid}_1 \qquad \text{base}_2$$

Hydrated hydrogen ion is the strongest acid that can exist in aqueous solution because any stronger acid will protonate water and produce H_3O^+. Hydroxide ion is the strongest base that can exist in aqueous solution, because any stronger base will deprotonate water and produce OH^-.

10-3 Strengths of Acids and Bases

The strength of an acid or base refers to its degree of ionization (acids and weak bases) or dissociation (strong bases) in aqueous solution. Those acids and bases that are essentially completely ionized are strong, while those that are sparingly ionized are weak. Lists of the common strong acids and the strong soluble bases in water were given in Table 4-1 of this book. Other acids can be assumed to be weak. Other hydroxide bases have low solubilities that preclude production of a very basic solution. The common weak bases are derivatives of ammonia.

Chapter 10 Reactions in Aqueous Solutions I: Acids, Bases, and Salts

Reviewing the Brønsted-Lowry concept reveals that because a strong acid nearly completely ionizes, its conjugate base must have essentially no tendency to accept a proton in aqueous solution. The generalizations derived from this are:

1. a stronger acid has a weaker conjugate base as its anion.
2. a stronger base has a weaker conjugate acid as its cation.

The following conjugate pairs illustrate this point:

$HCl > HF > HCN$
← Increasing Acid Strength

$Cl^- < F^- < CN^-$
→ Increasing Base Strength

$NaOH > CH_3NH_2 > NH_3$
← Increasing Base Strength

$Na^+ < CH_3NH_3^+ < CN^-$
→ Increasing Acid Strength

As discussed in detail in Section 10-5 of the text, binary hydride strengths generally **increase** down a column of the periodic chart as electronegativities decrease and radii substantially increase. From left to right across a period, radii are similar and binary acid strengths **increase** as the higher electronegativity of the central atom causes the bond to be more polar. As acid strengths increase (more likely to give up a proton), base strengths decrease (less likely to accept a proton). Examples of these trends are below.

$H_2Te > H_2Se > H_2S > H_2O$ as an acid

$NH_3 > PH_3 > AsH_3 > SbH_3$ as a base

For analogous ternary acids in the same family, acid strengths generally **increase** with increasing electronegativity of the central atom because a more electronegative element pulls electron density away from the O-H bond, weakening it. For the same reason, in analogous acids with the same central atom, acid strengths **increase** as more oxygen atoms, which are highly electronegative, are bonded to the central atom. The following acids illustrate these trends.

$HClO_2 > HBrO_2 > HIO_2$ as an acid (electronegativity of central atom decreases)

$HClO_4 > HClO_3 > HClO_2 > HClO$ as an acid (number of oxygen atoms decrease)

Example 10-1 Ternary acid Strength

Rank the following in order of increasing acid strength: HNO_3, HNO, HNO_2.

Plan

We recognize that these acids have the same central atom. More oxygen atoms are characteristic of a stronger acid.

Solution

$$\boxed{HNO_3 > HNO_2 > HNO \text{ as an acid}}$$

Example 10-2 Conjugate Base Strength

Rank the following in order of increasing base strength: HSO_3^- and HSO_4^-.

Plan

These are the conjugate bases of H_2SO_4 and H_2SO_3. H_2SO_4 has more oxygen atoms and is a stronger acid than H_2SO_3. A stronger acid has a weaker conjugate base.

Solution

$$\boxed{HSO_3^- > HSO_4^- \text{ as a base}}$$

10-4 Reactions of Acids and Bases

Review the material in Section 4-3 of this book that discusses molecular, total ionic, and net ionic equations. To correctly write the latter two types of equation, it is necessary to know the following about each acid, base, and salt of interest: (1) whether it is soluble or insoluble in water; and, (2) if it is soluble, whether it is a strong or weak electrolyte. It is essential to review the lists of strong acids and strong soluble bases in Table 4-1 of this book. Remember that the solubility rules for common salts are given in Section 4-2.5 of the text. It may be assumed unless otherwise specified that water-soluble salts are predominantly dissociated in aqueous solution.

Example 10-3 Equations for Acid-Base Reactions

Write (a) molecular; (b) total ionic; and (c) net ionic equations for each of the following acid-base reactions in dilute aqueous solution. Assume that only normal salts, in which all acidic hydrogens and all hydroxide ions have reacted, are formed.

1. $HF + Ca(OH)_2$
2. $Ba(OH)_2 + HI$
3. $HCl + Ni(OH)_2$
4. $NH_3 + H_2S$

Plan

We will consider both strengths and solubilities and apply the rules for writing equations discussed in Chapter 4. The systems are:

1. weak soluble acid with strong soluble base; soluble salt
2. strong soluble base with strong soluble acid; soluble salt
3. strong soluble acid with insoluble base; soluble salt
4. weak soluble base with weak soluble acid; soluble salt

Solution

1. (a) $2HF(aq) + Ca(OH)_2(aq) \rightarrow CaF_2(aq) + 2H_2O(\ell)$
   (b) $2HF(aq) + [Ca^{2+}(aq) + 2OH^-(aq)] \rightarrow [Ca^{2+}(aq) + 2F^-(aq)] + 2H_2O(\ell)$
   (c) $HF(aq) + OH^-(aq) \rightarrow F^-(aq) + H_2O(\ell)$

2. (a) $Ba(OH)_2(aq) + 2HI(aq) \rightarrow BaI_2(aq) + 2H_2O(\ell)$
   (b) $[Ba^{2+}(aq) + 2OH^-(aq)] + 2[H^+(aq) + I^-(aq)] \rightarrow [Ba^{2+}(aq) + 2I^-(aq)] + 2H_2O(\ell)$
   (c) $OH^-(aq) + H^+(aq) \rightarrow H_2O(\ell)$

3. (a) $2HCl(aq) + Ni(OH)_2(s) \rightarrow NiCl_2(aq) + 2H_2O(\ell)$
   (b) $2[H^+(aq) + Cl^-(aq)] + Ni(OH)_2(s) \rightarrow [Ni^{2+}(aq) + 2Cl^-(aq)] + 2H_2O(\ell)$
   (c) $2H^+(aq) + Ni(OH)_2(s) \rightarrow Ni^{2+}(aq) + 2H_2O(\ell)$

4. (a) $2NH_3(aq) + H_2S(aq) \rightarrow (NH_4)_2S(aq)$
   (b) $2NH_3(aq) + H_2S(aq) \rightarrow [2NH_4^+(aq) + S^{2-}(aq)]$
   (c) $2NH_3(aq) + H_2S(aq) \rightarrow 2NH_4^+(aq) + S^{2-}(aq)$

Example 10-4 Equations for Acid-Base Reactions

Write the same type of equations as in Example 10-3 for acid-base reactions to produce the following soluble salts. 1. K_2S 2. $NaHCO_3$ 3. $LiBr$ (The salt $NaHCO_3$ is an **acidic salt**, because it can react with additional base to yield the normal salt. It has one unreacted acidic hydrogen.)

Plan

We inspect the formulas to decide which acids and bases are needed to make the salts. We then follow the rules for writing the equations discussed in Chapter 4.
1. K_2S made from KOH, a strong soluble base, and H_2S, a weak soluble acid.
2. $NaHCO_3$ made from NaOH, a strong soluble base, and H_2CO_3, a weak soluble acid.
3. LiBr made from LiOH, a strong soluble base and HBr, a strong soluble acid.

Solution

1. (a) $2KOH(aq) + H_2S(aq) \rightarrow K_2S(aq) + 2H_2O(\ell)$
   (b) $2[K^+(aq) + OH^-(aq)] + H_2S(aq)] \rightarrow [2K^+(aq) + S^{2-}(aq)] + 2H_2O(\ell)$
   (c) $2OH^-(aq) + H_2S(aq) \rightarrow S^{2-}(aq) + 2H_2O(\ell)$

2. (a) $NaOH(aq) + H_2CO_3(aq) \rightarrow NaHCO_3(aq) + H_2O(\ell)$
   (b) $[Na^+(aq) + OH^-(aq)] + H_2CO_3(aq) \rightarrow [Na^+(aq) + HCO_3^-(aq)] + H_2O(\ell)$
   (c) $OH^-(aq) + H_2CO_3(aq) \rightarrow HCO_3^-(aq) + H_2O(\ell)$

3. (a) $LiOH(aq) + HBr(aq) \rightarrow LiBr(aq) + H_2O(\ell)$
   (b) $[Li^+(aq) + OH^-(aq)] + [H^+(aq) + Br^-(aq)] \rightarrow [Li^+(aq) + Br^-(aq)] + H_2O(\ell)$
   (c) $OH^-(aq) + H^+(aq) \rightarrow H_2O(\ell)$

10-5 The Lewis Theory

The basic ideas of the most comprehensive classification of acids and bases were stated by G. N. Lewis. A **Lewis acid** is any species that accepts a share in an electron pair. A **Lewis base** is any species which "donates" an electron pair to form a coordinate covalent bond. **Neutralization** is the formation of a coordinate covalent bond.

Chapter 10 Reactions in Aqueous Solutions I: Acids, Bases, and Salts

In the example below, the NH₃ provides the electron pair from which the coordinate covalent bond is made; therefore, NH₃ is the base. The HBr provides the H⁺ that accepts a share in the electron pair; therefore, HBr is the acid.

$$H\text{-}\ddot{\underset{..}{Br}}\text{:}(g) + H\text{-}\underset{\underset{H}{|}}{\overset{..}{N}}\text{-}H (g) \rightarrow H\text{-}\underset{\underset{H}{|}}{\overset{\overset{H}{|}}{N}}\text{-}H\ ^{+}\ :\ddot{\underset{..}{Br}}\text{:}^{-} (s)$$

acid   base

In the example below, the NH₃ provides the electron pair from which the coordinate covalent bond is made; therefore, NH₃ is the base. The H₂O provides the H⁺ that accepts a share in the electron pair; therefore, the H₂O is the acid.

$$H\text{-}\underset{\underset{H}{|}}{\overset{..}{N}}\text{-}H (g) + H\text{-}\underset{\underset{H}{|}}{\overset{..}{O}}\text{:}(\ell) \rightleftharpoons H\text{-}\underset{\underset{H}{|}}{\overset{\overset{H}{|}}{N}}\text{-}H\ ^{+} (aq) + H\text{-}\ddot{\underset{..}{O}}\text{:}^{-}(aq)$$

base   acid   acid   base

In the example below, the Cl⁻ provides the electron pair from which the coordinate covalent bond is made; therefore, KCl is the base. The AlCl₃ accepts a share in the electron pair; therefore, the AlCl₃ is the acid.

$$\underset{\underset{:\ddot{Cl}:}{|}}{\overset{\overset{:\ddot{Cl}:}{|}}{Al}}\text{-}\ddot{\underset{..}{Cl}}\text{:}(s) + K^{+}\ :\ddot{\underset{..}{Cl}}\text{:}^{-}(s) \rightarrow K^{+}\ \ :\ddot{Cl}\text{-}\underset{\underset{:\ddot{Cl}:}{|}}{\overset{\overset{:\ddot{Cl}:}{|}}{Al}}\text{-}\ddot{\underset{..}{Cl}}\text{:}^{-} (s)$$

acid   base

This third reaction is not an acid-base reaction by the previous definitions, but is one by the Lewis definition.

Example 10-5 Acid-Base Definitions

Write the equations below with Lewis structures and tell under what definitions they would be classified as acid-base reactions.

1. $HCl + KOH \rightarrow KCl + H_2O$
2. $HI + H_2O \rightarrow H_3O^+ + I^-$
3. $CH_3NH_2 + H_2O \rightleftharpoons CH_3NH_3^+ + OH^-$
4. $BCl_3 + CH_3NH_2 \rightarrow CH_3NH_2BCl_3$
5. $SiF_4 + 2\ F^- \rightarrow SiF_6^{2-}$

Plan
We will use the rules from Chapter 7 to write the Lewis structures. We then analyze the systems according to the acid-base definitions.

Solution

1. $H-\ddot{\underset{..}{Cl}}:$ + K^+ $:\ddot{O}-H^-$ → K^+ $:\ddot{\underset{..}{Cl}}:^-$ + $H-\ddot{O}:$
   $|$
   H

   Arrhenius and Brønsted-Lowry and Lewis

2. $H-\ddot{\underset{..}{I}}:$ + $H-\ddot{O}:$ → K^+ $:\ddot{\underset{..}{I}}:^-$ + $H-\ddot{O}-H^+$
   $||$
   HH

   Brønsted-Lowry and Lewis

3. $H-\underset{\underset{H}{|}}{\overset{\overset{H}{|}}{C}}-\underset{\underset{H}{|}}{\overset{\overset{H}{|}}{N}}:$ + $H-\ddot{O}:$ ⇌ $H-\underset{\underset{H}{|}}{\overset{\overset{H}{|}}{C}}-\underset{\underset{H}{|}}{\overset{\overset{H}{|}}{N}}-H^+$ + $:\ddot{O}-H^-$
   $|$
   H

   Brønsted-Lowry and Lewis

4. $H-\underset{\underset{H}{|}}{\overset{\overset{H}{|}}{C}}-\underset{\underset{H}{|}}{\overset{\overset{H}{|}}{N}}:$ + $\underset{\underset{:\ddot{Cl}:}{|}}{\overset{\overset{:\ddot{Cl}:}{|}}{B}}-\ddot{\underset{..}{Cl}}:$ → $H-\underset{\underset{H}{|}}{\overset{\overset{H}{|}}{C}}-\underset{\underset{H}{|}}{\overset{\overset{H}{|}}{N}}-\underset{\underset{:\ddot{Cl}:}{|}}{\overset{\overset{:\ddot{Cl}:}{|}}{B}}-\ddot{\underset{..}{Cl}}:$

   Lewis only

5. $\underset{\underset{:\ddot{F}:}{|}}{\overset{\overset{:\ddot{F}:}{|}}{:\ddot{F}-Si-\ddot{F}:}}$ + $2 :\ddot{\underset{..}{F}}:^-$ → $[:\ddot{F}-Si(\ddot{F})_5:]^{2-}$

   Lewis only

Chapter 10 Reactions in Aqueous Solutions I: Acids, Bases, and Salts

EXERCISES

Acid-Base Definitions

Draw Lewis structures for reactants and products in the equations below. If a structure exists as a mixture of resonance structures show only one of the forms. Classify the reactions according to one or more of the definitions in the chapter. Identify the acids and bases and, for those which are Brønsted-Lowry acid-base reactions, indicate conjugate acid-base pairs.

1. $HF(aq) + N_2H_4(aq) \rightleftharpoons N_2H_5F(aq)$
2. $SnF_4(\ell) + 2LiF(s) \rightarrow Li_2SnF_6(s)$
3. $2HI(aq) + Mg(OH)_2(s) \rightarrow MgI_2(aq) + 2H_2O(\ell)$
4. $OCl^-(aq) + H_2O(\ell) \rightleftharpoons HClO(aq) + OH^-(aq)$
5. $MgO(s) + SO_2(g) \rightarrow MgSO_3(s)$

Acid and Base Strengths

6. Rank the following in terms of increasing acid strengths.
   (a) H_2CO_3, H_3BO_3, HNO_3
   (b) H_2SeO_4, H_2SeO_3, H_2SO_4
   (c) PH_3, H_2S, SiH_4, HCl
   (d) $HBrO_2$, $HClO_2$, HIO, $HClO_3$, HIO_2
   (e) $HClO_4$, H_3PO_3, H_2SO_4, H_3PO_4

7. Rank the following in order of increasing base strengths.
   (a) H_2O, H_2Te, H_2Se, H_2S
   (b) Cl^-, Br^-, I^-, F^-
   (c) BrO_2^-, ClO_2^-, IO^-, ClO_3^-
   (d) NO_3^-, NO_2^-
   (e) HSO_4^-, $HTeO_3^-$, $HSeO_3^-$, $HTeO_2^-$

Writing Acid-Base Reactions

Write the (a) molecular, (b) total ionic and (c) net ionic equations for the acid-base reactions indicated below. All solutions are aqusous. Classify each acid and base as soluble, insoluble, strong, weak, etc. In each case, assume the product is a normal salt.

8. $H_2CO_3 + RbOH$

9. $HBrO + (CH_3)_3N(aq)$

10. $HIO + KOH$

11. $Cr(OH)_3 + HClO_4$

12. $HBr + Ba(OH)_2$

13. $CH_3NH_2(aq) + HCl$

Write (a) molecular, (b) total ionic and (c) net ionic equations for acid-base reactions that produce the following salts in aqueous solution.

14. AlI_3

15. Cs_2S

16. $Co(NO_3)_2$

17. $Mg(MnO_4)_2$

18. $Ca(H_2PO_2)_2$ (The salt is soluble and H_3PO_2 is a soluble weak acid.)

… 171

ANSWERS TO EXERCISES

1. 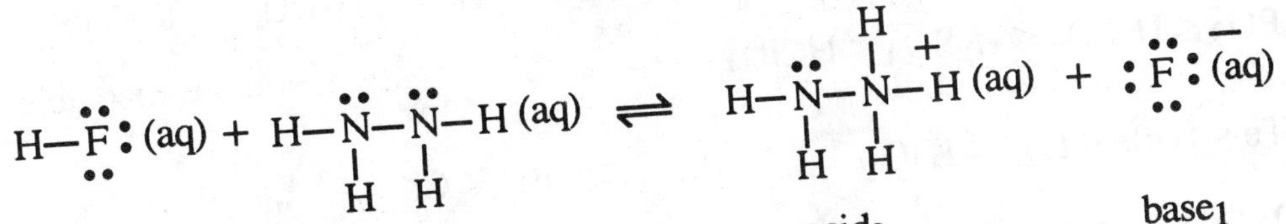 $H-\ddot{F}:$ (aq) $+ H-\underset{H}{\overset{..}{N}}-\underset{H}{\overset{..}{N}}-H$ (aq) $\rightleftharpoons H-\underset{H}{\overset{H}{N}}-\underset{H}{\overset{+}{N}}-H$ (aq) $+ :\ddot{F}:^{-}$ (aq)

   acid$_1$ base$_2$ acid$_2$ base$_1$

   (Brønsted-Lowry and Lewis)

2. 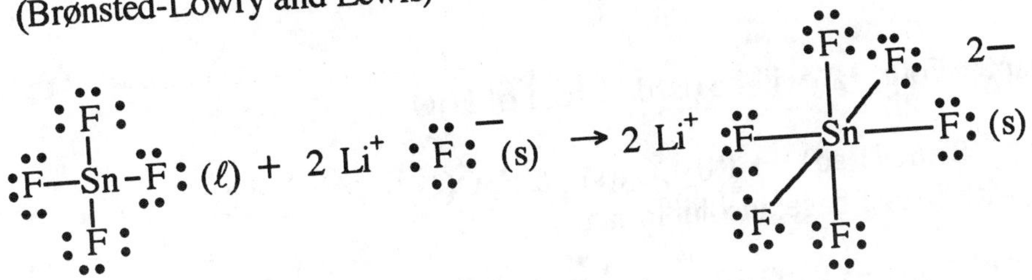 $:\!\ddot{F}\!-\!Sn\!-\!\ddot{F}\!:$ (ℓ) $+ \; 2 \; Li^{+} \; :\!\ddot{F}\!:^{-}$ (s) $\rightarrow 2 \; Li^{+} \; [SnF_6]^{2-}$ (s)

   acid   base

   (Lewis only)

3. 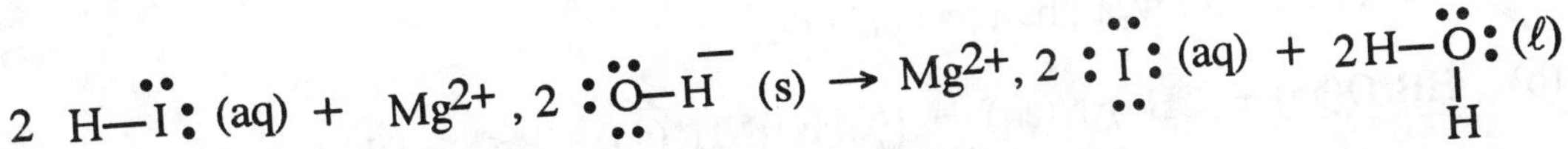 $2 \; H-\ddot{I}:$ (aq) $+ \; Mg^{2+}, \; 2 \; :\!\ddot{O}\!-\!H^{-}$ (s) $\rightarrow Mg^{2+}, \; 2 \; :\!\ddot{I}\!:^{-}$ (aq) $+ \; 2 H-\ddot{O}:$ (ℓ)

   acid$_1$ base$_2$ base$_1$ acid$_2$

   (Arrhenius, Brønsted-Lowry and Lewis)

4. 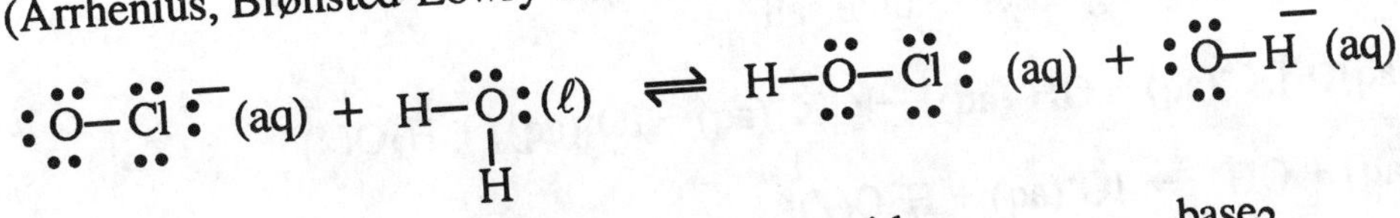 $:\!\ddot{O}\!-\!\ddot{C}l\!:^{-}$ (aq) $+ \; H-\ddot{O}:$ (ℓ) $\rightleftharpoons H-\ddot{O}-\ddot{C}l:$ (aq) $+ \; :\!\ddot{O}\!-\!H^{-}$ (aq)

   base$_1$ acid$_2$ acid$_1$ base$_2$

   (Brønsted-Lowry and Lewis)

5. 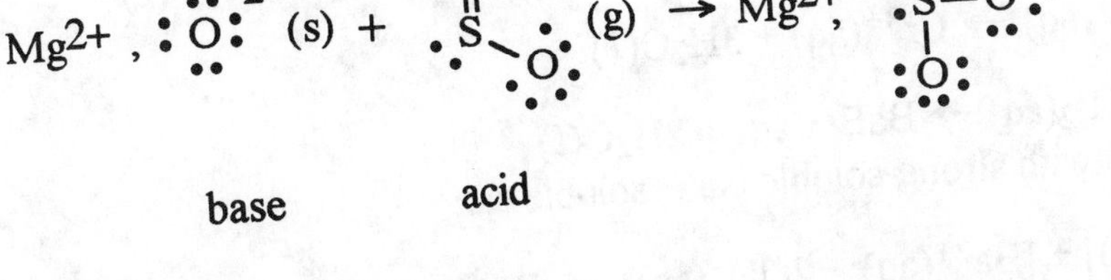 $Mg^{2+}, \; :\!\ddot{O}\!:^{2-}$ (s) $+ \; SO_3$ (g) $\rightarrow Mg^{2+}, \; SO_4^{2-}$

   base   acid

   (Lewis only)

6. (a) H_3BO_3 $H_2CO_3 < HNO_3$
   (b) $H_2SeO_3 < H_2SeO_4 < H_2SO_4$
   (c) $SiH_4 < PH_3 < H_2S < HCl$
   (d) $HIO < HIO_2 < HBrO_2 < HClO_2 < HClO_3$
   (e) $H_3PO_3 < H_3PO_4 < H_2SO_4 < HClO_4$

7. (a) $H_2Te < H_2Se < H_2S < H_2O$
   (b) $I^- < Br^- < Cl^- < F^-$
   (c) $ClO_3^- < ClO_2^- < BrO_2^- < IO^-$
   (d) $NO_3^- < NO_2^-$
   (e) $HSO_4^- < HSeO_3^- < HTeO_3^- < HTeO_2^-$

For brevity, as in Chapter Four, H^+ will be used instead of H_3O^+.

8. (a) $H_2CO_3(aq) + 2RbOH(aq) \rightarrow Rb_2CO_3(s) + 2H_2O(\ell)$
   weak acid with strong base; insoluble salt
   (b) $H_2CO_3 + 2[Rb^+(aq) + OH^-(aq)] \rightarrow Rb_2CO_3(s) + 2H_2O(\ell)$
   (c) $H_2CO_3(aq) + 2Rb^+(aq) + 2OH^-(aq) \rightarrow Rb_2CO_3(s) + 2H_2O(\ell)$

9. (a) $HBrO(aq) + (CH_3)_3N(aq) \rightarrow (CH_3)_3NHBrO(aq)$
   weak acid with weak base; soluble salt
   (b) $HBrO(aq) + (CH_3)_3N(aq) \rightarrow [(CH_3)_3NH^+(aq) + BrO^-(aq)]$
   (c) $HBrO(aq) + (CH_3)_3N(aq) \rightarrow (CH_3)_3NH^+(aq) + BrO^-(aq)$

10. (a) $HIO(aq) + KOH(aq) \rightarrow KIO(aq) + H_2O(\ell)$
    weak acid with strong base; soluble salt
    (b) $HIO(aq) + [K^+(aq) + OH^-(aq)] \rightarrow [K^+(aq) + IO^-(aq)] + H_2O(\ell)$
    (c) $HIO(aq) + OH^- \rightarrow IO^-(aq) + H_2O(\ell)$

11. (a) $Cr(OH)_3(s) + 3HClO_4(aq) \rightarrow Cr(ClO_4)_3(aq) + 3H_2O(\ell)$
    insoluble base with strong acid; soluble salt
    (b) $Cr(OH)_3(s) + 3[H^+(aq) + ClO_4^-(aq)] \rightarrow [Cr^{3+}(aq) + 3ClO_4^-] + 3H_2O(\ell)$
    (c) $Cr(OH)_3(s) + 3H^+(aq) \rightarrow Cr^{3+}(aq) + 3H_2O(\ell)$

12. (a) $2HBr(aq) + Ba(OH)_2(aq) \rightarrow BaBr_2(aq) + 2H_2O(\ell)$
    strong soluble acid with strong soluble base; soluble salt
    (b) $2[H^+(aq) + Br^-(aq)] + [Ba^{2+}(aq) + 2OH^-(aq)] \rightarrow [Ba^{2+}(aq) + 2Br^-(aq)] + 2H_2O(\ell)$
    (c) $H^+(aq) + OH^-(aq) \rightarrow H_2O(\ell)$

Chapter 10 Reactions in Aqueous Solutions I: Acids, Bases, and Salts

13. (a) $CH_3NH_2(aq) + HCl(aq) \rightarrow CH_3NH_3Cl(aq)$
    weak soluble base with strong acid; soluble salt

    (b) $CH_3NH_2(aq) + [H^+(aq) + Cl^-(aq)] \rightarrow [CH_3NH_3^+(aq) + Cl^-(aq)]$

    (c) $CH_3NH_2(aq) + H^+(aq) \rightarrow CH_3NH_3^+(aq)$

14. (a) $Al(OH)_3(s) + 3HI(aq) \rightarrow AlI_3(aq) + 3H_2O(\ell)$
    insoluble base with strong soluble acid; soluble salt

    (b) $Al(OH)_3(s) + 3[H^+(aq) + I^-(aq)] \rightarrow [Al^{3+}(aq) + 3I^-(aq)] + 3H_2O(\ell)$

    (c) $Al(OH)_3(s) + 3H^+(aq) \rightarrow Al^{3+}(aq) + 3H_2O(\ell)$

15. (a) $2CsOH(aq) + H_2S(aq) \rightarrow Cs_2S(aq) + 2H_2O(\ell)$
    strong soluble base with weak acid; soluble salt

    (b) $2[Cs^+(aq) + OH^-(aq)] + H_2S(aq) \rightarrow [2Cs^+(aq) + S^{2-}(aq)] + 2H_2O(\ell)$

    (c) $2OH^-(aq) + H_2S(aq) \rightarrow S^{2-}(aq) + 2H_2O(\ell)$

16. (a) $Co(OH)_2(s) + 2HNO_3(aq) \rightarrow Co(NO_3)_2(aq) + 2H_2O(\ell)$
    insoluble base with strong acid; soluble salt

    (b) $Co(OH)_2(s) + 2[H^+(aq) + NO_3^-(aq)] \rightarrow [Co^{2+}(aq) + 2NO_3^-(aq)] + 2H_2O(\ell)$

    (c) $Co(OH)_2(s) + 2H^+(aq) \rightarrow Co^{2+}(aq) + 2H_2O(\ell)$

17. (a) $Mg(OH)_2(s) + 2HMnO_4(aq) \rightarrow Mg(MnO_4)_2(aq) + 2H_2O(\ell)$
    insoluble base with weak acid; soluble salt

    (b) $Mg(OH)_2(s) + 2HMnO_4(aq) \rightarrow [Mg^{2+}(aq) + 2MnO_4^-(aq)] + 2H_2O(\ell)$

    (c) $Mg(OH)_2(s) + 2HMnO_4(aq) \rightarrow Mg^{2+}(aq) + 2MnO_4^-(aq) + 2H_2O(\ell)$

18. (a) $Ca(OH)_2(aq) + 2H_3PO_2(aq) \rightarrow Ca(H_2PO_2)_2(aq) + 2H_2O(\ell)$
    strong soluble base with weak acid; soluble salt

    (b) $[Ca^{2+}(aq) + 2OH^-(aq)] + 2H_3PO_2(aq) \rightarrow [Ca^{2+}(aq) + 2H_2PO_2^-(aq)] + 2H_2O(\ell)$

    (c) $2OH^-(aq) + 2H_3PO_2(aq) \rightarrow 2H_2PO_2^-(aq) + 2H_2O(\ell)$

Chapter 11

Reactions in Aqueous Solutions II: Calculations

11-1 Calculations Involving Molarity

The meaning of molarity and the use of molarity in stoichiometric calculations were discussed in Chapter Three. When working with stoichiometry problems, remember to use balanced equations to determine proper reaction ratios (either moles or millimoles).

Example 11-1 Acid-Base Reactions

A 25.00 mL sample of H_2Se requires 38.27 mL of 0.2118 M NaOH for complete neutralization. What is the molarity of the H_2Se solution?

Plan

We must (a) write the balanced equation for this acid-base reaction, and (b) use the stoichiometric techniques discussed in Chapter Three to solve the problem.

mL NaOH solution → mmol NaOH → mmol H_2Se → $M\, H_2Se$ solution

Solution

The neutralization reaction produces a normal salt and water:

$$2NaOH + H_2Se \rightarrow Na_2Se + 2H_2O$$

$$?\text{ mol } H_2Se = 38.27 \text{ mL NaOH} \times \frac{0.1059 \text{ mol NaOH}}{1 \text{ L NaOH}} \times \frac{1 \text{ mmol } H_2Se}{2 \text{ mmol NaOH}} = 4.053 \text{ mmol } H_2Se$$

$$?\, M\, H_2Se = \frac{4.053 \text{ mmol } H_2Se}{25.00 \text{ mL}} = \boxed{0.1621\, M\, H_2Se}$$

Example 11-2 Acid-Base Reactions

What are the molarities of all dissolved species in the solution that results from the reaction in Example 11-1?

Plan

Because the reaction is a complete neutralization and there is no limiting reactant, the number of moles of Na_2Se that forms can be found by considering the number of millimoles of either reactant. To find the molarity of the salt we need to divide that number of millimoles by the volume. The final volume is the sum of the two reacting volumes–25.00 mL + 38.27 mL = 63.27 mL. Finally, we take into account the dissociation of the salt.

Solution

$$? \text{ mmol } Na_2Se = 4.053 \text{ mmol } H_2Se \times \frac{1 \text{ mmol } Na_2Se}{1 \text{ mmol } H_2Se} = 4.053 \text{ mmol } Na_2Se$$

$$? M \, Na_2Se = \frac{4.053 \text{ mmol } Na_2Se}{63.27 \text{ mL solution}} = 0.06406 \, M \, Na_2Se$$

The Na_2Se is a strong electrolyte and will be totally dissociated; the solution will contain:

$$? M \, Na^+ = \frac{0.06406 \text{ mol } Na_2Se}{1 \text{ L solution}} \times \frac{2 \text{ mol } Na^+}{1 \text{ mol } Na_2Se} = \boxed{0.1281 \, M \, Na^+}$$

$$? M \, Se^{2-} = \frac{0.06406 \text{ mol } Na_2Se}{1 \text{ L solution}} \times \frac{1 \text{ mol } Se^{2-}}{1 \text{ mol } Na_2Se} = \boxed{0.06406 \, M \, Se^{2-}}$$

Example 11-3 Acid Base Reactions

What are the molarities of all dissolved species after 20.0 mL of $0.100M$ HBr is reacted with 15.0 mL of $0.100M$ KOH?

Plan

We first determine the initial number of millimoles of reactants. We then write the balanced equation and construct a reaction summary.

Solution

$$\text{initial millimoles of HBr} = 20.0 \text{ mL} \times \frac{0.100 \text{ mmol}}{\text{mL}} = 2.00 \text{ mmol HBr}$$

initial millimoles of KOH = 15.0 mL × $\dfrac{0.100 \text{ mmol}}{\text{mL}}$ = 1.50 mmol KOH

	KOH +	HBr	→	KBr +	H$_2$O
Initial	1.50 mmol	2.00 mmol		0.00 mmol	
Change	−1.50 mmol	−1.50 mmol		+1.50 mmol	
Final	0.00 mmol	0.50 mmol		1.50 mmol	

KOH is the limiting reactant and will be completely consumed. The total volume after the reaction is 15.0 mL + 20.0 mL = 35.0 mL and the molarities are:

$$? M \text{ HBr} = \dfrac{0.50 \text{ mmol HBr}}{35.0 \text{ mL}} = 0.014 \, M \text{ HBr}$$

$$? M \text{ KBr} = \dfrac{1.50 \text{ mmol KBr}}{35.0 \text{ mL}} = 0.0429 \, M \text{ KBr}$$

Both HBr and KBr are strong electrolytes and will be totally dissociated. The molarities of ions in the solution are $\boxed{0.014 \, M \text{ H}^+}$ and $\boxed{0.0429 \, M \text{ K}^+}$. The total bromide concentration is $0.014 \, M \text{ Br}^- + 0.0429 \, M \text{ Br}^- = \boxed{0.057 \, M \text{ Br}^-}$.

Example 11-4 Volume of Acid to Neutralize a Base

What volume of $0.315 M$ HClO$_3$ solution neutralizes 30.0 mL of $0.200 \, M$ Ba(OH)$_2$ solution?

Plan

We must first determine the neutralization reaction. The volume can then be found using stoichiometric techniques.

mL Ba(OH)$_2$ solution → mol Ba(OH)$_2$ → mol HClO$_3$ → mL HClO$_3$

Solution

The neutralization equation is: 2HClO$_3$ + Ba(OH)$_2$ → Ba(NO$_3$)$_2$ + 2H$_2$O.

$$? \text{ mL HClO}_3 = 30.0 \text{ mL Ba(OH)}_2 \times \dfrac{0.200 \text{ mol Ba(OH)}_2}{1000 \text{ mL Ba(OH)}_2} \times \dfrac{2 \text{ mol HClO}_3}{1 \text{ mol Ba(OH)}_2}$$

$$\times \dfrac{1000 \text{ mL HClO}_3}{0.315 \text{ mol HClO}_3} = \boxed{38.1 \text{ mL HClO}_3}$$

11-2 Standardization and Acid-Base Titrations

Standardization of acids or bases is the process by which the concentration of a solution of an acid or base is determined precisely by allowing a known volume to react with an accurately determined mass of a primary standard base or acid, or with an accurately determined volume of base or acid of known concentration. A variety of methods using either moles or millimoles is possible. Treatment of titrations is subclassified into two main methods as follow.

11-3 The Mole Method and Molarity

Example 11-1 is one example of a typical calculation of using the results of a titration to determine molarity. The technique is the one employed in Chapter Three. Several more examples follow to further illustrate the concept.

Example 11-5 Standardization of a Base Solution

Potassium hydrogen phthalate (abbreviated KHP) is a monoprotic acid with formula $KC_6H_4(COO)COOH$. KHP is often used as a primary standard for determining base concentration. If a 28.11 mL sample of a $Ca(OH)_2$ solution is required to neutralize 0.1714 g of KHP, what is the molarity of the $Ca(OH)_2$ solution?

Plan

First, we must write the neutralization reaction. We then use stoichiometric techniques as in previous examples to determine the molarity of the base.

$$g\ KHP \rightarrow mol\ KHP \rightarrow mol\ Ca(OH)_2 \rightarrow M\ Ca(OH)_2$$

Solution

The neutralization reaction is: $Ca(OH)_2 + 2KHP \rightarrow Ba(KP)_2 + 2H_2O$.

$$?\ mol\ Ca(OH)_2 = 0.1714\ g\ KHP \times \frac{1\ mol\ KHP}{204.2\ g\ KHP} \times \frac{1\ mol\ Ca(OH)_2}{2\ mol\ KHP}$$

$$= 4.197 \times 10^{-4}\ mol\ Ca(OH)_2$$

$$?\ M\ Ca(OH)_2 = \frac{4.197 \times 10^{-4}\ mol\ Ca(OH)_2}{0.02811\ L\ Ca(OH)_2} = \boxed{0.01493\ M\ Ca(OH)_2}$$

Example 11-6 Determination of Percent Acid

During "spoiling", fruit juices often oxidize to create "vinegar", which is a dilute acetic acid (CH_3COOH) solution. A sample of the beverage with a mass of 2.578 grams requires 18.62 milliliters of 0.09465 M NaOH for neutralization. What is the mass percent acetic acid in the sample, assuming that no other acidic substances are present?

Plan

We first must write the neutralization equation and then follow the normal stoichiometric techniques to find the mass of acetic acid. From the mass, we use the definition of mass percent to find the requested information.

L NaOH → mol NaOH → mol CH_3COOH → g CH_3COOH

Solution

The neutralization equation is: $CH_3COOH + NaOH \rightarrow NaCH_3COO + H_2O$

$$? \text{ mol g } CH_3COOH = 0.01862 \text{ L NaOH} \times \frac{0.09465 \text{ mol NaOH}}{1 \text{ L solution}} \times \frac{1 \text{ mol } CH_3COOH}{1 \text{ mol NaOH}}$$

$$\times \frac{60.06 \text{ g } CH_3COOH}{1 \text{ mol } CH_3COOH} = 0.1059 \text{ g } CH_3COOH$$

$$\% \, CH_3COOH = \frac{0.1059 \text{ g } CH_3COOH}{2.578 \text{ g sample}} \times 100 = \boxed{4.108\%}$$

11-4 Equivalent Weights and Normality

The concept of the "equivalent" finds application mainly in the study of acid-base reactions. One **equivalent weight (eq)** of an acid is the mass of that acid that produces one mole of hydronium ions or reacts with 1 mole of hydroxide ions. One equivalent weight of a base is, conversely, the mass of a base that provides one mole of hydroxide ions or reacts with one mole of hydronium ions. One equivalent weight of an acid exactly reacts with one equivalent weight of a base. The equivalent weight of an acid or base cannot be determined by just looking at the formula of the acid or base unless complete neutralization is assumed. Instead, the equivalent weight depends upon the reaction in which the compound is used. The following reactions illustrate this concept.

1. $2HBr$ + $Ca(OH)_2$ → $CaBr_2 + 2H_2O$
    2 mol   1 mol
    161.8 g   74.1 g
    2 eq   2 eq

One mole of HBr supplies one mole of H^+ but one mole of $Ca(OH)_2$ supplies two moles of OH^-. This indicates that 1 mol HBr = 1 equivalent = 80.9 g HBr; however, 1/2 mol $Ca(OH)_2$ = 1 equivalent = 37.0 g $Ca(OH)_2$.

2. H_3PO_4 + KOH → KH_2PO_4 + H_2O
   1 mol   1 mol
   98.0 g   56.1 g
   1 eq   1 eq

In this reaction, the H_3PO_4 reacts to form the acidic salt in which only one of the possible hydrogen ions is neutralized. Therefore, 1 mol H_3PO_4 = 1 equivalent = 98.0 g H_3PO_4, but 1 mol KOH = 1 equivalent = 56.1 g KOH.

3. H_3PO_4 + 3KOH → K_3PO_4 + $3H_2O$
   1 mol   3 mol
   98.0 g   168.3 g
   3 eq   3 eq

Contrast this with the previous example. In this case, all three possible hydrogen ions are neutralized. In this case, 1/3 mol H_3PO_4 = 1 equivalent = 98.0/3 = 32.7 g H_3PO_4; and, 1 mol KOH = 1 equivalent = 56.1 g KOH.

Note that in all these cases, the number of equivalents of acid equals the number of equivalents of base even though the number of moles may not be equal. This is true for any acid-base reaction. A concentration unit called **normality**, N, is designed to take advantage of this fact. The normality of a solute is the number of equivalents of solute per liter of solution (or milliequivalents per milliliter of solution). Mathematically:

$$N = \frac{\text{equivalents of solute}}{\text{L of solution}} = \frac{\text{millequivalents of solute}}{\text{mL of solution}}$$

The relationship between molarity and normality depends upon the particular acid-base reaction:

$$N = M \times \frac{\text{number of equivalents of solute}}{1 \text{ mol of solute}}$$

Since the number of equivalents of acid must equal the number of equivalents of base, for any acid-base reaction in which normality is employed:

$$V_{acid} \times N_{acid} = V_{base} \times N_{base}$$

Example 11-7 Concentration of a Solution

What are the normality and the molarity of a sulfuric acid, H_2SO_4, solution if 30.00 mL of it are required to completely neutralize 24.00 mL of 0.2000 N NaOH?

Plan

To find the normality, we know three of the four variables in the relationship above so we can solve for the fourth. To find the molarity from the normality, we must first write the neutralization reaction and use the relationship between normality and molarity.

Solution

$mL_{acid} \times N_{acid} = mL_{base} \times N_{base}$, therefore:

$$? N\, H_2SO_4 = \frac{mL_{base} \times N_{base}}{mL_{acid}} = \frac{24.00 \text{ mL} \times 0.2000 N}{30.00 \text{ mL}} = \boxed{0.1600\, N\, H_2SO_4}$$

The neutralization reaction is: $H_2SO_4 + 2\, NaOH \rightarrow Na_2SO_4 + 2\, H_2O$
One mole of H_2SO_4 reacts with two moles of hydroxide ion, therefore there are 2 equivalents per one mole of H_2SO_4.

$$? M\, H_2SO_4 = \frac{0.1600 \text{ equivalents}}{L} \times \frac{1 \text{ mol } H_2SO_4}{2 \text{ equivalents}} = \boxed{0.08000\, M\, H_2SO_4}$$

Example 11-8 Standardization of an Acid Solution

Anhydrous sodium carbonate is often used for the standardization of acidic solutions. Carbonate ion reacts with two hydrogen ions, producing CO_2 and water in the process. If 21.28 mL of HNO_3 is required to react with 0.1648 g of Na_2CO_3, what is the normality of the HNO_3 solution?

Plan

Because one mole of Na_2CO_3 can react with two moles of hydrogen ion, there are two equivalents in each mole of Na_2CO_3. We can find the equivalent weight of the base and then use the knowledge that the number of equivalents of acid is equal to the number of equivalents of base. Finally, we use the definition of normality as equivalents per liter.

Solution

$$\text{? eq Na}_2\text{CO}_3 = 0.1648 \text{ g Na}_2\text{CO}_3 \times \frac{1 \text{ mol Na}_2\text{CO}_3}{106.0 \text{ g Na}_2\text{CO}_3} \times \frac{2 \text{ eq Na}_2\text{CO}_3}{1 \text{ mol Na}_2\text{CO}_3}$$

$$= 0.003109 \text{ eq Na}_2\text{CO}_3 = 0.003109 \text{ eq HNO}_3$$

$$\text{? } N \text{ HNO}_3 = \frac{0.003109 \text{ eq HNO}_3}{0.02128 \text{ L}} = \boxed{0.1461 \text{ } N \text{ HNO}_3}$$

Example 11-9 Percent Purity of an Acid

A 0.5374 gram sample of impure diprotic tartaric acid, $(C_2H_4O_2)(COOH)_2$ required 17.68 mL of 0.3748 N NaOH for complete neutralization. If no other acidic components are in the sample, what is the percent purity of the sample?

Plan

We can find the number of equivalents of base and we know that the number of equivalents of acid must equal the number of equivalents of base. We also know that because the acid is diprotic, its equivalent weight is one half of its molecular weight of 150.10 g/mol. We can use this information to find the mass and the percent of acid in the sample.

Solution

$$\text{? eq NaOH} = 0.01768 \text{ L} \times \frac{0.3748 \text{ eq NaOH}}{\text{L}} = 0.006626 \text{ eq NaOH}$$

The number of equivalents of acid is also 0.006626 eq.

$$\text{? g } (C_2H_4O_2)(COOH)_2 = 0.006626 \text{ eq acid} \times \frac{75.05 \text{ g acid}}{1 \text{ eq acid}} = 0.4973 \text{ g acid}$$

$$\text{? \% } (C_2H_4O_2)(COOH)_2 = \frac{0.4973 \text{ g acid}}{0.5374 \text{ g sample}} \times 100 = \boxed{92.54\% \text{ } (C_2H_4O_2)(COOH)_2}$$

11-5 The Change-in-Oxidation-Number Method

The **change-in-oxidation-number (CON) method** for balancing redox equations is based on equalizing the total increases and decreases in oxidation numbers. The general procedure is:

1. Write as much of the overall unbalanced reaction as possible.

2. Use the rules in Chapter 4 to assign oxidation numbers to find those elements that undergo oxidation and reduction.
3. Draw a bracket to connect atoms of the element being oxidized. Show the increase in oxidation number per atom. Repeat with the element being reduced and show the decrease in oxidation number per atom.
4. Insert coefficients into the equation to make the total increase and decrease in oxidation numbers equal.
5. Balance remaining atoms and charges by inspection.

Example 11-10 Balancing Redox Equations (CON method)

Balance the redox equation: $Mn + FeCl_3 + KOH \rightarrow Mn(OH)_2 + FeCl_2 + KCl$

Plan

A formula unit equation is already provided. We then follow the five step procedure to balance the equation.

Solution

$$\overset{0}{Mn} + \overset{+3\ -1}{FeCl_3} + \overset{+1\ -2+1}{KOH} \rightarrow \overset{+2\ -2\ +1}{Mn(OH)_2} + \overset{+2\ -1}{FeCl_2} + \overset{+1\ -1}{KCl}$$

Oxidation Numbers	Change/Atom	Equalizing Changes Gives
Mn = 0 → Mn = +2	+2	1(+2) = +2
Fe = +3 → Fe = +2	−1	2(−1) = −2

The coefficients needed to equalize changes are transferred back to the original equation.

$$Mn + 2FeCl_3 + KOH \rightarrow Mn(OH)_2 + 2FeCl_2 + KCl$$

Finally, the K, Cl, O and H are balanced by inspection. The KCl must be multiplied by 2 to equalize the Cl and multiplication of the KOH by 2 provides the 2 K atoms and the 2 OH− ions needed on the right. The completed equation is:

$$\boxed{Mn + 2\ FeCl_3 + 2KOH \rightarrow Mn(OH)_2 + 2FeCl_2 + 2KCl}$$

Atom balance: 2 Fe, 6 Cl, 1 Mn, 2 O, 2 K   Charge balance: 0

11-6 Addition of H^+ or OH^- and/or Water

Often, only enough information is provided to construct an incomplete, as well as unbalanced equation. In these cases, atom and charge balances may be achieved by adding, as necessary, OH^- and/or H_2O (but not H^+) in reactions in **basic** solution or H^+ and/or H_2O (but not OH^-) for reactions in **acidic** solution. To decide where to add these species, the following scheme is useful:

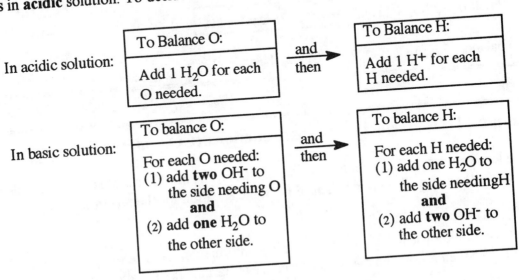

Example 11-11 Balancing Redox Equations (CON method)

Chlorous acid, $HClO_2$, oxidizes sulfur dioxide, SO_2, to sulfate ions and is reduced to chloride ions in acidic solution. Write the balanced net ionic equation for the reaction.

Plan

We use the information provided and our knowledge of the other ion formulas to write the partial ionic equation. We then follow the five step procedure adding H^+ and/or H_2O as needed because this is an acidic solution.

Solution

The initial information yields the partial ionic equation:

$$\overset{+1\ +3\ -2}{HClO_2} + \overset{+4\ -2}{SO_2} \rightarrow \overset{-1}{Cl^-} + \overset{+6\ -2}{SO_4^{2-}}$$

Oxidation Numbers	Change/Atom	Equalizing Changes Gives
$Cl = +3 \rightarrow Cl = -1$	-4	$1(-4) = -4$
$S = +4 \rightarrow S = +6$	$+2$	$2(+2) = +4$

The coefficients needed to equalize changes are transferred back to the equation.

$$HClO_2 + 2SO_2 \rightarrow Cl^- + 2SO_4^{2-}$$

Because this is an acidic solution, the oxygen atoms can be balanced by adding $2H_2O$ molecules to the left side, followed by adding 5 H^+ to the right side to balance both the hydrogen atoms and the charge. The completed equation is:

$$\boxed{HClO_2 + 2SO_2 + 2H_2O \rightarrow Cl^- + 2SO_4^{2-} + 5H^+}$$

Atom balance: 5 H, 1 Cl, 8 O, 2 S   Charge balance: 0

Example 11-12 Balancing Redox Equations (CON method)

Nitrate ions can oxidize sulfite ions in acidic solution. The products are nitrogen dioxide, NO_2, and sulfate ions. Write the balanced net ionic equation for the reaction.

Plan

We can use our knowledge of ions and the given information to write the partial ionic equation. We then follow the five step procedure to balance the equation.

Solution

The partial ionic equation is:

$$\overset{+5\ -2}{NO_3^-} + \overset{+4\ -2}{SO_3^{2-}} \rightarrow \overset{+4\ -2}{NO_2} + \overset{+6\ -2}{SO_4^{2-}}$$

Oxidation Numbers	Change/Atom	Equalizing Changes Gives
N = +5 → N = +4	-1	2(-1) = -2
S = +4 → S = +6	+2	1(+2) = +2

The corrected coefficients are inserted into the equation to give:

$$2NO_3^- + SO_3^{2-} \rightarrow 2NO_2 + SO_4^{2-}$$

To balance oxygen atoms 1 H_2O is added to the right. To balance H atoms and charge 2H^+ are added to the left. The completed net ionic equation is:

$$\boxed{2NO_3^- + SO_3^{2-} + 2H^+ \rightarrow 2NO_2 + SO_4^{2-} + H_2O}$$

Atom balance: 2 N, 1 S, 9 O, 2 H   Charge balance: -2

Example 11-13 Balancing Redox Equations (CON method)

In basic solution, chlorate, ClO_3^-, ions can oxidize dinitrogen tetroxide, N_2O_4, to nitrate ions, forming chloride ions as the other product. Write the balanced net ionic equation for this reaction.

Plan
We can use our knowledge of ions and the given information to write the partial ionic equation. We then follow the five step procedure to balance the equation.

Solution
The partial ionic equation is:

$$\overset{+5\,-2}{ClO_3^-} + \overset{+4\,-2}{N_2O_4} \rightarrow \overset{+5\,-2}{NO_3^-} + \overset{-1}{Cl^-}$$

Oxidation Numbers	Change/Atom	Equalizing Changes Gives
$Cl = +5 \rightarrow Cl = -1$	-6	$1(-6) = -6$
$N = +4 \rightarrow N = +5$	$+1$	$6(+1) = +6$

Since nitrogen atoms appear in groups of 2 in dinitrogen tetroxide, only half that many N_2O_4 molecules are needed. The corrected coefficients are inserted into the equation to give:

$$ClO_3^- + 3N_2O_4 \rightarrow 6NO_3^- + Cl^-$$

The total charge on the left is -1 and on the right is -7. We can balance charge by adding $6OH^-$ to the left hand side and finish balancing the oxygen and hydrogen by adding $3H_2O$ to the right hand side. The completed net ionic equation is:

$$\boxed{ClO_3^- + 3N_2O_4 + 6OH^- \rightarrow 6NO_3^- + Cl^- + 3H_2O}$$

Atom balance: 1 Cl, 2 N, 21 O, 6 H   Charge balance: -7

11-7 The Half-Reaction Method

According to the **half-reaction (HR) method** for balancing redox equations, the oxidation and reduction half-reactions are balanced separately and completely. Then the number of electrons gained and lost are made equal and the resulting half-reactions are added to give the overall balanced equation. The general procedure is:

1. Write as much of the overall unbalanced equation as possible.

2. Construct unbalanced oxidation and reduction half-reactions using the complete formulas of polyatomic ions and molecules.
3. Balance all the atoms except H and O by inspection. Then use the chart in Section 4-10 of the text to balance the H and O in each half-reaction.
4. Balance the charge on each half-reaction by adding electrons as reactants or products.
5. Balance the electron transfer by multiplying the balanced half-reactions by appropriate integers.
6. Add the resulting half-reactions and eliminate any common terms to obtain the balanced equation.

Example 11-14 Balancing Redox Equations (HR method)

Nitrate ions oxidize metallic copper to copper(II) ions, Cu^{2+}, and are reduced to nitrogen oxide, NO, in acidic solution. Write the balanced net ionic equation for this reaction.

Plan
We can use our knowledge of ions and the given information to write the partial ionic equation. We then follow the six step procedure to balance the equation.

Solution
The initial ionic equation is:

$$NO_3^- + Cu \rightarrow NO + Cu^{2+}$$

The steps above are then followed to give balanced half reactions:

$NO_3^- \rightarrow NO$ (reduction)
$NO_3^- + 4H^+ \rightarrow NO + 2H_2O$ (acidic solution)
$NO_3^- + 4H^+ + 3e^- \rightarrow NO + 2H_2O$ (balance charge)

$Cu \rightarrow Cu^{2+}$ (oxidation)
$Cu \rightarrow Cu^{2+} + 2e^-$ (balance charge)

The reduction half-reaction must be multiplied by 2 and the oxidation half-reaction by 3 to make electron transfer equal. The two half-reactions are added and electrons cancelled.

$2 [NO_3^- + 4H^+ + 3e^- \rightarrow NO + 2H_2O]$
$3[Cu \rightarrow Cu^{2+} + 2e^-]$

$$\boxed{2NO_3^- + 8H^+ + 3Cu \rightarrow 2NO + 3Cu^{2+} + 4H_2O}$$

Atom balance: 2 N, 6 O, 8 H, 3 Cu   Charge balance: +6

Example 11-15 Balancing Redox Equations (HR method)

Permanganate, MnO_4^-, ions can oxidize hydrogen sulfide, H_2S, to elemental sulfur, S_8, in acidic solution, while being reduced to Mn^{2+} ions. Write the balanced net ionic equation for this reaction.

Plan
We can use our knowledge of ions and the given information to write the partial ionic equation. We then follow the six step procedure to balance the equation.

Solution
The initial ionic equation is:
$$MnO_4^- + H_2S \rightarrow S_8 + Mn^{2+}$$

The steps are then followed as above.

$MnO_4^- \rightarrow Mn^{2+}$ (reduction)
$MnO_4^- + 8H^+ \rightarrow Mn^{2+} + 4H_2O$ (acidic solution)
$MnO_4^- + 8H^+ + 5e^- \rightarrow Mn^{2+} + 4H_2O$ (balance charge)

$8H_2S \rightarrow S_8$ (oxidation)
$8H_2S \rightarrow S_8 + 16H^+$ (acidic solution)
$8H_2S \rightarrow S_8 + 16H^+ + 16e^-$ (balance charge)

The reduction half-reaction must be multiplied by 16 and the oxidation half-reaction by 5 to make electron transfer equal. The two half-reactions are added and like terms cancelled.

$16[MnO_4^- + 8H^+ + 5e^- \rightarrow Mn^{2+} \, 4H_2O]$
$5[8H_2S \rightarrow S_8 + 16H^+ + 16e^-]$
$$\boxed{16MnO_4^- + 48H^+ + 40H_2S \rightarrow 16Mn^{2+} + 5S_8 + 64H_2O}$$

Atom balance: 16 Mn, 40 S, 128 H, 64 O   Charge balance: +32

Example 11-16 Balancing Redox Equations (HR method)

Hydrogen peroxide, H_2O_2, can be oxidized to oxygen gas by $AuCl_4^-$ ions in acidic solutions. Metallic gold and chloride ions are formed in the process. Write the balanced net ionic equation for this reaction.

Plan
We can use our knowledge of ions and the given information to write the partial ionic equation. We then follow the six step procedure to balance the equation.

Solution
The initial ionic equation is:

$$H_2O_2 + AuCl_4^- \rightarrow O_2 + Au + Cl^-$$

The steps are then followed as above.

$H_2O_2 \rightarrow O_2$
$H_2O_2 \rightarrow O_2 + 2H^+$ (acidic solution)   (oxidation)
$H_2O_2 \rightarrow O_2 + 2H^+ + 2e^-$ (balance charge)

$AuCl_4^- \rightarrow Au + Cl^-$ (reduction)
$AuCl_4^- \rightarrow Au + 4Cl^-$ (balance atoms)
$AuCl_4^- + 3e^- \rightarrow Au + 4Cl^-$ (balance charge)

The reduction half-reaction must be multiplied by 2 and the oxidation half-reaction by 3 to make electron transfer equal. The two half-reactions are added and like terms cancelled.

$3[H_2O_2 \rightarrow O_2 + 2H^+ + 2e^-]$
$2[AuCl_4^- + 3e^- \rightarrow Au + 4Cl^-]$

$$\boxed{3H_2O_2 + 2AuCl_4^- \rightarrow 3O_2 + 6H^+ + 2Au + 8Cl^-}$$

Atom balance: 6 H, 6 O, 2 Au, 8 Cl   Charge balance: −2

Example 11-17 Balancing Redox Equations (HR method)

In basic solution, nitrite, NO_2^-, ion can oxidize metallic aluminum to the $Al(OH)_4^-$ ion as it is reduced to ammonia, NH_3. Write the balanced net ionic equation for this reaction.

Plan
We can use our knowledge of ions and the given information to write the partial ionic equation. We then follow the six step procedure to balance the equation.

Solution
The initial ionic equation is:
$$NO_2^- + Al \rightarrow NH_3 + Al(OH)_4^-$$

The steps are then followed as above.

$Al \rightarrow Al(OH)_4^-$ (oxidation)
$Al + 4OH^- \rightarrow Al(OH)_4^-$ (basic solution)
$Al + 4OH^- \rightarrow Al(OH)_4^- + 3e^-$ (balance charge)

$5NO_2^- \rightarrow NH_3$ (reduction)
$5H_2O + NO_2^- \rightarrow NH_3 + 7OH^-$ (basic solution)
$5H_2O + NO_2^- + 6e^- \rightarrow NH_3 + 7OH^-$ (balance charge)

The reduction half-reaction must be multiplied by 1 and the oxidation half-reaction by 2 to make electron transfer equal. The two half-reactions are added and like terms cancelled.

$2[Al + 4OH^- \rightarrow Al(OH)_4^- + 3e^-]$
$1[5H_2O + NO_2^- + 6e^- \rightarrow NH_3 + 7OH^-]$
$$\boxed{2Al + 8OH^- + 5H_2O + NO_2^- \rightarrow 2Al(OH)_4^- + NH_3 + 7OH^-}$$

Atom balance: 2 Al, 15 O, 18 H, 1 N   Charge Balance: -9

11-8 The Stoichiometry of Oxidation-Reduction Reactions

One method of analyzing substances that are capable of being oxidized or reduced is to perform a titration. In a **redox** (oxidation-reduction) titration, one measures the volume of a standard solution of an oxidizing agent or reducing agent required to react with a specific amount, either volume or mass, of reducing agent or oxidizing agent. Problems can be worked in a manner completely analagous to that used in acid-base titrations.

Example 11-18 Redox Titration

In acidic solution, the following reaction occurs. What volume of 0.150 M Mn(NO$_3$)$_2$ is needed to react with 60.0 mL of 0.0500 M NaBiO$_3$?

$$5BiO_3^- + 14 H^+ + 2 Mn^{2+} \rightarrow 5 Bi^{3+} + 7 H_2O + 2 MnO_4^-$$

Plan

We realize that the Mn^{2+} is provided by the Mn(NO$_3$)$_2$ solution and the BiO$_3^-$ is provided by the NaBiO$_3$. The nitrate and sodium ions will be spectator ions in this process. We then follow our normal stoichiometric analysis.

mL NaBiO$_3$ → mol NaBiO$_3$ → mol Mn(NO$_3$)$_2$ → mL Mn(NO$_3$)$_2$

Solution

? mol NaBiO$_3$ = 0.0600 L NaBiO$_3$ × $\dfrac{0.0500 \text{ mol NaBiO}_3}{1 \text{ L NaBiO}_3}$ = 0.00300 mol NaBiO$_3$

? mol Mn(NO$_3$)$_2$ = 0.00300 mol NaBiO$_3$ × $\dfrac{2 \text{ mol Mn(NO}_3)_2}{5 \text{ mol NaBiO}_3}$

= 0.00120 mol Mn(NO$_3$)$_2$

? mL Mn(NO$_3$)$_2$ = 0.00120 mol Mn(NO$_3$)$_2$ × $\dfrac{1 \text{ L Mn(NO}_3)_2}{0.150 \text{ mol Mn(NO}_3)_2}$

= 0.00800 L Mn(NO$_3$)$_2$ = $\boxed{8.00 \text{ mL Mn(NO}_3)_2}$

Example 11-19 Redox Titration

A 0.1578 g sample of FeSO$_4$ is oxidized by 34.20 mL of a solution of K$_2$Cr$_2$O$_7$ according to the equation below. What is the molarity of the K$_2$Cr$_2$O$_7$ solution?

$$Cr_2O_7^{2-} + 6Fe^{2+} + 14H^+ \rightarrow 6Fe^{3+} + 2Cr^{3+} + 7H_2O$$

Plan

We recognize that the K$_2$Cr$_2$O$_7$ provides the Cr$_2$O$_7^{2-}$ and the FeSO$_4$ provides the Fe^{2+}. K$^+$ and SO$_4^{2-}$ are spectator ions. We then use our normal stoichiometric techniques.

g FeSO$_4$ → mol FeSO$_4$ → mol K$_2$Cr$_2$O$_7$ → M K$_2$Cr$_2$O$_7$

Solution

$$? \text{ mol } K_2Cr_2O_7 = 0.1578 \text{ g FeSO}_4 \times \frac{1 \text{ mol FeSO}_4}{151.9 \text{ g FeSO}_4} \times \frac{1 \text{ mol } K_2Cr_2O_7}{6 \text{ mol FeSO}_4}$$

$$= 1.731 \times 10^{-4} \text{ mol } K_2Cr_2O_7$$

$$? M \, K_2Cr_2O_7 = \frac{1.731 \times 10^{-4} \text{ mol } K_2Cr_2O_7}{0.03420 \text{ L } K_2Cr_2O_7} = \boxed{0.005061 \, M \, K_2Cr_2O_7}$$

Example 11-20 Redox Titration

Exactly 25.0 mL of a $KMnO_4$ solution are required for complete reaction with 40.0 mL of 0.200 M $FeSO_4$ solution according to the equation below. What is the molarity of the $FeSO_4$ solution?

$$MnO_4^- + 8H^+ + 5Fe^{2+} \rightarrow Mn^{2+} + 5Fe^{3+} + 4H_2O$$

Plan

We recognize that the $KMnO_4$ provides the MnO_4^- and the $FeSO_4$ provides the Fe^{2+}. K^+ and MnO_4^- are spectator ions. We then use our normal stoichiometric techniques.

mL $FeSO_4$ → mmol $FeSO_4$ → mmol $KMnO_4$ → M $KMnO_4$

Solution

$$? \text{ mmol } KMnO_4 = 40.0 \text{ mL FeSO}_4 \times \frac{0.200 \text{ mol FeSO}_4}{1 \text{ L solution}} \times \frac{1 \text{ mol } KMnO_4}{5 \text{ mol FeSO}_4}$$

$$= 1.60 \text{ mmol } KMnO_4$$

$$? M \, KMnO_4 = \frac{1.60 \text{ mm } KMnO_4}{25.0 \text{ mL } KMnO_4} = \boxed{0.0640 \, M \, KMnO_4}$$

Example 11-21 Redox Titration

A 0.987 gram sample of Mo_2O_3 reacts with 30.0 mL of a $KMnO_4$ solution according to the following equation. What is the molarity of the $KMnO_4$ solution?

$$3MnO_4^- + 5Mo^{3+} + 8H_2O \rightarrow 3Mn^{2+} + 5MoO_4^{2-} + 16H^+$$

Plan

We see that **one** mole of Mo_2O_3 provides **two** moles of Mo^{3+} ion and that $KMnO_4$ provides the MnO_4^- ion. K^+ is a spectator ion. We then follow the normal stoichiometric procedure.

g Mo_2O_3 → mol Mo_2O_3 → mol $KMnO_4$ → M $KMnO_4$

Solution

$$? \text{ mol } KMnO_4 = 0.987 \text{ g } Mo_2O_3 \times \frac{1 \text{ mol } Mo_2O_3}{239.9 \text{ g } Mo_2O_3} \times \frac{2 \text{ mol } Mo^{3+}}{1 \text{ mol } Mo_2O_3} \times \frac{3 \text{ mol } KMnO_4}{5 \text{ mol } Mo^{3+}}$$

$$= 0.00494 \text{ mol } KMnO_4$$

$$? M \text{ } KMnO_4 = \frac{0.00494 \text{ mol } KMnO_4}{0.0300 \text{ L } KMnO_4} = \boxed{0.165 \text{ } M \text{ } KMnO_4}$$

Example 11-22 Redox Titration

A 1.517 gram sample containing some Mo_2O_3 reacts with 21.37 mL of 0.1026 M $KMnO_4$ solution. How many grams of Mo_2O_3 are contained in the sample and what is the percentage of Mo_2O_3 in the sample? Assume that no other components react with $KMnO_4$. Use the equation in Example 11-21 for this example.

Plan

We must first use our normal stoichiometric techniques to determine the mass of Mo_2O_3 in the sample. After calculating this mass, we use the definition of mass percent to find the requested information.

Solution

$$? \text{ g } Mo_2O_3 = 0.02137 \text{ L } KMnO_4 \times \frac{0.1026 \text{ mol } KMnO_4}{1 \text{ L solution}} \times \frac{5 \text{ mol } Mo^{3+}}{3 \text{ mol } KMnO_4}$$

$$\times \frac{1 \text{ mol } Mo_2O_3}{2 \text{ mol } Mo^{3+}} \times \frac{239.9 \text{ g } Mo_2O_3}{1 \text{ mol } Mo_2O_3} = \boxed{0.4383 \text{ g } Mo_2O_3}$$

$$? \% \text{ } Mo_2O_3 = \frac{0.4382 \text{ g } Mo_2O_3}{1.517 \text{ g sample}} \times 100 = \boxed{28.89\% \text{ } Mo_2O_3}$$

EXERCISES

Molarity and Moles in Acid-Base Titrations

1. What are the molarities of the ions in the solutions prepared by mixing of the following pairs of solutions? Assume that only normal salts are formed.

   (a) 300 mL of 1.00 M HBr and 150 mL of 1.00 M Ba(OH)$_2$

   (b) 250 mL of 0.150 M HClO$_3$ and 250 mL of 0.0750 M Ba(OH)$_2$

   (c) 500 mL of 2.00 M KOH and 100 mL of 3.33 M H$_3$PO$_4$

2. What are the molarities of the ions in the solutions prepared by mixing the following solutions? Assume that only normal salts are formed.

   (a) 22.0 mL of 0.220 M LiOH and 28.0 mL of 0.220 M HClO$_4$

   (b) 20.0 mL of 0.165 M RbOH and 40.0 mL of 0.165 M HNO$_3$

   (c) 300 mL of 0.200 M KOH and 100 mL of 0.100 M H$_3$PO$_4$

   (d) 15.0 mL of 0.175 M CsOH and 15.0 mL of 0.400 M HBr

3. What is the molarity of an HCl solution if 15.83 milliliters of it are needed to dissolve and completely neutralize a 0.1078 gram sample of Cr(OH)$_3$?

4. How many grams of NaOH are required to completely neutralize 175 mL of 0.345 M H$_2$SO$_4$ solution?

5. What mass of KHP (MW = 204.2 g/mol) is required to completely neutralize 20.00 mL of 0.03145 M Ba(OH)$_2$?

6. Titration of a 0.5240 gram sample of impure solid Na$_2$CO$_3$ requires 43.60 mL of 0.1077 M H$_2$SO$_4$ according to the equation below. How many grams of Na$_2$CO$_3$ does the sample contain and what is the percent purity of the sample? Assume that the impurities do not react with H$_2$SO$_4$.

$$H_2SO_4 + Na_2CO_3 \rightarrow Na_2SO_4 + CO_2 + H_2O$$

7. Wine has a distinct flavor partially due to certain acids present in the fruits from which it is made. Acid is often reported as if only tartaric acid, C$_2$H$_4$O$_2$(COOH)$_2$ were present. A 10.0 gram sample of white wine (d = 1.00 g/mL) was titrated with 0.1124 M KOH and required 8.18 mL of the KOH for the reaction below:

$$C_2H_4O_2(COOH)_2 + 2KOH \rightarrow 2H_2O + C_2H_4O_2(COOK)_2$$

Calculate both the molarity of the tartaric acid in the wine and the number of grams of tartaric acid per 100 mL of wine.

8. Most commercially available vinegars are about 5% acetic acid, CH_3COOH. A 1.399 gram sample of vinegar was titrated with 0.1038 M NaOH solution and required 11.48 mL to the equivalence point. What was the percent by mass of acetic acid in this vinegar sample? Assume that the acetic acid is the only acidic component in the sample.

9. NaOH is a poor primary standard both because it is hygroscopic (absorbs moisture) and reacts with atmospheric CO_2. A 100.0 mL solution was made from 10.16 grams of NaOH pellets. Titration of 3.750 grams of benzoic acid, $C_6H_5CO_2H$, a good monoprotic primary standard, required 13.62 mL of the NaOH solution for complete neutralization. What was the molarity of NaOH in the prepared solution and what was the percent purity of the NaOH in the pellets?

Equivalent Weights and Normality in Acid-Base Titrations

10. Calculate the equivalent weight of each acid and base in the following equations.

    (a) $2HNO_3 + Sr(OH)_2 \rightarrow Sr(NO_3)_2 + 2H_2O$

    (b) $H_2TeO_4 + RbOH \rightarrow RbHSO_4 + H_2O$

    (c) $H_2SeO_3 + 2LiOH \rightarrow Li_2SeO_4 + 2H_2O$

    (d) $2HO(CH_2COOH)_2COOH + 3Sr(OH)_2 \rightarrow Sr_3[HO(CH_2COO)_2COO]_2 + 6H_2O$

    (e) $HO(CH_2COOH)_2COOH + 2KOH \rightarrow K_2[HO(CH_2COO)_2COOH] + 2H_2O$

    (f) $H_3PO_4 + Ga(OH)_3 \rightarrow GaPO_4 + 3H_2O$

    (g) $H_3PO_4 + NH_3 \rightarrow (NH_4)H_2PO_4$

11. What are the normalities of the solutions made by dissolving the following amounts of material if the substances are to be used in the reactions specified?

    (a) 2.741 g $Sr(OH)_2$ in 500.0 mL solution for reaction 10(a)

    (b) 5.028 g RbOH in 2.00 L solution for reaction 10(b)

    (c) 0.121 g H_2SeO_3 in 25.00 mL solution for reaction 10(c)

    (d) 2.50 g $HO(CH_2COOH)_2COOH$ in 125 mL solution for reaction 10(d)

    (e) 2.50 g $HO(CH_2COOH)_2COOH$ in 125 mL solution for reaction 10(e)

    (f) 1.44 mol $CaOH)_2$ in 12.0 L solution for reaction 10(d)

    (g) 0.0250 mole H_3PO_4 in 50.0 mL solution for reaction 10(f)

    (h) 0.0250 mole H_3PO_4 in 50.0 mL solution for reaction 10(g)

12. Calculate the volume of 6.00 M H_3PO_4 required to prepare 500 milliliters of 0.200 N H_3PO_4 for the reactions:

    (a) $H_3PO_4 + NaOH \rightarrow NaH_2PO_4 + H_2O$

    (b) $H_3PO_4 + 2NaOH \rightarrow Na_2HPO_4 + 2H_2O$

    (c) $H_3PO_4 + 3NaOH \rightarrow Na_3PO_4 + 3H_2O$

13. What volume of 0.140 M H_3PO_4 would be required to react with 35.0 mL of 0.112 N NaOH solution according to the equation in 12(c)?

14. A 25.00 mL sample of an aqueous ammonia, NH_3, reacts with 30.00 mL of 0.1684 N HCl. What is the normality of the ammonia solution?

15. A 0.2168 gram sample of $Ga(OH)_3$ requires 22.67 mL of HBr solution for complete neutralization. What is the normality of the HBr solution?

16. A 0.9225g sample of impure Na_2CO_3 is completely neutralized by 26.37 mL of 0.1432 N HCl solution. No other basic impurities are present to react with the HCl. What is the percent Na_2CO_3 in the sample?

17. What is the normality of an RbOH solution if 35.38 mL of it are needed to completely neutralize 25.00 mL of 0.1376 M H_2SO_4 solution?

Change in Oxidation Number Method

18. Balance the following unbalanced formula unit equations.

    (a) $HI(g) + H_2SO_4(\ell) \rightarrow I_2(s) + H_2S(g) + H_2O(\ell)$

    (b) $H_2O(\ell) + As_4O_6(aq) + HI_3(aq) \rightarrow As_4O_{10}(s) + HI(aq)$

    (c) $Cl_2(g) + I_2(s) + H_2O(\ell) \rightarrow HCl(aq) + HIO_3(aq)$

    (d) $C(s) + HNO_3(aq) \rightarrow NO_2(g) + CO_2(g) + H_2O(\ell)$

    (e) $CuSO_4(aq) + KCN(aq) \rightarrow CuCN(s) + K_2SO_4 + C_2N_2$

    (f) $I_2(s) + HNO_3(aq) \rightarrow HIO_3(aq) + NO(g) + H_2O(\ell)$

    (g) $Pb(s) + PbO_2(s) + H_2SO_4(aq) \rightarrow PbSO_4(s) + H_2O(\ell)$

    (h) $As_2S_3(s) + H_2O_2(\ell) \rightarrow H_3AsO_4(aq) + H_2SO_4(aq) + H_2O(\ell)$

    (i) $HI(g) + H_2SO_4(aq) \rightarrow I_2(s) + H_2S(g) + H_2O(\ell)$

19. Balance the following ionic equations.

    (a) $Cr_2O_7^{2-}(aq) + SO_3^{2-}(aq) + H^+(aq) \rightarrow Cr^{3+}(aq) + SO_4^{2-}(aq) + H_2O(\ell)$

(b) $C_2H_6O(\ell) + Cr_2O_7^{2-}(aq) + H^+(aq) \to C_2H_4O_2(aq) + Cr^{3+}(aq) + H_2O(\ell)$

(c) $Cr(OH)_3(s) + OH^-(aq) + Cl_2(g) \to CrO_4^{2-}(aq) + Cl^-(aq) + H_2O(\ell)$

(d) $P_4(s) + OH^-(aq) + H_2O(\ell) \to PH_3(aq) + HPO_3^{2-}(aq)$

20. Balance the following ionic equations. $H^+(aq)$ and/or $H_2O(\ell)$ may be added for acidic solutions while $OH^-(aq)$ and/or $H_2O(\ell)$ may be added for basic solutions.

(a) $HClO(aq) + Hg(\ell) + Br^-(aq) \to Cl^-(aq) + HgBr_4^{2-}(aq)$ (acidic)

(b) $S_2O_6^{2-}(aq) + Br_2(\ell) \to H_2SO_3(aq) + BrO_3^-(aq)$ (acidic)

(c) $UF_6^-(aq) + H_2O_2(aq) \to UO_2^{2+}(aq) + HF(aq)$ (acidic)

(d) $Zn(s) + NO_3^-(aq) \to Zn^{2+}(aq) + NH_4^+(aq)$ (acidic)

(e) $Te(s) + NO_3^-(aq) \to TeO_2(s) + NO(g)$ (acidic)

(f) $Br_2(\ell) \to BrO_3^-(aq) + Br^-(aq)$ (basic)

(g) $NO_2^-(aq) + MnO_4^-(aq) \to MnO_2(s) + NO_3^-(aq)$ (basic)

(h) $HNO_2(aq) + Cl_2(g) \to ClO_4^-(aq) + N_2O(g)$ (acidic)

Half-Reaction Method

21. Balance the following ionic equations.

(a) $H_2O_2(aq) + MnO_4^-(aq) + H^+(aq) \to Mn^{2+}(aq) + O_2(g) + H_2O(\ell)$

(b) $I_3^-(aq) + H_2O(\ell) \to HIO_3(aq) + I^-(aq) + H^+(aq)$

(c) $CN^-(aq) + CrO_4^{2-}(aq) + H_2O(\ell) \to CNO^-(aq) + Cr(OH)_4^-(aq) + OH^-(aq)$

(d) $Cr_2O_7^{2-}(aq) + Si(s) + OH^-(aq) \to Cr(s) + SiO_3^{2-}(aq) + H_2O(\ell)$

22. Balance the following ionic equations. $H^+(aq)$ and/or $H_2O(\ell)$ may be added for acidic solutions while $OH^-(aq)$ and/or $H_2O(\ell)$ may be added for basic solutions.

(a) $As_2O_3(s) + NO_3^-(aq) \to AsO_4^{3-}(aq) + NO(g)$ (acidic)

(b) $C_3H_8O_3(aq) + MnO_4^-(aq) \to CO_3^{2-}(aq) + MnO_4^{2-}(aq)$ (acidic)

(c) $CrO_4^{2-}(aq) + Si(s) \to Cr(OH)_3(s) + SiO_3^{2-}(aq)$ (basic)

(d) $P_4(s) + HClO(aq) \to H_3PO_4(aq) + Cl^-(aq)$ (basic)

(e) $HClO_2(aq) + NH_4^+(aq) \rightarrow Cl^-(aq) + N_2(g)$ (acidic)

(f) $Cr_2O_7^{2-}(aq) + NH_4^+(aq) \rightarrow Cr^{3+}(aq) + N_2H_5^+(aq)$ (acidic)

(g) $ClO_3^-(aq) + N_2H_4(aq) \rightarrow Cl^-(aq) + N_2(g)$ (basic)

(h) $S_8(s) + Cl_2(g) \rightarrow H_2S(aq) + ClO_4^-(aq)$ (acidic)

(i) $H_2O_2(aq) + NO(g) \rightarrow NO_3^-(aq)$ (acidic)

Molarity and Moles in Redox Titrations

23. A standardized solution is $0.0404\ M$ in sodium thiosulfate, $Na_2S_2O_3$. A 20.0 mL sample of this solution reacts with 33.4 mL of a solution of potassium triiodide, KI_3, (which was prepared by dissolving I_2 in an excess of aqueous KI). What is the molarity of the KI_3?

$$2S_2O_3^{2-}(aq) + I_3^-(aq) \rightarrow 3\ I^-(aq) + S_4O_6^{2-}(aq)$$

24. What was the mass of iodine, I_2, that dissolved in 500 mL of excess aqueous KI in Exercise 18? Assume that the reaction below went to completion and that no significant volume change occurred on addition of the solid KI.

$$I_2(s) + KI(aq) \rightarrow KI_3(aq)$$

25. How many grams of $KMnO_4$ are required to oxidize 50.00 mL of a $0.2078\ M$ solution of $FeSO_4$? The products are Mn^{2+} and Fe^{3+}.

26. In photographic developers silver, Ag^+, ion oxidizes hydroquinone, $C_6H_6O_2$, to quinone, C_6H_4O, and is reduced to silver metal. How many mL of $0.02181\ M\ C_6H_6O$ are needed to reduce 5.422 milligrams of silver ion?

27. Oxalate salts in acidic solution can be titrated with permanganate ion, yielding CO_2 and Mn^{2+} ions as products. A 0.6461 gram sample containing some $Na_2C_2O_4$ requires 15.55 mL of $0.02296\ M\ KMnO_4$ for oxidation. What are the mass and the percent $Na_2C_2O_4$ in the sample? Assume that the $Na_2C_2O_4$ is the only component that will react with the permanganate ion.

28. Copper can be dissolved by concentrated nitric acid ($16M\ HNO_3$) according to the reaction below. How many mL of $16\ M\ HNO_3$ are needed to dissolve 3.85 grams of copper?

$$Cu(s) + 4HNO_3(aq) \rightarrow Cu(NO_3)_2(aq) + 2NO_2(g) + 2H_2O(\ell)$$

29. What is the molarity of a $K_2Cr_2O_7$ solution if 14.68 mL of it are required to react with 25.00 mL of $0.3421\ M\ Na_2SO_3$ solution? The reaction is:

$$8H^+ + Cr_2O_7^{2-} + 3SO_3^{2-} \rightarrow 2Cr^{3+} + 3SO_4^{2-} + 4H_2O$$

30. Iron(II) ions can be titrated using dichromate ions in acidic solution. Reaction products are iron(III) and chromium(III) ions. A 0.5184 gram sample containing some iron is dissolved in 3 M H_2SO_4. The resultant solution is titrated with 0.01612 M $K_2Cr_2O_7$ and requires 10.87 mL of that solution. What are the mass and the percentage of iron in the sample?

Miscellaneous Exercises

31. How many mL of $Ba(OH)_2$ solution are required to completely react with 44.78 mL of 0.2098 M NH_4Cl solution according to the acid-base reaction below?

$$2NH_4Cl(aq) + Ba(OH)_2(aq) \rightarrow 2NH_3(aq) + BaCl_2(aq) + 2 H_2O(\ell)$$

32. What are the concentrations of all species after 30.0 mL of 0.35 M solution of HNO_3 is added to 30.0 mL of a 0.55 M solution of the weak base C_5H_5N?

33. What is the percent $KMnO_4$ in an impure sample if 0.1614 grams of the sample require 22.58 mL of 0.1058 M $H_2C_2O_4$ for complete reaction according to the equation below. No other reducible species are present.

$$2\ MnO_4^- + 6H^+ + 5H_2C_2O_4 \rightarrow 2Mn^{2+} + 10CO_2 + 8H_2O$$

34. How many mL of 0.180 M HCl solution are required to completely neutralize 15.00 mL of 0.250 M $Ba(OH)_2$ solution?

35. How many mL of 0.180 M HCl solution are required to completely neutralize 15.00 mL of 0.250 N $Ba(OH)_2$ solution?

36. A 0.617 gram sample of impure $Al(OH)_3$ is completely neutralized by 75.0 mL of 0.276 M HCl. How many grams of $Al(OH)_3$ are in the sample? What is its percent purity?

37. One method of determining the oxidizing power of a household bleach (which usually contains NaOCl) is to add an unmeasured excess of iodide ion, I^-, in acidic solution. The iodine, I_2, formed can then be titrated with thiosulfate ion, $S_2O_3^{2-}$, using a starch indicator. The pertinent equations are:

$$HOCl + H^+ + 2I^- \rightarrow I_2 + Cl^- + H_2O$$

$$I_2 + 2S_2O_3^{2-} \rightarrow 2I^- + S_4O_6^{2-}$$

A 0.918 g sample of liquid household cleanser was dissolved in water and excess KI was added. It required 16.83 mL of 0.108 N $Na_2S_2O_3$ to titrate the I_2 produced.

(a) How much KI would need to be added to be sure that the entire sample would react if it were 100% NaOCl?

(b) How many grams of NaOCl were in the sample and what was the percent NaOCl in the cleanser?

Chapter 11 Reactions in Aqueous Solutions II: Calculations

38. What are the molarities of all species after 20.0 mL of 0.350 M NaOH solution is added to 15.0 mL of 0.400 M HClO₃ solution?

39. How many grams of zinc can be oxidized to Zn^{2+} by the action of 48.61 mL of 0.02234 M KMnO₄ in basic solution? The other product is MnO_2.

40. An impure sample of Fe(NH₄)₂(SO₄)₂·6H₂O weighing 13.118 grams required 42.46 mL of 0.1033 M K₂Cr₂O₇ for complete reaction according to the equation of Exercise 30. Assuming no other oxidizable species were present in the sample, how many grams of Fe(NH₄)₂(SO₄)₂ were contained in the sample and what was its percent purity?

41. A 1.427 gram sample of impure As₂O₃ is dissolved in concentrated NaOH solution to form NaAsO₂, which is in turn reacted with HCl to produce H₃AsO₃. The resulting solution reacts with 32.60 mL of 0.749 N Ce(SO₄)₂. What were the mass of As₂O₃ and percent purity in the original sample?

$$H_3AsO_3 + 2Ce^{4+} + H_2O \rightarrow H_3AsO_4 + 2Ce^{3+} + 2H^+$$

42. Write balanced net ionic equations for the following redox reactions.

   (a) In a neutral solution gold(III) ions oxidize iodide ions to solid iodine and are reduced to solid gold(I) iodide.

   (b) In basic solution, potassium permanganate oxidizes potassium nitrite to potassium nitrate and is reduced to manganese(IV) oxide.

   (c) Iron(II) ion reduces nitrite ion to nitrogen oxide and is oxidized to iron(III) ion in acidic solution.

   (d) In acidic solution, uranium(IV) ion reduces permanganate ion to manganese(II) ion and is oxidized to UO_2^+ ion.

43. Balance the following ionic equations. H^+(aq) and/or $H_2O(\ell)$ may be added for acidic solutions while OH^-(aq) and/or $H_2O(\ell)$ may be added for basic solutions. Tell what is the oxidizing and reducing agent in each case.

   (a) $Au^{3+}(aq) + I_2(aq) \rightarrow Au(s) + IO_3^-(aq)$ (acidic)

   (b) $OsO_4(s) + C_4H_8(OH)_2(aq) \rightarrow Os^{4+}(aq) + C_4H_8O_2(aq)$ (basic)

   (c) $I^-(aq) + NO_2^-(aq) \rightarrow NO(g) + I_2(aq)$ (acidic)

   (d) $MnO_4^-(aq) + SCN^-(aq) \rightarrow Mn^{2+}(aq) + SO_4^{2-}(aq) + CO_2(g) + N_2(g)$ (acidic)

   (e) $ClO_3^-(aq) + I_2(aq) \rightarrow IO_3^-(aq) + Cl^-(aq)$ (acidic)

   (f) $HXeO_4^-(aq) + Pb(s) \rightarrow Xe(g) + HPbO_2^-(aq)$ (basic)

ANSWERS TO EXERCISES

1. (a) 0.333 M Ba^{2+}, 0.667 M Br^-
   (b) 0.0375 M Ba^{2+}, 0.0750 M ClO_3^-
   (c) 0.555 M PO_4^{3-}, 1.67 M K^+

2. (a) 0.0264 M H^+, 0.123 M ClO_4^-, 0.0968 M Li^+
   (b) 0.110 M NO_3^-, 0.0550 M Rb^+, 0.0550 M H^+
   (c) 0.150 M K^+, 0.0250 M PO_4^{3-}, 0.0750 M OH^-
   (d) 0.200 M Br^-, 0.112 M H^+, 0.0875 M Cs^+

3. 0.1984 M HCl   4. 4.83 g NaOH   5. 0.2569 g KHP

6. 0.4977 g Na_2CO_3, 94.98% Na_2CO_3

7. 0.0460 M $C_2H_4O_2(COOH)_2$, 0.690 g/100 mL   8. 5.111% CH_3COOH

9. 0.2256 M NaOH, 9.024 g NaOH, 88.82% pure

10. (a) 63.01 g HNO_3/eq, 60.82 g $Sr(OH)_2$/eq
    (b) 193.62 g H_2TeO_4/eq, 102.5 g RbOH/eq
    (c) 64.49 g H_2SeO_3/eq, 23.95 g LiOH/eq
    (d) 64.05 g $HO(CH_2COOH)_2COOH$/eq, 37.00 g $Ca(OH)_2$/eq
    (e) 96.07 g $HO(CH_2COOH)_2COOH$/eq, 56.1 g KOH/eq
    (f) 32.67 g H_3PO_4/eq, 40.25 g $Ga(OH)_3$/eq
    (g) 98.01 g H_3PO_4/eq, 17.04 g NH_3/eq

11. (a) 0.09023 N $Sr(OH)_2$
    (c) 0.0751 N H_2SeO_3
    (e) 0.208 N $HO(CH_2COOH)_2COOH$
    (g) 1.50 N H_3PO_4
    (b) 0.02453 N RbOH
    (d) 0.312 N $HO(CH_2COOH)_2COOH$
    (f) 0.240 N $Ca(OH)_2$
    (h) 0.500 N H_3PO_4

12. (a) 16.7 mL H_3PO_4 (b) 8.33 mL H_3PO_4 (c) 5.56 mL H_3PO_4

13. 28.0 mL H_3PO_4 14. 0.2021 N NH_3 15. 0.2376 N HBr

16. 86.78% Na_2CO_3 17. 0.1945 N RbOH

18. (a) $8HI(g) + H_2SO_4(\ell) \rightarrow 4I_2(s) + H_2S(g) + 4H_2O(\ell)$
    (b) $4H_2O + As_4O_6(aq) + 4HI_3(aq) \rightarrow As_4O_{10}(s) + 12HI(aq)$

(c) $5Cl_2(g) + I_2(s) + 6H_2O(\ell) \rightarrow 10HCl(aq) + 2HIO_3(aq)$

(d) $C(s) + 4HNO_3(aq) \rightarrow 4NO_2(g) + CO_2(g) + 2H_2O(\ell)$

(e) $2CuSO_4(aq) + 4KCN(aq) \rightarrow 2CuCN(s) + C_2N_2(g) + 2K_2SO_4(aq)$

(f) $3I_2(s) + 10HNO_3(aq) \rightarrow 6HIO_3(aq) + 2H_2O(\ell) + 10NO(g)$

(g) $Pb(s) + PbO_2(s) + 2H_2SO_4(aq) \rightarrow 2PbSO_4(s) + 2H_2O(\ell)$

(h) $As_2S_3(s) + 14H_2O_2(aq) \rightarrow 2H_3AsO_4(aq) + 3H_2SO_4(aq) + 8H_2O(\ell)$

(i) $8HI(g) + H_2SO_4(aq) \rightarrow 4I_2(s) + H_2S(g) + 4H_2O(\ell)$

19. (a) $Cr_2O_7^{2-}(aq) + 3SO_3^{2-} + 8H^+(aq) \rightarrow 2Cr^{3+}(aq) + 3SO_4^{2-}(aq) + 4H_2O(\ell)$

(b) $3C_2H_6O(\ell) + 2Cr_2O_7^{2-}(aq) + 16H^+(aq) \rightarrow 3C_2H_4O_2(aq) + 4Cr^{3+}(aq) + 11H_2O(\ell)$

(c) $2Cr(OH)_3(s) + 10OH^-(aq) + 3Cl_2(g) \rightarrow 2CrO_4^{2-}(aq) + 6Cl^-(aq) + 8H_2O(\ell)$

(d) $P_4(s) + 4OH^-(aq) + 2H_2O(\ell) \rightarrow 2PH_3(aq) + 2HPO_3^{2-}(aq)$

20. (a) $HClO(aq) + Hg(\ell) + 4Br^-(aq) + H^+(aq) \rightarrow Cl^-(aq) + HgBr_4^{2-}(aq) + H_2O(\ell)$

(b) $5S_2O_6^{2-}(aq) + Br_2(\ell) + 8H^+(aq) + 6H_2O(\ell) \rightarrow 10H_2SO_3(aq) + 2BrO_3^-(aq)$

(c) $2UF_6^-(aq) + H_2O_2(aq) + 6H^+(aq) + 2H_2O(\ell) \rightarrow 2UO_2^{2+}(aq) + 12HF(aq)$

(d) $4Zn(s) + NO_3^-(aq) + 10H^+(aq) \rightarrow 4Zn^{2+}(aq) + NH_4^+(aq) + 3H_2O(\ell)$

(e) $3Te(s) + 4NO_3^-(aq) + 4H^+(aq) \rightarrow 3TeO_2(s) + 4NO(g) + 2H_2O(\ell)$

(f) $3Br_2(\ell) + 6OH^-(aq) \rightarrow BrO_3^-(aq) + 5Br^-(aq) + 3H_2O(\ell)$

(g) $3NO_2^-(aq) + 2MnO_4^-(aq) + H_2O(\ell) \rightarrow 2MnO_2(s) + 3NO_3^-(aq) + 2OH^-(aq)$

(h) $14HNO_2(aq) + 2Cl_2(g) \rightarrow 7N_2O(g) + 5H_2O(\ell) + 4ClO_4^-(aq) + 4H^+(aq)$

21. (a) $5H_2O_2(aq) + 2MnO_4^-(aq) + 6H^+(aq) \rightarrow 2Mn^{2+}(aq) + 5O_2(g) + 8H_2O(\ell)$

(b) $3I_3^-(aq) + 3H_2O(\ell) \rightarrow HIO_3 + 8I^-(aq) + 5H^+(aq)$

(c) $3CN^-(aq) + 2CrO_4^{2-}(aq) + 5H_2O(\ell) \rightarrow 3CNO^-(aq) + 2Cr(OH)_4^-(aq) + 2OH^-(aq)$

(d) $Cr_2O_7^{2-}(aq) + 3Si(s) + 4OH^-(aq) \rightarrow 2Cr(s) + 3SiO_3^{2-}(aq) + 2H_2O(\ell)$

22. (a) $3As_2O_3(s) + 4NO_3^-(aq) + 7H_2O(\ell) \rightarrow 6AsO_4^{3-}(aq) + 14H^+(aq) + 4NO(g)$

(b) $C_3H_8O_3(aq) + 14MnO_4^-(aq) + 20OH^-(aq) \rightarrow 3CO_3^{2-}(aq) + 14MnO_4^{2-}(aq) + 14H_2O(\ell)$

(c) $4CrO_4^{2-}(aq) + 3Si(s) + 7H_2O(\ell) \rightarrow 4Cr(OH)_3(s) + 3SiO_3^{2-}(aq) + 2OH^-(aq)$

(d) $P_4(s) + 10HClO(aq) + 6H_2O(\ell) \rightarrow 4H_3PO_4(aq) + 10H^+(aq) + 10Cl^-(aq)$

(e) $3HClO_2(aq) + 4NH_4^+(aq) \rightarrow 3Cl^-(aq) + 2N_2(g) + 6H_2O(\ell) + 7H^+(aq)$

(f) $Cr_2O_7^{2-}(aq) + 6NH_4^+(aq) + 5H^+(aq) \rightarrow 2Cr^{3+}(aq) + 3N_2H_5^+(aq) + 7H_2O(\ell)$

(g) $2ClO_3^-(aq) + 3N_2H_4(aq) \rightarrow 2Cl^-(aq) + 3N_2(g) + 6H_2O(\ell)$

(h) $7S_8(s) + 8Cl_2(g) + 64H_2O(\ell) \rightarrow 56H_2S(g) + 16ClO_4^-(aq) + 16H^+(aq)$

(i) $3H_2O_2(aq) + 2NO(g) \rightarrow 2NO_3^-(aq) + 2H^+(aq) + 2H_2O(\ell)$

23. $0.0121\ M\ KI_3$ 24. $1.54\ g\ I_2$ 25. $3.284\ g\ KMnO_4$ 26. $1.152\ mL\ C_6H_6O_2$

27. $0.1196\ g\ Na_2C_2O_4$, $18.51\%\ Na_2C_2O_4$

28. $15\ mL\ HNO_3$ 29. $0.1942\ M\ K_2Cr_2O_7$ 30. $0.05871\ g\ Fe$, $11.33\%\ Fe$

31. $31.86\ mL\ Ba(OH)_2$ 32. $0.10\ M\ C_5H_5N$, $0.18\ M\ C_5H_5NH^+$, $0.18\ M\ NO_3^-$

33. $93.57\%\ KMnO_4$ 34. $41.7\ mL\ HCl$ 35. $20.8\ mL\ HCl$

36. $0.538\ g\ Al(OH)_3$, 87.2% 37. (a) $4.09\ g\ KI$ (b) $0.0677\ g\ NaOCl$, $7.38\%\ NaOCl$

38. $0.200\ M\ Na^+$, $0.171M\ ClO_3^-$, $0.029\ M\ OH^-$ 39. $0.1065\ g\ Zn$

40. $7.427\ g\ Fe(NH_4)_2(SO_4)_2 \cdot 6H_2O$, 56.62%

41. $1.21\ g\ As_2O_3$, $84.8\%\ As_2O_3$

42. (a) $Au^{3+}(aq) + 3I^-(aq) \rightarrow AuI(s) + I_2(s)$

(b) $2MnO_4^-(aq) + 3NO_2^-(aq) + H_2O(\ell) \rightarrow 3NO_3^-(aq) + 2MnO_2(s) + 2OH^-(aq)$

(c) $2H^+(aq) + NO_2^-(aq) + Fe^{2+}(aq) \rightarrow Fe^{3+}(aq) + NO(g) + H_2O(\ell)$

(d) $6H_2O(\ell) + 5U^{4+}(aq) + MnO_4^-(aq) \rightarrow 5UO_2^+(aq) + Mn^{2+}(aq) + 12H^+(aq)$

43. (a) $10Au^{3+}(aq) + 3I_2(aq) + 18H_2O(\ell) \rightarrow 10Au(s) + 6IO_3^-(aq) + 36H^+(aq)$
Au^{3+} is the oxidizing agent; I_2 is the reducing agent.

(b) $OsO_4(s) + 2C_4H_8(OH)_2(aq) \rightarrow Os^{4+}(aq) + 2C_4H_8O_2(aq) + 4OH^-(aq)$
$C_4H_8(OH)_2$ is the reducing agent; OsO_4 is the oxidizing agent.

(c) $2I^-(aq) + 2NO_2^-(aq) + 4H^+(aq) \rightarrow 2NO(g) + I_2(aq) + 2H_2O(\ell)$
I^- is the reducing agent; NO_2^- is the oxidizing agent.

(d) $22MnO_4^-(aq) + 10SCN^-(aq) + 56H^+(aq) \rightarrow 22Mn^{2+}(aq) + 10SO_4^{2-}(aq) + 10CO_2(g) + 5N_2(g) + 28H_2O(\ell)$
MnO_4^- is the oxidizing agent; SCN^- is the reducing agent.

(e) $5ClO_3^-(aq) + 3I_2(aq) + 3H_2O(\ell) \rightarrow 6IO_3^-(aq) + 5Cl^-(aq) + 6H^+(aq)$
ClO_3^- is the oxidizing agent; I_2 is the reducing agent.

(f) $4OH^-(aq) + 2HXeO_4^-(aq) + 6Pb(s) \rightarrow 2Xe(g) + 6HPbO_2^-(aq)$
$HXeO_4^-$ is the oxidizing agent; Pb is the reducing agent.

Chapter 12: Gases and the Kinetic-Molecular Theory

12-1 Pressure

The **pressure** of a gas in a container is defined as the force it exerts per unit area on the walls of the container. Pressure is often measured by and, is proportional to, the height to which a column of mercury is supported. Pressure is expressed in terms of **millimeters of mercury (mm Hg)**. The reference point for pressure is set at exactly 760 mm Hg and is defined to be **one atmosphere (1 atm)** of pressure. The term torr is used interchangeably with the term mm Hg, so that **1 mm Hg = 1 torr**. In SI units, the pascal (Pa), which has units of kg m^{-1}s^{-2} or Nm^{-2} is used. The relationships between the units are shown in Appendix D of the text.

12-2 Boyle's Law

Boyle's Law states the the volume of a given mass of a gas at constant temperature is inversely proportional to its pressure. This means that as pressure increases, volume decreases and vice-versa. Algebraically, Boyle's Law is stated as follows.

$$P_1V_1 = P_2V_2 \text{ (Temperature and mass constant)}$$

Here, and throughout the chapter, the subscripts refer to initial, **1**, and final, **2**, conditions of the gas sample.

Example 12-1 Boyle's Law Calculation

A sample of Ne occupies a volume of 55.6 milliliters at 25°C and 625 mm Hg. At what pressure will the sample occupy a volume of 95.5 milliliters at 25°C?

Plan
We know an initial pressure and volume and want to find the pressure associated with a new volume. We can apply Boyle's law directly.

Solution
$P_1 = 625$ mm Hg, $P_2 = ?$, $V_1 = 55.6$ mL, $V_2 = 95.5$ mL

$$P_1V_1 = P_2V_2 \text{ so } P_2 = \frac{P_1V_1}{V_2} = \frac{625 \text{ mm Hg} \times 55.6 \text{ mL}}{95.5 \text{ mL}} = \boxed{364 \text{ mm Hg}}$$

12-3 Charles' Law and the Absolute Temperature Scale

Charles' Law states that the volume of a given mass of a particular gas, at constant pressure, is directly proportional to its **absolute** temperature. This means that as temperature is increased, volume also increases. Algebraically, Charles' Law is stated:

$$\frac{V_1}{T_1} = \frac{V_2}{T_2} \quad \text{(Pressure and mass constant)}$$

Example 12-2 Charles' Law Calculation

A gas at 1.00 atmospheres pressure has a volume of 25.0 liters at 25°C. At what temperature will the gas have a volume of 10.0 liters?

Plan
We know an initial volume and temperature and want to find the temperature associated with the new volume. We must convert the temperatures to kelvins and then we can directly apply the algebraic relationship above.

Solution
$T_1 = 273 + 25°C = 298K$, $T_2 = ?$, $V_1 = 25.0$ L, $V_2 = 10.0$ L

$$\frac{V_1}{T_1} = \frac{V_2}{T_2} \text{ so } T_2 = \frac{V_2T_1}{V_1} = \frac{10.0 \text{ L} \times 298K}{25.0 \text{ L}} = \boxed{119K} = \boxed{-154°C}$$

Because both the temperature and the pressure affect the volume of a gas and, thus, its density, it is useful to describe the volume at some standard set of conditions. The defined standard conditions for a gas (STP conditions) are 1 atmosphere pressure and 0°C (273 K).

Example 12-3 Charles' Law Calculation; STP

A gas sample occupies a volume of 38.0 milliliters at -85°C and 1.00 atmosphere. What would its volume be at standard temperature and pressure?

Plan

We recognize this as a Charles' Law problem and know that standard temperature is 273K and that standard pressure is 1 atmosphere. We can convert our initial temperature to kelvins and use the formula directly.

Solution

$T_1 = 273 + (-85°C) = 188K$, $T_2 = 273$ K, $V_1 = 38.0$ mL, $V_2 = ?$

$$\frac{V_1}{T_1} = \frac{V_2}{T_2} \text{ so } V_2 = \frac{V_1 \times T_2}{T_1} = \frac{38.0 \text{ mL} \times 273K}{188K} = \boxed{55.2 \text{ mL}}$$

12-4 The Combined Gas Law Equation

In instances in which only the mass of gas is held constant and the other parameters are allowed to vary, the **combined gas law** is used. It is so called because it is a combination of Boyle's and Charles' Laws:

$$\frac{P_1 V_1}{T_1} = \frac{P_2 V_2}{T_2} \quad \text{(Mass of gas constant)}$$

Example 12-4 Combined Gas Law Calculation

A sample of argon occupies 44.7 mL and exerts a pressure of 415 torr at 18°C. What volume would it occupy at STP?

Plan

We know the STP conditions and realize that all three parameters of the sample are changing. We tabulate what is known, converting the temperature to kelvins, and then solve for the unknown, substituting in known values.

Solution

$V_1 = 44.7$ mL, $V_2 = ?$, $P_1 = 415$ torr, $P_2 = 760$ torr $T_1 = 273 + 30°C = 303K$ $T_2 = 273K$

$$\frac{P_1V_1}{T_1} = \frac{P_2V_2}{T_2} \text{ so } V_2 = \frac{P_1V_1T_2}{P_2T_1} = \frac{(415 \text{ torr})(44.7 \text{ mL})(273K)}{(760 \text{ torr})(291K)} = \boxed{22.9 \text{ mL}}$$

Example 12-5 Combined Gas Law Calculation

A certain mass of methane occupies a volume of 10.0 liters at STP. What temperature is necessary for the same mass of methane to exert a pressure of 2.50 atmospheres in an 9.00 liter container?

Plan

The approach is the same as for the previous example, except that the temperature is the unknown.

Solution

$P_1 = 1.00$ atm, $P_2 = 2.50$ atm, $V_1 = 10.0$ L, $V_2 = 8.00$ L, $T_1 = 273K$, $T_2 = ?$

$$T_2 = \frac{P_2V_2T_1}{P_1V_1} = \frac{(2.50 \text{ atm})(9.00 \text{ L})(273K)}{(1.00 \text{ atm})(10.0 \text{ L})} = \boxed{614K} = \boxed{341°C}$$

12-5 Avogadro's Law and the Standard Molar Volume

Avogadro's Law states that at the same temperature and pressure, equal volumes of any gases contain equal numbers of molecules. At STP 1 mole (Avogadro's number of molecules) of an ideal gas occupies 22.4 liters. This is called the **standard molar volume** and is a good approximation for most real gases at pressures and temperatures near normal living conditions. Avogadro's Law can also be interpreted as stating that the number of moles (represented by n) of a gas is proportional to its volume at constant pressure and temperature.

Example 12-6 Standard Molar Volume Calculation

How many liters are occupied by a 25.0 gram sample of XeF_6 at STP?

Plan

We can use the molecular weight of XeF_6 to find the number of moles of gas. We can then use the fact that one mole of the gas occupies 22.4 L at STP.

Solution

$$? \text{ mol } XeF_6 = 25.0 \text{ g } XeF_6 \times \frac{1 \text{ mol } XeF_6}{245.3 \text{ g } XeF_6} = 0.102 \text{ mol } XeF_6$$

$$? \text{ L } XeF_6 = 0.102 \text{ mol } XeF_6 \times \frac{22.4 \text{ L } XeF_6}{1 \text{ mol } XeF_6} = \boxed{2.28 \text{ L } XeF_6}$$

For gases not at STP, the combined gas law and the standard molar volume can be used to convert data relating the volume at STP to any other desired conditions. This gives one method of finding the number of moles of a gas present when conditions other than STP are encountered.

Example 12-7 Combined Gas Law Calculation

What is the pressure exerted by 25.0 grams of XeF_6 in a 10.0 L container at 25°C?

Plan

We refer to Example 12-6 and note that this sample occupied a volume of 2.28 liters at STP. We use the combined gas law to solve for the pressure at the new temperature and volume.

Solution

$P_1 = 1.00$ atm, $P_2 = ?$, $V_1 = 2.28$ L, $V_2 = 10.0$ L, $T_1 = 273$K, $T_2 = 25°C + 273 = 298$K

$$P_2 = \frac{P_1 V_1 T_2}{V_2 T_1} = \frac{(1.00 \text{ atm})(2.28 \text{ L})(298\text{K})}{(10.0 \text{ L})(273\text{K})} = \boxed{0.249 \text{ atm}}$$

12-6 The Ideal Gas Equation

The previous gas laws give several relationships among variables P, V, n and T. Volume is related to all of the other parameters: $V \propto \frac{1}{P}$, $V \propto T$ and $V \propto n$. These can be combined to yield: $V \propto \frac{nT}{P}$. Because a proportionality can be written as an equality using a proportionality constant, in this case **R**, the universal gas constant, this can be changed to: $V = \f(RnT)$. This is more commonly expressed as **PV = nRT** and is called the **ideal gas equation**. R can be evaluated using the parameters at STP and the standard molar volume:

$$R = \frac{PV}{nT} = \frac{(1.00 \text{ atm})(22.4 \text{ L})}{(1.00 \text{ mol})(273 \text{K})} = 0.0821 \frac{\text{L} \cdot \text{atm}}{\text{mol} \cdot \text{K}}$$

To use this value of **R**, units **must** be converted to liters, atmospheres, moles, and kelvins. If different units are used, R has different numerical values, some of which are presented in Appendix D in the text. The ideal gas equation is usually used in preference to converting to STP conditions and using the standard molar volume.

Example 12-8 Ideal Gas Equation

Solve Example 12-7 using the ideal gas equation instead of the combined gas law.

Plan

We know the number of moles of xenon from Example 12-6. We can solve the ideal gas equation for pressure and evaluate using the known parameters.

Solution

$P = ?, V = 10.0 \text{ L}, T = 298\text{K}, n = 0.102 \text{ mol}, R = 0.0821 \frac{\text{L} \cdot \text{atm}}{\text{mol} \cdot \text{K}}$

$$PV = nRT \text{ so } P = \frac{nRT}{V} = \frac{(0.102 \text{ mol}) \left[0.0821 \frac{\text{L} \cdot \text{atm}}{\text{mol} \cdot \text{K}} \right](298\text{K})}{(10.0 \text{L})} = \boxed{0.250 \text{ atm}}$$

Example 12-9 Ideal Gas Equation

To what temperature must a 0.400 mole sample of argon be heated to exert a pressure of 7.50 atmospheres in a 3.50 liter container?

Plan

We know all the parameters needed for the ideal gas equation except for temperature. We can rearrange and solve the equation for T.

Solution

$P = 7.50 \text{ atm}, V = 3.50 \text{ L}, n = 0.400 \text{ mol}, R = 0.0821 \frac{\text{L} \cdot \text{atm}}{\text{mol} \cdot \text{K}}$

$$PV = nRT, \text{ so } T = \frac{PV}{nR} = \frac{(3.50 \text{ atm})(7.50\text{L})}{(0.400 \text{ mol})\left[0.0821 \frac{\text{L·atm}}{\text{mol·K}}\right]} = \boxed{799\text{K}} = \boxed{526°\text{C}}$$

Example 12-10 Ideal Gas Equation

What volume is occupied by 53.6 grams of NH_3 gas in a balloon at 19°C and 954 torr pressure?

Plan

We know the mass of NH_3 and, therefore, can find the number of moles. All the parameters needed for the ideal gas equation except for volume are then known. We can rearrange and solve the equation.

Solution

$T = 19°C + 273 = 292K$, $P = 954 \text{ torr} \times \frac{1 \text{ atm}}{760 \text{ torr}} = 1.26 \text{ atm}$, $R = 0.0821 \frac{\text{L·atm}}{\text{mol·K}}$

$? \text{ mol } NH_3 = n = 53.6 \text{ g } NH_3 \times \frac{1 \text{ mol } NH_3}{17.0 \text{ g } NH_3} = 3.15 \text{ mol } NH_3$

$$PV = nRT, \text{ so } V = \frac{nRT}{P} = \frac{(3.15 \text{ mol})\left[0.0821 \frac{\text{L·atm}}{\text{mol·K}}\right](292K)}{1.26 \text{ atm}} = \boxed{59.9 \text{ L}}$$

The density of a gas, usually expressed in g/L can be found using gas laws. To do this, we find the mass of a one liter sample of the gas.

Example 12-11 Ideal Gas Equation; Density of a Gas

What is the density of NH_3, in g/L, at -35°C and 666 torr?

Plan

We can find the density by assuming that we have exactly one liter of the gas. We can determine the mass in that volume from the ideal gas equation and the molar mass of NH_3.

Solution

$P = 666 \text{ torr} \times \dfrac{1 \text{ atm}}{760 \text{ torr}} = 0.876 \text{ atm} \qquad T = -35°C + 273 = 238 \text{ K} \qquad R = 0.0821 \dfrac{L \cdot atm}{mol \cdot K}$

Assume V = exactly 1 L, so infinite significant digits

$PV = nRT \text{ so } n = \dfrac{PV}{RT} = \dfrac{(0.876 \text{ atm})(1 \text{ L})}{(238 \text{ K})\left[0.0821 \dfrac{L \cdot atm}{mol \cdot K}\right]} = 0.0448 \text{ mol } NH_3$

$0.0448 \text{ mol } NH_3 \times \dfrac{17.04 \text{ g } NH_3}{1 \text{ mol } NH_3} = 0.764 \text{ g } NH_3$; therefore, density = $\boxed{0.764 \text{ g } NH_3/L}$

12-7 Determination of Molecular Weights and Molecular Formulas of Gaseous Substances

Gas laws are one method of determining molecular weights of materials that can be vaporized. If the temperature, pressure, and volume are known, the number of moles of gas present in a sample of known mass can be found. Dimensional analysis provides the molecular weight in g/mol. This can be combined with techniques for finding the empirical formula, discussed in Chapter Two to determine the formula of an unknown gas.

Example 12-12 Ideal Gas Equation; Molecular Formula Calculation

A 1.61 gram sample of a gas occupies 72.4 mL at 3.20 atmospheres pressure and 50°C. The gas is found to be 69.6% sulfur and 30.4% nitrogen. Assume that the gas behaves ideally. What is its molecular formula?

Plan
We find the simplest formula, using techniques discussed in Chapter Two. We then use dimensional analysis to find the molecular weight of the gas. We divide the molecular weight by the empirical weight to find the number of empirical units in one molecule of the gas.

Solution
First, assume that there are 100.0 g of the gas, then there would be 69.6 g S and 30.4 g N. We find the number of moles of each element.

$$? \text{ mol S} = 69.6 \text{ g S} \times \frac{1 \text{ mol S}}{32.06 \text{ g S}} = 2.17 \text{ mol S}$$

$$? \text{ mol N} = 30.4 \text{ g N} \times \frac{1 \text{ mol N}}{14.01 \text{ g N}} = 2.17 \text{ mol N}$$

We can see that the simplest formula of the gas is SN and the empirical weight of the gas is 14.01 + 32.06 or 46.07 g/mol.

Next, calculate the number of moles of the unknown:

$P = 3.20$ atm, $V = 72.4$ mL $= 0.0724$ L, $T = 50°C + 273 = 323$K, $R = 0.0821 \frac{L \cdot atm}{mol \cdot K}$

$$PV = nRT, \text{ so } n = \frac{PV}{RT} = \frac{(3.20 \text{ atm})(0.0724 \text{ L})}{\left[0.0821 \frac{L \cdot atm}{mol \cdot K}\right](323K)} = 0.00874 \text{ mol}$$

$$? \text{ g/mol} = \frac{1.61 \text{ g}}{0.00874 \text{ mol}} = 184 \text{ g/mol}$$

Finally, the molecular weight is divided by the empirical weight: $\frac{184 \text{ g/mol}}{46.07 \text{ g/mol}} = 3.99$

The molecular formula is, therefore, $\boxed{S_4N_4}$.

12-8 Dalton's Law of Partial Pressures

Dalton's Law of Partial Pressures states that in a mixture of gases, each gas exerts the pressure that it would exert if it occupied the volume alone. The total pressure in the system is the sum of the partial pressures of each of the gases involved. If A, B, C, ••• are different gases in a mixture then:

$$P_{total} = P_A + P_B + P_C + \cdots$$

This law can be derived from the fact that moles of nonreactive gases are additive:

$$n_{total} = n_A + n_B + n_C + \cdots \text{ and that } n = \frac{PV}{RT}$$

In the same container, the $\frac{V}{RT}$ term is a constant so the number of moles is directly proportional to the pressure. A useful extension of Dalton's Law is the relationship that the mole fraction, X, of a gas is related to the partial pressure of that gas and the total pressure.

$$X_A = \frac{n_A}{n_{total}} = \frac{P_A}{P_{total}} \text{ or } P_A = \frac{n_A}{n_{total}} P_{total} = X_A P_{total}$$

Example 12-13 Partial Pressure and Mole Fraction

What is the total pressure and the pressure due to the nitrogen in a gas mixture composed of 1.00 gram of N_2, 1.00 gram of H_2, and 1.00 gram of O_2 in a 1.00 liter container at 0°C?

Plan

First, we must find the number of moles of each component. We can find the total pressure using the total number of moles and the other known parameters in the ideal gas equation. Second, we can find the partial pressure of the nitrogen by using its number of moles in the ideal gas equation. Alternatively, we can use the mole fraction of nitrogen to determine its partial pressure from the total pressure.

Solution

$$? \text{ mol } N_2 = 1.00 \text{ g } N_2 \times \frac{1 \text{ mol } N_2}{28.02 \text{ g } N_2} = 0.0357 \text{ mol } N_2$$

$$? \text{ mol } H_2 = 1.00 \text{ g } H_2 \times \frac{1 \text{ mol } H_2}{2.02 \text{ g } H_2} = 0.495 \text{ mol } H_2$$

$$? \text{ mol } O_2 = 1.00 \text{ g } O_2 \times \frac{1 \text{ mol } O_2}{32.00 \text{ g } O_2} = 0.0313 \text{ mol } O_2$$

$$? \text{ moles total} = 0.0357 + 0.495 + 0.0313 \text{ mol} = 0.562 \text{ mol gas}$$

$$P_{total} = \frac{(n_{total})RT}{V} = \frac{(0.562 \text{ mol}) \left[0.0821 \frac{L \cdot atm}{mol \cdot K}\right](273K)}{1.00 \text{ L}} = \boxed{12.6 \text{ atm}}$$

$$P_{N_2} = \frac{(n_{N_2})RT}{V} = \frac{(0.0357 \text{ mol}) \left[0.0821 \frac{L \cdot atm}{mol \cdot K}\right](273K)}{1.00 \text{ L}} = \boxed{0.800 \text{ atm}}$$

Alternatively,

$$P_{N_2} = X_{N_2} P_{total} = \frac{n_{N_2}}{n_{total}}(P_{total}) = \frac{0.0357 \text{ mol } N_2}{0.562 \text{ mol total}} \times 12.6 \text{ atm} = \boxed{0.800 \text{ atm}}$$

12-9 Vapor Pressure

Gaseous products of chemical reactions are often collected over water. Water (or any other liqu has molecules in the vapor phase above the liquid surface. These vapor molecules exert a press that increases with increasing temperature. In doing calculations involving gases collected over water, Dalton's Law is used to correct for the pressure due to the water. Table 12-1 shows the vapor pressures of water near room temperature. A complete table for the common liquid range of water is shown in Appendix E of the text. Each substance has its own specific vapor pressur at different temperatures.

Table 12-1 Vapor Pressure of Water Near Room Temperature

Temperature (°C)	Vapor Pressure of Water (torr)	Temperature (°C)	Vapor Pressure of Water (torr)
20	17.54	26	25.21
21	18.65	27	26.74
22	19.83	28	28.35
23	21.07	29	30.04
24	22.38	30	31.82
25	23.76	31	33.69

Example 12-14 Gas Collected Over Water

Hydrogen can be generated by the reaction of an active metal with an acid. A 500 milliliter sample of H_2 was collected over water at 30°C at a barometric pressure of 742 torr. What mass of H_2 was collected?

Plan

We know that we can find the mass of hydrogen if we can find the number of moles of hydrogen. We know the volume, temperature, and total pressure of the hydrogen sample. If we use the vapor pressure of water at 30°C and Dalton's law, we can find the pressure due to the hydrogen. The ideal gas equation can then be used to find the number of moles of hydrogen. We finish the process by multiplying the number of moles by the molar mass.

Solution

$P_{total} = P_{water} + P_{hydrogen}$, so $P_{hydrogen} = P_{total} - P_{water}$

$$P_{hydrogen} = 742 \text{ torr} - 31.82 \text{ torr} = 710 \text{ torr} \times \frac{1 \text{ atm}}{760 \text{ torr}} = 0.934 \text{ atm}$$

$$?\ n_{H_2} = \frac{PV}{RT} = \frac{(0.934 \text{ atm})(0.500 \text{ L})}{\left[0.0821 \frac{\text{L} \cdot \text{atm}}{\text{mol} \cdot \text{K}}\right](303\text{K})} = 0.0188 \text{ mol H}_2$$

$$?\ g\ H_2 = 0.0188 \text{ mol H}_2 \times \frac{2.02 \text{ g H}_2}{1 \text{ mol H}_2} = \boxed{0.0380 \text{ g H}_2}$$

12-10 Diffusion and Effusion of Gases; Graham's Law

Graham's Law states that the rate of diffusion or effusion of a gas is inversely proportional to the square root of its density (or molecular weight). The rates of diffusion or effusion of two gases at the same temperature can be expressed as:

$$\frac{\text{rate}_1}{\text{rate}_2} = \sqrt{\frac{d_2}{d_1}} \quad \text{or} \quad \frac{\text{rate}_1}{\text{rate}_2} = \sqrt{\frac{MW_2}{MW_1}}$$

This law is derivable from the tenet of the kinetic molecular theory that states that the average kinetic energy of any two gases at the same temperature is the same. The kinetic energy is equal to $1/2\ mu^2$, where **m** is the mass, which is proportional to the molecular weight, and **u** is the velocity, which is proportional to the rate of diffusion or effusion.

Example 12-15 Molecular Speed

Hydrogen chloride and hydrogen bromide are both allowed to effuse out of a container with a small orifice at the same temperature and pressure. Which gas effuses at a faster rate and how much faster does it effuse than the other gas?

Plan
We can find the molecular weight of each gas and then use Graham's law.

Solution

$$\frac{\text{rate}_{HCl}}{\text{rate}_{HBr}} = \sqrt{\frac{MW_{HCl}}{MW_{HBr}}} = \sqrt{\frac{80.9 \text{ g/mol}}{36.5 \text{ g/mol}}} = \boxed{1.49}$$

The HCl effuses 1.49 times as fast as the HBr.

Example 12-16 Molecular Mass of a Gas

A gas diffuses only 0.500 times as fast as an O_2 molecule at the same temperature. What is the molecular weight of the gas?

Plan

We use the known molecular weight of O_2 and the fact that the rate of the other gas is only 0.500 rate$_{O_2}$ in the Graham's Law equation.

Solution

$$\frac{rate_x}{rate_{O_2}} = \frac{0.500 \, rate_{O_2}}{rate_{O_2}} = 0.500 = \sqrt{\frac{MW_{O_2}}{MW_x}} = \sqrt{\frac{32.00 \, amu}{MW_x}}$$

Squaring both sides of the equation yields:

$$0.250 = \frac{32.00 \, amu}{MW_x} \quad \text{or} \quad 0.250 MW_x = 32.00 \, amu$$

$$MW_x = \frac{32.00 \, amu}{0.250} = \boxed{128 \, amu}$$

12-11 Real Gases and Deviations from Ideality

Real gases have finite molecular volumes and attractive forces. At high pressures and low temperatures, where a gas is close to liquefying, these two factors can cause considerable deviation from properties calculated with the ideal gas equation. The best known equation for correcting for non-ideality is the van der Waals equation:

$$(P + \frac{n^2 a}{V^2})(V - nb) = nRT$$

The "a" term corrects for the attractive forces and the "b" term corrects for the volume occupied by the gas molecules themselves. Table 12-5 in the text gives some examples of "a" and "b" values for different molecules.

Example 12-17 van der Waal's Equation

Calculate the pressure exerted by 2.95 moles of CO_2 in a 600 mL container at 30.0°C. For carbon dioxide: $a = 3.59 \dfrac{L^2 \cdot atm}{mol^2}$, and $b = 0.0427$ L/mol. Compute the corrected pressure to that found if the ideal gas equation is used instead and compare the two answers.

Plan
We substitute the data directly into the van der Waal's equation solving for P. We then repeat the process with the ideal gas equation.

Solution
Using the van der Waal's equation:

$$\left(P + \dfrac{n^2 a}{V^2}\right)(V - nb) = nRT$$

$$\left[P + \dfrac{(2.95 \text{ mol})^2 (3.59 \dfrac{L^2 \cdot atm}{mol^2})}{(0.600 \text{ L})^2}\right][0.600 \text{ L} - (2.95 \text{ mol})(0.0427 \text{ L/mol})] =$$

$$(2.95 \text{ mol})\left[0.0821 \dfrac{L \cdot atm}{mol \cdot K}\right](303 K)$$

$(P + 86.8)(0.474) = 73.4$ so $0.474P + 41.1 = 73.4$

$0.474P = 32.3$

$P = \boxed{68.1 \text{ atm}}$

Using the ideal gas equation:

$$P = \dfrac{nRT}{V} = \dfrac{(2.95 \text{ mol})\left[0.0821 \dfrac{L \cdot atm}{mol \cdot K}\right](303 K)}{0.600 L} = \boxed{122 \text{ atm}}$$

The computed pressure is nearly **twice** as large as the more accurate van der Waals value. CO_2 differs greatly from ideality under these conditions.

12-12 Mass-Volume Relationships in Reactions Involving Gases

Gas law concepts can be integrated into the stoichiometric relationships discussed in Chapter Three. If at STP, the standard molar volume of 22.4 liters per mole of gas may be used. If not at STP, the ideal gas equation must be used.

Example 12-18 Mass-Volume Calculations

At 20°C and 2.50 atm pressure, 5.00 liters of sulfur dioxide were consumed in the following reaction. How many grams of oxygen were consumed and how many grams of sulfur trioxide were produced?

$$2SO_2(g) + O_2(g) \rightarrow 2SO_3(g)$$

Plan

We must use the ideal gas equation to find the number of moles of SO_2, because the system is not at STP. Once that amount has been found, we can find masses with our usual stoichiometric techniques as in Chapter Three.

Solution

First, we find the number of moles of SO_2 used.

$P = 2.50$ atm, $V = 5.00$ L, $T = 20°C + 273 = 293$K, $R = 0.0821 \frac{L \cdot atm}{mol \cdot K}$

$$? \text{ mol } SO_2 = \frac{PV}{RT} = \frac{(2.50 \text{ atm})(5.00 \text{ L})}{\left[0.0821 \frac{L \cdot atm}{mol \cdot K}\right](293K)} = 0.520 \text{ mol } SO_2$$

Now, we can use stoichiometric techniques to find masses of the other gases.

$$? \text{ g } O_2 = 0.520 \text{ mol } SO_2 \times \frac{1 \text{ mol } O_2}{2 \text{ mol } SO_2} \times \frac{32.00 \text{ g } O_2}{1 \text{ mol } O_2} = \boxed{8.32 \text{ g } O_2}$$

$$? \text{ g } SO_3 = 0.520 \text{ mol } SO_2 \times \frac{2 \text{ mol } SO_3}{2 \text{ mol } SO_2} \times \frac{80.06 \text{ g } SO_3}{1 \text{ mol } SO_3} = \boxed{41.6 \text{ g } SO_3}$$

Example 12-19 Mass-Volume Calculation

Oxygen can be prepared by heating solid potassium chlorate, $KClO_3$. An impure 58.2 gram sample containing potassium chlorate was heated until it yielded no more oxygen. The total volume of dry oxygen collected at 27°C and 741 torr was 15.0 liters. What was the percentage of potassium chlorate in the original sample?

$$2KClO_3(s) \rightarrow 2KCl(s) + 3O_2(g)$$

Plan
We must first use the ideal gas equation to find the number of moles of oxygen produced. We can then use normal stoichiometric techniques to find the mass of $KClO_3$ that was originally in the sample. Finally, we can find the percent $KClO_3$.

Solution
The number of moles of O_2 is found first.

$$P = 741 \text{ torr} \times \frac{1 \text{ atm}}{760 \text{ torr}} = 0.975 \text{ atm}, \quad V = 15.0 \text{ L}, \quad T = 27° + 273 = 300K, \quad R = 0.0821 \frac{L \cdot atm}{mol \cdot K}$$

$$? \; n = \frac{PV}{RT} = \frac{(0.975 \text{ atm})(15.0 \text{ L})}{\left[0.0821 \frac{L \cdot atm}{mol \cdot K}\right](300K)} = 0.594 \text{ mol } O_2$$

Now, the mass of $KClO_3$ can be found.

$$? \text{ g } KClO_3 = 0.594 \text{ mol } O_2 \times \frac{2 \text{ mol } KClO_3}{3 \text{ mol } O_2} \times \frac{122.6 \text{ g } KClO_3}{1 \text{ mol } KClO_3} = 48.5 \text{ g } KClO_3$$

Finally, the percentage of $KClO_3$ in the sample can be determined.

$$? \; \%KClO_3 = \frac{48.5 \text{ g } KClO_3}{58.2 \text{ g sample}} \times 100 = \boxed{83.3\% \; KClO_3}$$

EXERCISES

Boyle's Law

1. A sample of hydrogen sulfide, H_2S, occupies a volume of 14.0 liters and exerts a pressure of 56.0 atm at 28°C. If the volume expands to 55.0 liters with no temperature change, what pressure will the gas exert?

2. A gas occupies 33.8 mL and exerts a pressure of 366 torr. If the temperature remains constant, what volume must it occupy to exert a pressure of 285 torr?

3. A 20.0 liter tank of helium is at a pressure of 40.0 atm. How many 2.50 liter balloons at 1.00 atmosphere pressure can be filled from this tank? The helium remaining in the tank will also be at 1.00 atmosphere pressure.

Charles' Law

4. A sample of a gas occupies 82.3 mL at 14°C. If the pressure remains constant, what volume will it occupy at 78°C?

5. A sample of nitrogen occupies 25.0 mL at 0°C. At constant pressure, what temperature is needed to decrease the volume to 20.0 mL?

6. A rubber football is filled with 1.98 liters of air in a room at 25°C. A game is played outside at the same pressure. The volume of gas in the football when outside was 1.78 liters. There was no gas leakage. What was the temperature outside?

The Combined Gas Law and Standard Conditions

7. A gaseous sample occupies 22.8 mL at -45°C and 335 torr. What volume will it occupy at STP?

8. A sample of gas occupies 14.8 liters at 14°C and 2.28 atm. At what temperature will it occupy a volume of 35.2 liters at 1.04 atm?

9. A sample of gas is enclosed in a 350.0 mL container at 55°C and 745 torr. What pressure will it exert in a 750.0 mL container at 16°C?

10. A gas occupies 88.9 liters at STP. What will be its pressure if the volume is reduced to 18.6 liters while the temperature increases to 25°C?

Chapter 12 Gases and the Kinetic Molecular Theory

11. A tire is found to have a volume of 6.40 liters and a pressure of 32.0 lb/in² at 20.0°C. After the car has been driven, the volume is 6.48 liters and the pressure is 35.1 lb/in². What is the temperature of the gas in the tire?

Avogadro's Law and the Standard Molar Volume

12. What volume is occupied by 30.0 grams of radon at STP?

13. A sample of PH_3 occupies 650.0 mL at STP. What is the mass of PH_3 in the sample?

14. A 0.785 gram sample of a diatomic element occupies 248 mL at STP. What are the molecular weight and the formula of the element?

The Ideal gas equation and Dalton's Law

15. What volume will be occupied by 8.41 grams of SO_2 at 500°C and 5.40 atm pressure?

16. How many grams of H_2 are in a 300.0 mL sample collected over water at an atmospheric pressure of 758 torr in a room at 30°C?

17. A 10.0 mL sample of O_2 was collected over water at 24.0°C when the barometric pressure was 738 torr. How many grams of oxygen were present? What volume would the dry oxygen occupy at STP?

18. What are the partial pressures and the total pressure exerted by a mixture of 2.22 grams of neon and 4.78 grams of argon at 47°C in a container with a volume of 11.6 liters?

19. A gas is found to be 36.8% nitrogen and 63.2% oxygen. When a 1.25 gram sample of the gas is analyzed by the Dumas method, it is found to occupy 422 mL at 702 torr and 17°C. What are the molecular weight and molecular formula of the gas?

20. A sample of a noble gas with a mass of 25.6 grams occupies 3.85 liters at 1.24 atm and 25°C. What is the molecular weight and which noble gas is it?

21. How many grams of ethane, C_2H_6, are in a 25.0 L container at 23°C and 22.0 atm?

22. What pressure is exerted by 2.89 grams of xenon tetrafluoride in a 622 mL container at 29°C?

23. Nickel carbonyl is an extremely toxic chemical that is formed as a byproduct in the purification of nickel. It is also used as an agent in organic synthesis. Nickel carbonyl is 28.1% carbon, 34.4% nickel, and 37.5% oxygen. When 3.15 grams of the compound are vaporized at 52°C and 748 torr, the sample occupies a volume of 499 mL. What are the molar mass and the molecular formula of nickel carbonyl?

24. A sample of a gas contains 0.305 grams of carbon, 0.407 grams of oxygen and 1.805 grams of chlorine. A different sample of the same gas weighing 1.72 grams occupies 2.00 liters and exerts a pressure of 992 torr at 1560°C. What is the molecular formula of the gas?

25. What is the density of NO_2 gas in grams/liter at 18°C and 555 torr?

26. Cyclopropane is a gas that has been used as an anesthetic in surgery. It is found to be 85.7% carbon and 14.3% hydrogen. A 0.154 gram sample occupies 90.9 mL at 15.0°C and 750 torr? What is the molecular formula of cyclopropane?

27. Natural gas has no odor; therefore, a substance with a strong odor is added to natural gas samples to warn people of leaks in their gas lines. Ethyl mercaptan, which is 38.7% carbon, 9.74% hydrogen, and 51.6% sulfur, is a common odorant. A 1.27 gram sample of ethyl mercaptan occupies 520 mL at 742 torr and 31°C. What are the molar mass and the molecular formula of ethyl mercaptan?

28. Para-dichlorobenzene is used in moth balls. A 25.00 gram sample was found to be composed of 12.25 grams of carbon, 0.68 grams of hydrogen, and 12.07 grams of chlorine. When the entire sample was vaporized, it occupied a volume of 5.95 liters at 150°C and 755 torr? What is this substance's molecular formula?

Graham's Law

29. A xenon molecule has an average velocity of 252 m/s at 62°C. A different noble gas molecule, at the same temperature, has a velocity of 643 m/s. What is the molecular weight of the other noble gas? Which noble gas is it?

30. What is the molecular weight of a gas that diffuses 0.707 times as fast as nitrogen, N_2?

31. He gas has an average velocity of 1.36×10^3 m/s at 25°C. What is the average velocity of N_2O_4 at the same temperature?

32. Calculate the density of N_2 at STP. What is the density of a gas at STP that diffuses 0.741 times as fast as N_2?

Mass-Volume Relationships

33. A sample that contains 155 grams of argon has a pressure of 12.8 atm at some temperature. What is the total pressure in the container if 214 grams of xenon are added to the same container at the same temperature?

34. A sample of PH_3 with a mass of 18.9 grams has a volume of 22.3 liters at 22°C and some pressure. What is the volume of 8.26 grams of AsH_3 at 36°C and the same pressure?

35. Sulfur hexafluoride is used in tennis balls because of its low rate of effusion. It can be prepared from direct reaction of the elements according to the equation below.

$$S_8(s) + 24F_2(g) \rightarrow 8SF_6(g)$$

(a) How many grams of sulfur are required to react with 4.25 liters of fluorine at 44°C and 7.11 atm?

(b) When 57.4 grams of fluorine react with excess sulfur, the SF_6 formed is collected in a 2.00 liter container at 25°C. What is the pressure of the SF_6?

36. Sodium hypochlorite solution, NaClO, is often used as a disinfectant agent in swimming pools. It is decomposed into sodium chloride and oxygen by sunlight.

$$2NaClO(aq) \rightarrow 2NaCl(aq) + O_2(g)$$

(a) What volume of dry O_2 at STP is produced by the decomposition of 50.0 grams of NaClO?

(b) How many grams of NaCl are produced when 11.2 liters of O_2 at STP are generated by this reaction?

37. HCN, a poisonous gas used in gas chambers, can be produced by the reaction of acid on cyanide salts. How many liters of HCN at 1.25 atm and 21°C can be produced by reacting 206 grams of $Ca(CN)_2$ with excess acid?

$$Ca(CN)_2(s) + 2HCl(aq) \rightarrow CaCl_2(aq) + 2HCN(g)$$

38. Aluminum reacts with hydrochloric acid to yield aluminum chloride and hydrogen gas.

(a) Write the balanced equation for this reaction.

(b) How many grams of aluminum must react to produce 40.0 liters of dry hydrogen as measured at 30°C and 3.65 atm?

(c) How many milliliters of hydrogen gas, measured at 745 torr and 29°C, can be formed by the reaction of 0.136 grams of aluminum?

(d) What mass of aluminum chloride is prepared in the process in (c)?

39. Methanol, CH_3OH, which is also known as "wood alcohol," is used as a fuel for many race cars. It is commercially synthesized from carbon monoxide and hydrogen. How many grams of methanol could be produced from the reaction of 16.6 liters of CO with 38.5 liters of H_2 at STP? Assume 77.6% yield with respect to the limiting reactant.

$$CO(g) + 2H_2(g) \rightarrow CH_3OH(\ell)$$

40. Ammonia reacts with hot copper(II) oxide to form metallic copper, nitrogen and steam.

$$2NH_3(g) + 3CuO(s) \rightarrow N_2(g) + 3Cu(s) + 3H_2O(g)$$

Assuming 93.0% yield of metallic copper with respect to copper(II) oxide, what volume of gaseous ammonia reacts with 58.6 grams of copper(II) oxide at 1000°C and 5.00 atm pressure?

Miscellaneous Exercises

41. How many liters of dry H_2 at 26°C and 754 torr atmospheric pressure are produced by the reaction of 1.75 grams Na? The hydrogen is originally collected over water.

$$2Na(s) + 2H_2O(\ell) \rightarrow 2NaOH(aq) + H_2(g)$$

42. Butane, C_4H_{10}, a common hydrocarbon fuel, burns in the presence of excess oxygen to produce carbon dioxide and water vapor.

$$2C_4H_{10}(\ell) + 13O_2(g) \rightarrow 8CO_2(g) + 10H_2O(g)$$

(a) How many total liters of gas at 30.0°C and 783 torr can be released in the combustion of 50.0 grams of butane?

(b) How many liters of O_2 are needed for the combustion under the conditions in part (a)?

(c) How many grams of butane would be needed to react with 27.0 liters of oxygen at 65.0 atm pressure and 40.0°C?

43. Naphthalene is another compound used in moth balls. A 50.0 gram sample was found to be composed of 46.8 grams of carbon and 3.2 grams of hydrogen. When a 21.79 gram sample of naphthalene was vaporized, it occupied a volume of 5.95 liters at 150°C and 755 torr. What is this substance's molecular formula?

44. The total pressure of a gas mixture at 0°C is 3.52 atm. If the gas mixture is composed of 3.55 grams of neon, 12.8 grams of chlorine, and 3.64 grams of oxygen, what are (a) the volume of the container and (b) the partial pressure of each gas?

45. When an impure 42.0 gram sample containing some Na_2CO_3 is treated with excess hydrochloric acid, 350.0 milliliters of dry CO_2 with a pressure of 18.3 atm at 22°C were obtained. What was the percent Na_2CO_3 in the sample? Assume that the impurities do not release oxygen.

$$Na_2CO_3(s) + 2HCl(aq) \rightarrow 2NaCl(aq) + H_2O(\ell) + CO_2(g)$$

46. A sample of helium has a pressure of 3.99 atm in a 12.5 liter container at 86°C. At what temperature would the sample have a pressure of 7.66 atm in a 5.08 liter container?

47. An fluorine sample has a pressure of 2.88 atm in a 2.55 liter container. A chlorine sample at the same temperature has a pressure of 4.66 atm in a 3.66 liter container. If both samples are transferred into a 12.0 liter container at the same temperature, what is the partial pressure of each gas and the total pressure in the container?

48. What is the density of SF_6 at 566 torr and 45°C?

49. A chlorine monofluoride sample in a 6.57 liter flask has a pressure of 8.33 atm at 39°C. What is the new pressure if the sample is transferred to a 21.0 liter container at 8°C?

50. A carbon monoxide molecule has a velocity of 5.50×10^2 m/s at some temperature. What is the velocity of a carbon dioxide molecule at the same temperature?

51. Use the van der Waal's equation to evaluate the pressure exerted by a 12.5 mol sample of NH_3 in a 2.00 L container at 0°C. For NH_3, $a = 4.17 \frac{L^2 \cdot atm}{mol^2}$ and $b = 0.0371$ L/mol. Repeat the calculation using the ideal gas equation.

ANSWERS TO EXERCISES

1. 14.3 atm
2. 43.4 mL
3. 312 balloons
4. 101 mL
5. 109K or -164°C
6. 218 K or -55°C
7. 0.585 L
8. 311K or 38°C
9. 306 torr
10. 5.22 atm
11. 325 K or 52°C
12. 3.03 L
13. 0.987 g PH_3
14. 70.9 g/mol; Cl_2
15. 1.54 L
16. 0.0233 g H_2
17. 0.0124 grams O_2; 9.41 mL at STP
18. 0.521 atm total; 0.249 atm Ne; 0.272 atm Ar
19. 76.0 g/mol; N_2O_3
20. 131 g/mol; Xe
21. 679 g C_2H_6
22. 0.554 atm
23. 171 g/mol; $Ni(CO)_4$
24. $COCl_2$
25. 1.41 g/L
26. C_3H_6
27. 62.1 g/mol; C_2H_6S
28. $C_6H_4Cl_2$
29. 20.2 g/mol; Ne
30. 56.0 g/mol
31. 284 m/s
32. N_2 = 1.25 g/L, other gas = 2.28 g/L
33. 18.2 atm
34. 4.45 L AsH_3
35. (a) 12.4 g S_8 (b) 6.17 atm
36. (a) 7.50 L O_2 (b) 58.5 g NaCl
37. 86.2 L HCN
38. (a) $2Al(s) + 6HCl(aq) \rightarrow 2AlCl_3(aq) + 3H_2(g)$
    (b) 106 g Al (c) 191 mL H_2 (d) 0.672 g $AlCl_3$
39. CO limiting; 18.4 g CH_3OH
40. 9.54 L NH_3
41. 0.974 L H_2
42. (a) 187 L gas (b) 135 L O_2 (c) 609 g C_4H_{10}
43. $C_{10}H_8$
44. (a) 3.00 L (b) P_{O_2} = 0.852 atm; P_{Ne} = 1.32 atm; P_{Cl_2} = 1.35 atm
45. 66.7% Na_2CO_3
46. 280K; 7°C
47. P_{F_2} = 0.612 atm, P_{Cl_2} = 1.42 atm, total pressure = 2.03 atm
48. 4.17 g/L
49. 2.35 atm
50. 4.38×10^2 m/s
51. van der Waal's pressure = 19.0 atm, ideal gas equation pressure = 140 atm

Chapter 13: Liquids and Solids

13-1 Heat Transfer

Heat is absorbed by an object when its temperature increases or when it is converted to a phase with higher potential energy, i.e., solid to liquid, solid to gas or liquid to gas. The reverse processes are accompanied by the release of heat of equal magnitude. In each case, heat is transferred between a substance and its surroundings.

13-2 Specific Heat and Heat Capacity

As discussed in Chapter One, the specific heat of a substance is the amount of heat required to raise the temperature of one gram of the substance by 1°C, with no phase change. For example, the specific heat of liquid water is 4.184 J/g•°C. Specific heats of the same substance in different phases are different. Solid water (ice) has a specific heat of 2.09 J/g•°C, while gaseous water (steam) has a specific heat of 2.03 J/g•°C. The molar heat capacity is expressed in J/mol•°C or and is the amount of energy needed to raise the temperature of one mole of the substance by 1°C. The molar heat capacity is equal to the molar mass times the specific heat. Specific heats and molar heat capacities for some common substances are shown in Appendix E in the text.

When substances at different temperatures are mixed, heat is transferred from the warmer substance to the cooler substance. The magnitude of the heat transferred is the same but the sign is different. If the substance gives off heat, the sign of heat is negative. If the substance absorbs heat, the sign of heat is positive. The final temperature of the system is intermediate between the original temperatures of the two substances.

Example 13-1 Heat Transfer

If 500.0 g sample of lead at 200.0°C is placed in 100.0 g of water at 20.0°C in an insulated container, what will the final temperature be? The specific heat of water is 4.184 J/g·°C.

Plan

The heat gained by the water equals the heat lost by the lead. Let **t** be the final temperature of both. The water will get hotter, so **t - 20.0** (Δt of H_2O) will be positive. The lead will get colder, so for Pb **200.0 - t** (Δt of Pb) will be positive.

Solution

$$(100.0 \text{ g } H_2O) \times (4.184 \text{ J/g } H_2O \cdot °C) \times (t°C - 20.0°C) =$$

$$(500.0 \text{ g Pb}) \times (0.159 \text{ J/g Pb} \cdot °C) \times (200.0°C - t°C)$$

$418.4 \text{ t} - 8368 = 15900 - 79.5 \text{ t}$

$497.9 \text{ t} = 24,268$

$t = \boxed{48.7°C}$ (48.740711)

13-3 Heats of Transformation–Phase Changes

When substances undergo phase changes (changes in state) with no concurrent temperature changes, the heat transferred is called a **heat of transformation**. The **heat of fusion** is the amount of heat that must be absorbed by one gram of a solid at its melting point to convert it to liquid with no change in temperature. The units are usually J/g or cal/g. The **molar heat of fusion**, ΔH_{fus}, is usually expressed in kJ/mol. The **heat of solidification** and **molar heat of solidification**, ΔH_{sol}, apply to the reverse process of freezing. They are equal in magnitude, but opposite in sign, to the corresponding heats of fusion. (The sign is positive for heat of fusion in which energy is required and negative for heat of solidification in which energy is released.) For example, one gram of ice absorbs 334 J as it melts at 0°C at one atmosphere pressure. One gram of water releases 334 J as it freezes at 0°C.

$$H_2O(s) \underset{\text{solidification}}{\overset{\text{fusion}}{\rightleftharpoons}} H_2O(\ell)$$

This corresponds to 6.02 kJ/mol as seen from dimensional analysis:

$$\frac{334 \text{ J}}{1 \text{ g}} \times \frac{18.02 \text{ g}}{1 \text{ mol}} \times \frac{1 \text{ kJ}}{1000 \text{ J}} = 6.02 \text{ kJ/mol}$$

This amount of heat is a measure of the strength of interactions holding particles together in the solid state compared to those binding the particles in the liquid state.

The **heat of vaporization** and the **heat of condensation** and the molar heats of vaporization, ΔH_{vap}, and condensation, ΔH_{cond} are defined similarly for transformations between one gram or one mole of liquid and gas at the boiling point of the liquid. The normal boiling point of a substance is the temperature at which the vapor pressure equals 1 atmosphere. The heat of vaporization and molar heat of vaporization of liquid water at 100°C are 2260 J/g and 40.7 kJ/mol, respectively. Condensation of a gas is just the reverse of vaporization of a liquid.

$$H_2O(\ell) \underset{\text{condensation}}{\overset{\text{vaporization}}{\rightleftharpoons}} H_2O(g)$$

Heats of condensation are equal in magnitude, but opposite in sign, to heats of vaporization. Heats of vaporization are related to the magnitude of the forces of attraction among liquid particles relative to the much smaller attractive forces among gaseous particles. Because of this large change in magnitude of attractive forces, for the same substance, $\Delta H_{vap} \gg \Delta H_{fus}$. For water, ΔH_{vap} is 40.7 kJ/mol, which is much larger than the ΔH_{fus} of 6.02 kJ/mol calculated earlier.

Under the appropriate conditions of pressure and temperature, some solids can be converted directly to gases, and vice-versa, without passing through the liquid state. The first process is called sublimation and the reverse process is called deposition. For example, CO_2 can be directly converted from its solid form, "dry ice," to its vapor at one atmosphere pressure.

$$CO_2(s) \underset{\text{deposition}}{\overset{\text{sublimation}}{\rightleftharpoons}} CO_2(g)$$

The heats of sublimation and deposition and molar heats of sublimation and deposition, ΔH_{sub} and ΔH_{dep}, are associated with these changes and are equal in magnitude, but opposite in sign.

Example 13-2 Heat Transfer with State Change

Calculate the amount of heat in kilojoules released when 50.0 grams of steam at 125.0°C are converted to 50.0 grams of ice at -25.0°C.

Plan

We see that there are five steps involved in this process: (1) cooling the 50.0 g of steam from 125.0°C to 100.0°C, which involves the specific heat of steam; (2) condensing the 50.0 g of steam, which involves the heat of condensation; (3) cooling the 50.0 g of liquid water from 100.0°C to 0.0°C, which involves the specific heat of liquid water; (4) freezing the 50.0 g of water, which involves the heat of solidification; and (5) cooling the 50.0 g of ice from 0.0°C to -25.0°C, which involves the specific heat of ice. We calculate the energy involved in each of these steps in turn and then add the results. (Figure 13-15 in the text illustrates a typical heating curve. This cooling curve would be reversed in appearance.)

Solution

(1) $50.0 \text{ g} \times \left[2.03 \dfrac{J}{g \cdot °C}\right] \times (100.0°C - 125.0°C) \times \dfrac{1 \text{ kJ}}{1000 \text{ J}} = -2.54 \text{ kJ}$

(2) $50.0 \text{ g} \times \dfrac{-2260 \text{ J}}{1 \text{ g}} \times \dfrac{1 \text{ kJ}}{1000 \text{ J}} = -113 \text{ kJ}$

(3) $50.0 \text{ g} \times \left[4.18 \dfrac{J}{g \cdot °C}\right] \times (0.0°C - 100.0°C) \times \dfrac{1 \text{ kJ}}{1000 \text{ J}} = -20.9 \text{ kJ}$

(4) $50.0 \text{ g} \times \dfrac{-333 \text{ J}}{1 \text{ g}} \times \dfrac{1 \text{ kJ}}{1000 \text{ J}} = -16.6 \text{ kJ}$

(5) $50.0 \text{ g} \times \left[2.09 \dfrac{J}{g \cdot °C}\right] \times ([-25.0°C] - 0.0°C) \times \dfrac{1 \text{ kJ}}{1000 \text{ J}} = -2.61 \text{ kJ}$

Total heat released = $\boxed{-156 \text{ kJ}}$

Example 13-3 Heat Transfer with State Change

If a 100 gram sample of ice at -20.0°C is dropped into 5.0 grams of water at 5.0°C in an insulated container, will the water freeze by the time equilibrium is established?

Plan
We can calculate the amount of heat absorbed in warming the 100.0 g of ice from -20.0°C to 0.0°C. This represents the maximum amount of energy that the ice could absorb without melting. We can then compare this to the amount of heat released in cooling the water to 0.0°C and then freezing it.

Solution
The amount of heat which the ice can absorb is calculated first:

$$? J = 100.0 \text{ g} \times \left[2.09 \frac{J}{g \cdot °C}\right] \times (0.0°C - [-20.0°C]) = 4.18 \times 10^3 \text{ J}$$

Now the heat released in cooling and then freezing the water is calculated:

$$? J = 5.00 \text{ g} \times \left[4.18 \frac{J}{g \cdot °C}\right] \times (0.0°C - 5.0°C) = -1.0 \times 10^2 \text{ J}$$

$$? J = 5.00 \text{ g} \times \frac{-333 \text{ J}}{1 \text{ g}} = -1.7 \times 10^3 \text{ J}$$

Total heat released by cooling and freezing water $= -1.8 \times 10^3$ J

Because the ice can absorb all the heat the water must release to freeze and still not melt, yes-the water will freeze.

Example 13-4 Heat Transfer with State Change

What will be the final temperature of the liquid water resulting from the mixing of 10.0 grams of steam at 130.0°C with 40.0 grams of ice at -10.0°C?

Plan
We know that the heat gained by the ice will be equal in magnitude to the heat lost by the steam but opposite in sign. We will need to heat the ice to the melting point, melt the ice, and then raise the temperature of the liquid water formed. We will need to cool the steam to the

boiling point, condense the steam, and then cool the liquid water to the final temperature. Let t equal the final temperature of the water.

Solution

Heat gained by ice:

$$\left\{40.0\text{g} \times \left[2.09 \frac{\text{J}}{\text{g}\cdot°\text{C}}\right] \times (0.0°\text{C} - [-10.0°\text{C}])\right\} + \left\{40.0\text{ g} \times \frac{334\text{ J}}{\text{g}}\right\}$$
$$+ \left\{40.0\text{ g} \times \left[4.18 \frac{\text{J}}{\text{g}\cdot°\text{C}}\right](t - 0.0°\text{C})\right\} = 1.42 \times 10^4 \text{ J} + 167t$$

Heat lost by steam:

$$\left\{10.0\text{ g} \times \left[2.03 \frac{\text{J}}{\text{g}\cdot°\text{C}}\right] \times (100.0°\text{C} - 130.0°\text{C})\right\} + \left\{10.0\text{ g} \times \frac{-2260\text{ J}}{\text{g}}\right\} +$$
$$\left\{10.0\text{ g} \times \left[4.18 \frac{\text{J}}{\text{g}\cdot°\text{C}}\right](t - 100.0°\text{C})\right\} = -2.74 \times 10^4 \text{ J} + 41.8t$$

Setting them equal, but opposite in sign yields:

$$1.42 \times 10^4 \text{ J} + 167t = -(-2.74 \times 10^4 \text{ J} + 41.8t)$$

$$209t = 1.32 \times 10^4$$

$$T = \frac{1.32 \times 10^4}{209} = \boxed{63.1°\text{C}}$$

13-4 The Clausius-Clapeyron Equation

Boiling points vary with atmospheric pressure because the boiling point is the defined as the temperature at which the vapor pressure is equal to the atmospheric pressure. When the temperature of a liquid is changed from T_1 to T_2 there is a concurrent change in the vapor pressure from P_1 to P_2. These changes can be approximated using the molar heat of vaporization, ΔH_{vap} and the Clausius-Clapeyron equation:

$$\ln\left[\frac{P_2}{P_1}\right] = \frac{\Delta H_{vap}}{R}\left[\frac{1}{T_1} - \frac{1}{T_2}\right] \text{ or } \log\left[\frac{P_2}{P_1}\right] = \frac{\Delta H_{vap}}{2.303R}\left[\frac{1}{T_1} - \frac{1}{T_2}\right]$$

Although ΔH_{vap} does change with temperature, for most purposes it can be approximated by using the ΔH_{vap} at the boiling point. Heat of vaporization data are tabulated for several substances in Appendix E. The units of R must agree with the units of the ΔH_{vap}. This equation is used to find vapor pressures at other temperatures or to approximate boiling points at other pressures.

Example 13-5 Estimation of Boiling Point

The boiling point of benzene, C_6H_6, is 80.1°C at 760 torr and the ΔH_{vap} is 30.8 kJ/mol. Estimate the boiling point at 300 torr.

Plan

We let our known boiling point and pressure have subscripts 1, so $P_1 = 760$ torr, $T_1 = 80.1°C + 273.2 = 353.3K$. We let our unknown temperature be T_2, therefore, $P_2 = 300$ torr. We use the Clausius-Clapeyron equation to solve for T_2. R must have units which agree with ΔH_{vap}. In this case, the appropriate unit is 8.314×10^{-3} kJ/mol·K.

Solution

$$\ln\left[\frac{300 \text{ torr}}{760 \text{ torr}}\right] = \frac{30.8 \text{ kJ/mol}}{8.314 \times 10^{-3} \text{ kJ/molK}}\left[\frac{1}{353.3K} - \frac{1}{T_2}\right]$$

$$\ln[0.395] = -0.930 = (3.70 \times 10^3)(2.83 \times 10^{-3} - 1/T_2)$$

$$-2.51 \times 10^{-4} = 2.83 \times 10^{-3} - 1/T_2$$

$$-1/T_2 = (-2.51 \times 10^{-4}) - (2.83 \times 10^{-3}) = -3.08 \times 10^{-3}$$

$$T_2 = \frac{-1}{-3.08 \times 10^{-3}} = \boxed{325K} = \boxed{52°C}$$

13-5 The Structures of Crystals

The structures of many solids have been determined very accurately by means of x-ray diffraction. Each crystal structure is described by a **unit cell**, which is the smallest repeating unit in the crystal. Each unit cell is characterized by its edge lengths, a, b, and c, and the angles between the edges, α, β, and γ as shown in Figure 13-21 in the text. Differences and similarities in these six parameters allow each of the fourteen types of unit cells to be placed into one of seven fundamental systems as shown in the text in Table 13-9.

Within the cubic system there are three subclasses: simple cubic, body-centered cubic, and face-centered cubic. These are shown in Figure 13-1 for homonuclear species. Attention here will be focused on these three types of crystal structure as well as the hexagonal close-packed structure, which is discussed in Section 13-16 in the text.

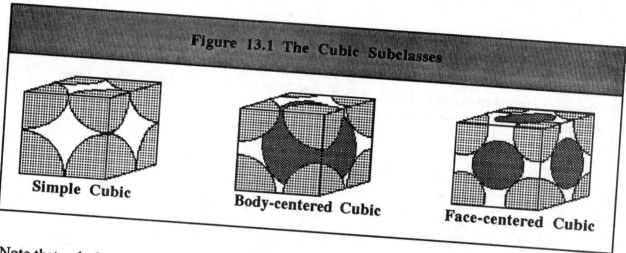

Note that only fractions of particles are included in each unit cell except for the center atom in a body-centered cubic structure. The contributions to the unit cell are summarized in Table 13-1. In the discussion which follows, attention will be paid to the edge length, **a**, the face diagonal, **h**, and the body diagonal, **j**. These are illustrated in the cube below.

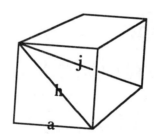

Table 13-1 Contribution to the Unit Cell by Position

Location of Particle	Contribution to One Unit (Cubic) Cell
Corner	1/8
Edge	1/4
Face-centered	1/2
Body-centered	1

The hexagonal close-packed and the face-centered cubic (also called cubic close-packed) structures represent maximum efficiency when identical spheres are considered. For both, 74% of the total volume is due to particles and only 26% is empty space. Each particle is in contact with twelve nearest neighbors and is said to have a coordination number of twelve. Many metals have close-packed structures.

Solids can also be classed into four other categories: (1) molecular solids, such as O_2 and Xe; (2) covalent solids such as SiO_2 and diamond; (3) ionic solids such as NaCl and KBr; and, (4) metallic solids such as Ni and Ag. In each case, the individual units comprising the compound occupy the lattice points that describe the unit cell. The general physical properties of each type of solid are consistent with the structures and interparticle attractions. A summary of the general physical characteristics of the types of solids appears in the text in Table 13-10.

Example 13-6 Simple Cubic Lattice

Below -219°C oxygen crystallizes as a molecular solid with O_2 molecules occupying the corners of a simple cubic lattice. How many O_2 molecules are there per unit cell?

Plan
We know that in the simple cubic structure all molecules are located in the corners of the cell. We consult Table 13-1 to see the contribution of corner to the unit cell.

Solution
8 corners(1/8 atom/corner) = $\boxed{1\ O_2 \text{ molecule per unit cell}}$

Example 13-7 Face-Centered Cubic Lattice

Xenon crystallizes in a face-centered cubic lattice when cooled below -112°C. How many Xe atoms are there per unit cell?

Plan
In a face-centered cubic structure there are eight corner atoms and six face-centered atoms per unit cell. We use Table 13-1 to determine the contribution of each position to the unit cell.

Solution
8 corners(1/8 atom/corner) + 6 faces(1/2 atom/face) = $\boxed{4 \text{ atoms/unit cell}}$

Example 13-8 Body-Centered Cubic Lattice

Chromium crystallizes in a metallic body-centered cubic lattice. How many chromium atoms are there in one unit cell?

Plan
We know that there are eight corner atoms and one body-centered atom per unit cell. We consult Table 13-1 to determine the contribution of each position to the unit cell.

Solution

8 corners(1/8 atom/corner) + 1(1 body-centered atom) = $\boxed{2 \text{ atoms/unit cell}}$

Example 13-9 Determination of Lattice Form

The α form of iron crystallizes in a cubic lattice with a unit cell edge length of a = 2.8665 Å. The density of iron is 7.86 g/cm³. Determine which cubic lattice corresponds to iron.

Plan
Since the various cubic lattices have one, two, or four particles per cell, we must determine the the number of iron atoms in the unit cell. We must first find the mass of one unit cell. We find the volume of the unit cell from the edge length, a, cubed, remembering from Chapter Four that 1 Å = 1 x 10⁻⁸ cm. We can multiply the volume by the density to get the mass of the cell. We can use Avogadro's number and the molar mass of iron to determine the number of iron atoms in the cell. Finally, we consult Table 13-1.

Solution

$$? \frac{\text{g Fe}}{\text{unit cell}} = (2.8665 \times 10^{-8} \text{ cm})^3 \times \frac{7.85 \text{ g Fe}}{1 \text{ cm}^3} = \frac{1.85 \times 10^{-22} \text{ g Fe}}{\text{unit cell}}$$

$$? \frac{\text{atoms Fe}}{\text{unit cell}} = \frac{1.85 \times 10^{-22} \text{ g Fe}}{\text{unit cell}} \times \frac{6.022 \times 10^{23} \text{ Fe atoms}}{55.85 \text{ g Fe}} = 1.99 \frac{\text{Fe atoms}}{\text{unit cell}}$$

Table 13-1 indicates that this is consistent with a $\boxed{\text{body-centered cubic crystal lattice}}$.

Example 13-10 Atomic Radius and Volume

Beryllium crystallizes in a hexagonal close-packed lattice and has a specific gravity of 1.85. Assume that this is an ideal close-packed lattice and determine the volume and radius of a beryllium atom in this lattice.

Plan
We can find the total volume occupied by one mole of Be. We can then find the part of the volume due to the atoms themselves (74%), not the empty space. From this we can calculate the volume of a single Be atom. Finally, using the fact that the volume of a sphere is equal to $4/3\pi r^3$, we can find the radius of a Be atom.

Solution

$$? \frac{cm^3}{mol\ Be\ atoms} = \frac{9.012\ g\ Be}{1\ mol\ Be\ atoms} \times \frac{1\ cm^3}{1.85\ g\ Be} = \frac{4.87\ cm^3}{1\ mol\ Be\ atoms}$$

The volume due to the atoms themselves, not empty space is:

$$\frac{4.87\ cm^3}{1\ mol\ Be\ atoms} \times 0.74 = \frac{3.6\ cm^3}{1\ mol\ Be\ atoms}$$

$$? \frac{cm^3}{Be\ atom} = \frac{3.6\ cm^3}{1\ mol\ Be\ atoms} \times \frac{1\ mol\ Be\ atoms}{6.022 \times 10^{23}\ Be\ atoms} = \boxed{\frac{6.0 \times 10^{-24}\ cm^3}{Be\ atom}}$$

Finally, the radius of the Be atom can be found from:

$$r = \sqrt[3]{\frac{V}{4\pi/3}} = \sqrt[3]{\frac{6.0 \times 10^{-24}\ cm^3}{4\pi/3}} = \boxed{1.1 \times 10^{-8}\ cm} \times \frac{1\ \text{Å}}{10^{-8}\ cm} = \boxed{1.1\ \text{Å}}$$

This value agrees well with the tabulated value of 1.11 Å.

Example 13-11 Atomic Radius

The γ form of nickel crystallizes in a face-centered cubic lattice with a = 3.525 Å. Calculate the radius of a Ni atom.

Plan

We can visualize one face of the face-centered cube as shown in Figure 13-1, assuming contact between the corner atoms and the face-centered atom. The length of the diagonal, **h**, is related to the edge length, **a**, by the Pythagorean theorem ($h^2 = a^2 + a^2$). We can also see from the diagram that **h** equals four times the radius of a Ni atom. We can use these relationships directly to solve the problem.

Solution

$$h^2 = a^2 + a^2 = 2a^2 \text{ so } h = a\sqrt{2}$$

$$h = 3.525 \text{ Å}\sqrt{2} = 3.525 \text{ Å}(1.414) = 4.985 \text{ Å}$$

The radius is equal to $\dfrac{h}{4} = \dfrac{4.985 \text{ Å}}{4} = \boxed{1.246 \text{ Å}}$ (Tabulated value 1.24 Å.)

Example 13-12 Atomic Radius

Rubidium crystallizes in a body-centered cubic lattice with a = 5.63 Å. Calculate the radius of a rubidium atom. Assume atom to atom contact through the body-centered atom.

Plan

We can inspect the body-centered cubic lattice as shown in Figure 11-1. The length of diagonal **h** was related to edge length, **a**, in Example 13-14. The body diagonal, **j**, is also the hypotenuse of a right triangle and is related to each **a** and **h** as follows:

$$j^2 = a^2 + h^2 = a^2 + 2a^2 = 3a^2; \text{ therefore } j = a\sqrt{3}$$

We can see that four radii comprise the body diagonal and use this piece on information with the evaluation of **j** to determine the atomic radius.

Solution

$$j = a\sqrt{3} = 5.63 \text{ Å}\sqrt{3} = 5.63 \text{ Å}(1.732) = 9.75 \text{ Å}$$

Four atomic radii comprise the body diagonal, so:

$$r_{Rb} = \frac{j}{4} = \frac{9.75 \text{ Å}}{4} = \boxed{2.44 \text{ Å}} \text{ (Tabulated value = 2.44 Å.)}$$

Ionic lattices are not quite as simple as metallic lattices. The unit cell can be defined by one kind of ion. Other ions occupy spaces among the first type of ions.

Example 13-13 Ionic Radius versus Crystal Data

Cesium chloride, CsCl, crystallizes in a lattice in which there are chloride ions at each corner of a cube and a cesium ion in a body-centered position. The edge length, **a**, is 4.121 Å and the radius of a chloride ion is 1.81 Å. Assuming anion-cation contact along the body diagonal, **j**, determine the ionic radius of Cs^+.

Plan

We know from Example 13-15 that the body diagonal, j, is equal to $a\sqrt{3}$. We also know from inspecting Figure 13-1 that the body diagonal is composed of two chloride radii and two cesium radii. We can find the length of the body diagonal and then determine the contribution of the cesium radius to that diagonal.

Solution

$$? \text{ body diagonal} = a\sqrt{3} = 4.121 \text{ Å}\sqrt{3} = 4.121 \text{ Å}(1.732) = 7.138 \text{Å}$$

$$2r_{Cs^+} + 2r_{Cl^-} = 7.138 \text{ Å} \quad \text{so} \quad 2r_{Cs^+} = 7.138 \text{ Å} - 2r_{Cl^-} = 7.138\text{Å} - 2(1.81 \text{ Å})$$

$$2r_{Cs^+} = 3.52 \text{ Å}, \text{ each } Cs^+ \text{ radius} = \frac{3.52 \text{ Å}}{2} = \boxed{1.76 \text{ Å}}$$

The tabulated value of a Cs^+ radius is 1.69 Å. There is probably not contact along the body diagonal.

EXERCISES

Consult Appendix E and this chapter for specific heats, heats of transformation, etc., to be used in solving these problems.

Heat Transfer and Heats of Transformation

1. How many kilojoules must be absorbed to raise the temperature of:
   (a) 35.0 grams of liquid ethanol, C_2H_5OH, from 15.0°C to 70.0°C?
   (b) 35.0 g of liquid water from 15.0°C to 70.0°C?
   (c) 35.0 g of copper from 15.0°C to 70.0°C?

2. What will be the temperature of 200.0 grams of the following substances, initially at 80.0°C, after the removal of 5.00 kJ of heat?
   (a) water   (b) iron   (c) aluminum

3. How many kilojoules of heat must be absorbed or released to change the temperature of each of the following substances as indicated?
   (a) 40.0 g of water at 50.0°C to steam at 120.0°C
   (b) 35.0 g of liquid aluminum at its melting point of 558°C to solid at 10.0°C
   (c) 75.0 g of gaseous ethanol, C_2H_5OH, at 100.0°C to liquid at -20.0°C
   (d) 56.8 g of solid diethyl ether, $(C_2H_6)_2$, at its melting pint of -116.0°C to vapor at the boiling point of 35.0°C
   (e) 55.0 g of liquid mercury at 55° to vapor at the boiling point of 357°C

4. How many kilojoules of heat must be removed in cooling:
   (a) 75.0 g of steam at 120°C to ice at -50.0°C?
   (b) 75.0 g of benzene at 120°C to solid benzene at 5.48°C?
   (c) 75.0 g water at 25.0°C to ice at -15°C?

5. Calculate the amount of heat in kilojoules absorbed when 7.50 moles of ice at 0.0°C are converted to steam at 170°C.

6. Will 10.0 grams of water at 12.0°C freeze when mixed with one kg of ice originally at -30.0°C in an insulated container?

7. Will a sample containing 45.0 grams of water at 20.0°C boil when 125 grams of iron at 230.0°C are added to it?

The Clausius-Clapeyron Equation

8. What is the vapor pressure of benzene at 0.0°C?

9. At what pressure does ethanol boil at 0.0°C?

10. What is the boiling point of mercury at a pressure of 225 torr?

The Structures of Crystals

11. The unit cell of the cesium chloride structure was discussed in Example 13-13. Calculate the number of formula units of CsCl in one unit cell.

12. The atomic radius of lead is 1.750 Å. Lead crystallizes in a face-centered cubic lattice. What is the density of lead?

13. Aluminum crystallizes in a face-centered cubic lattice. An aluminum atom has a radius of 1.43 Å. Calculate the edge length of the unit cell and the density of aluminum.

14. Calcium crystallizes in a face-centered cubic lattice with a unit cell edge length of 5.56Å. Calculate the radius of a calcium atom and the density of calcium.

15. Potassium bromide crystallizes in the sodium chloride type structure (Figure 13-27 in the text). The ionic radii of K^+ and Br^- are 1.33Å and 1.95 Å, respectively. Calculate the edge length of the unit cell, assuming anion-cation contact along the edge.

16. Refer to Exercise 15. Calculate the density of KBr.

17. An unknown metal crystallizes in a body-centered cubic lattice with a unit cell edge length of 3.301Å. Its density is 8.569 g/cm^3 at 20°C. Identify the metal.

18. What is the atomic radius of the metal in Exercise 17?

19. As mentioned in Example 13-8, chromium crystallizes in a body-centered cubic lattice. The radius of a chromium atom is 1.25Å. Calculate the edge length of the unit cell and the density of chromium.

20. Thallium(I) bromide is isomorphous with CsCl. The unit cell edge length is 3.98 Å. The ionic radius of Br⁻ is 1.95 Å. Calculate the ionic radius of Tl⁺ assuming anion-cation contact along the body diagonal.

21. Refer to Exercise 20. What is the density of TlBr?

Miscellaneous Exercises

22. What will be the final equilibrium temperature of the substances present in an insulated container if:

    (a) 25.0 g of Al at 150.0°C are placed in 125 g of H_2O at 10.0°C?

    (b) 20.0 g of ethanol at 65.0°C are mixed with 20.0 g H_2O at 10.0°C?

23. Vanadium crystallizes in a body-centered cubic lattice. The unit cell has an edge length of 3.011 Å. What is the density of vanadium?

24. What will be the final temperature of the substances present if 150 g of aluminum at 200°C are placed in an insulated container with 30.0 g of ice originally at -20.0°C?

25. An unknown metal crystallizes in a body-centered cubic lattice with a unit cell edge length of 5.344 Å. Its density is 0.862 g/cm³ at 20°C. Identify the metal.

26. How many kilojoules of energy are required to convert 75.0 grams of solid sodium at 18.5°C to liquid at its melting point of 97.8°C? For Na, ΔH_{vap} = 2.60 kJ/mol and the specific heat of the solid is 1.23 J/g·°C.

27. When 67.41 grams of gold at 89.50°C are placed into 75.00 grams of water at 21.30°C, the final temperature of the system is 23.17°C. Determine the specific heat of gold.

28. Use the Clausius-Clapeyron equation to estimate the atmospheric pressure required for a sample of diethyl ether to boil at -15°C.

29. How many kilojoules are needed to convert 56.4 grams of solid Cs from solid at the melting point of 28.7°C to vapor at the boiling point of 690.0°C? For Cs, the ΔH_{fus} = 278 kJ/mol and ΔH_{vap} = 9.07 x 10³ kJ/mol. The specific heat of the liquid cesium is 0.252 J/g·°C.

30. Use the Clausius-Clapeyron equation to estimate the vapor pressure of water at 65°C. Compare your answer to that in the table in Appendix E.

31. Copper has a specific gravity of 8.93 and crystallizes in a face-centered cubic lattice. Use this information to determine the edge length of the unit cell and the radius of a copper atom.

ANSWERS TO EXERCISES

1. (a) 4.74 kJ (b) 8.05 kJ (c) 0.741 kJ

2. (a) 74.0°C (b) 23.7°C (c) 52.2°C

3. (a) 100.3 kJ absorbed (b) 31.1 kJ released (c) 83.8 kJ released
   (d) 45.9 kJ absorbed   (e) 18.4 kJ absorbed

4. (a) 236 kJ (b) 52 kJ (c) 35.7 kJ   5. 426 kJ

6. Yes, the ice can absorb more than enough heat to freeze the water without any ice melting

7. No, the amount of heat required to drop the temperature of the iron below water's boiling point is less than the amount of heat required to raise the temperature of the water to its boiling point.

8. 35.3 torr   9. 16.2 torr   10. 568K or 295°C   11. one

12. 11.35 g/cm^3 13. a = 4.04Å; d = 2.71 g/cm^3

14. r = 1.96 Å; d = 1.55 g/cm^3 15. 6.56 Å   16. 2.80 g/cm^3 17. Nb

18. 1.429 Å   19. a = 2.89Å; d = 7.14 g/cm^3 20. 1.50 Å   21. 7.49 g/cm^3

22. (a) 15.8°C (b) 30.4°C   23. 6.198 g/cm^3 24. 60.5°C

25. K (potassium)   26. 15.80 kJ   27. 0.131 J/g·°C   28. 106 torr

29. 3.97 x 10^3 kJ   30. Calculated = 196 torr; table 187.5 torr

31. a = 3.62Å; r = 1.28Å or 1.28 x 10^{-8} cm

Chapter 14 Solutions

14-1 Concentrations of Solutions

Review the concepts of solutions that were discussed in Chapters Three, Four, Nine, Ten, and Eleven. This includes the three means of expressing concentrations: (1) percent by mass (mass percent); (2) molarity (M); and, (3) normality (N). Percent by mass and molarity were discussed in Chapter Three, while molarity and normality were both used in Chapter Eleven. Mole fraction was discussed in conjunction with gas mixtures and Dalton's law in Chapter Nine. The additional concentration unit of molality is included in this chapter and mole fraction is reviewed in the context of liquid, rather than gaseous, solutions. Density, which was presented in Chapter One, is useful in converting between concentration units. When performing conversions between concentration units, you should assume that a fixed amount of solution or solvent is present initially and find the related necessary quantities.

When the original concentration is:
- Mass percent
- Mole Fraction
- Molarity
- molality

Start with:
- 100 g solution
- 1 mol (solute + solvent)
- 1.00 liter = 1000 mL solution
- 1000 g solvent = 1 kg solvent

14-2 Molality

The **molality** (m) of a solute is defined as the number of moles of solute per kilogram of **solvent** (**not** solution).

$$\text{molality} = m = \frac{\text{mol solute}}{\text{kg solvent}}$$

This is the only concentration unit that we use in which the denominator depends on an amount of solvent instead of an amount of solution.

Example 14-1 Molality

A 7.58 gram sample of benzoic acid, $C_6H_5CO_2H$, is dissolved in 100.0 mL of benzene, C_6H_6. The density of benzene is 0.879 g/mL. What is the molality of the benzoic acid?

Plan

We convert the mass of the solute, $C_6H_5CO_2H$, to moles. Using the density, we convert the volume of solvent, C_6H_6, to mass in grams and then to kilograms. We then apply the definition of the molality.

Solution

$$? \text{ mol } C_6H_5CO_2H = 7.58 \text{ g } C_6H_5CO_2H \times \frac{1 \text{ mol } C_6H_5CO_2H}{122.13 \text{ g } C_6H_5CO_2H}$$

$$= 0.0621 \text{ mol } C_6H_5CO_2H$$

$$? \text{ kg } C_6H_6 = 100.0 \text{ mL } C_6H_6 \times \frac{0.879 \text{ g}}{\text{mL}} \times \frac{1 \text{ kg}}{1000 \text{ g}} = 0.0879 \text{ kg } C_6H_6$$

$$? \, m \, C_6H_5CO_2H = \frac{0.0621 \text{ mol } C_6H_5CO_2H}{0.0879 \text{ kg } C_6H_6} = \boxed{0.706 \, m \, C_6H_5CO_2H}$$

Example 14-2 Molality; Mass of Solvent

How many grams of water must be added to 35.0 grams of glucose, $C_6H_{12}O_6$, to make a $0.750 \, m$ glucose solution?

Plan

We convert the mass of solute to number of moles. We then use the definition of molality and dimensional analysis to find the mass of water.

Solution

$$? \text{ mol } C_6H_{12}O_6 = 35.0 \text{ g } C_6H_{12}O_6 \times \frac{1 \text{ mol } C_6H_{12}O_6}{180.18 \text{ g } C_6H_{12}O_6} = 0.194 \text{ mol } C_6H_{12}O_6$$

$$? \text{ g } H_2O = 0.194 \text{ mol } C_6H_{12}O_6 \times \frac{1 \text{ kg } H_2O}{0.759 \text{ mol } C_6H_{12}O_6} \times \frac{1000 \text{ g}}{1 \text{ kg}} = \boxed{256 \text{ g } H_2O}$$

Example 14-3 Molality from Mass Percent

What is the molality of ammonium sulfate in an aqueous solution that is 18.0% by mass ammonium sulfate, $(NH_4)_2SO_4$?

Plan

Because the mass percent is given we assume that we have 100 grams of solution. Then we know the mass of solute is 18.0 grams and we can find the mass of water by subtraction. We can convert the mass of solute to number of moles, the mass of water to kilograms, and then apply the definition of molality.

Solution

$$? \text{ mol } (NH_4)_2SO_4 = 18.0 \text{ g } (NH_4)_2SO_4 \times \frac{1 \text{ mol } (NH_4)_2SO_4}{132.1 \text{ g } (NH_4)_2SO_4} = 0.136 \text{ mol } (NH_4)_2SO_4$$

$$? \text{ kg } H_2O = (100.0 \text{ g solution} - 18.0 \text{ g } (NH_4)_2SO_4) \times \frac{1 \text{ kg}}{1000 \text{ g}} = 0.0820 \text{ kg } H_2O$$

$$? \, m \, (NH_4)_2SO_4 = \frac{0.136 \text{ mol } (NH_4)_2SO_4}{0.0820 \text{ kg } H_2O} = \boxed{1.66 \, m \, (NH_4)_2SO_4}$$

14-3 Mole Fraction

The **mole fraction** (X_A) of component A in a two component system composed of A and B is:

$$X_A = \frac{\text{number of moles of A}}{\text{number of moles of A} + \text{number of moles of B}}$$

The mole fraction of B would be similarly defined. Note that mole fractions are dimensionless quantities and that $X_A + X_B = 1$.

Example 14-4 Mole Fraction

Determine the mole fraction of $(NH_4)_2SO_4$ in the solution described in Example 14-3.

Plan

We found the number of moles of $(NH_4)_2SO_4$ in the previous example. We now find the number of moles of water from the mass of water and apply the definition of mole fraction.

Solution

$$? \text{ mol } H_2O = 82.0 \text{ g } H_2O \times \frac{1 \text{ mol } H_2O}{18.02 \text{ g } H_2O} = 4.55 \text{ mol } H_2O$$

$$? X_{(NH_4)_2SO_4} = \frac{0.136 \text{ mol } (NH_4)_2SO_4}{0.136 \text{ mol}(NH_4)_2SO_4 + 4.55 \text{ mol } H_2O} = \boxed{0.0290}$$

Example 14-5 Mole Fraction

What is the mole fraction of each component in a solution prepared from 100.0 g of glucose, $C_6H_{12}O_6$, dissolved in 50.0 g of water?

Plan

We find the number of moles of each component and the total number of moles of the two components in the solution. We then apply the definition of mole fraction.

Solution

$$? \text{ mol } C_6H_{12}O_6 = 100.0 \text{ g } C_6H_{12}O_6 \times \frac{1 \text{ mol } C_6H_{12}O_6}{180.18 \text{ g } C_6H_{12}O_6} = 0.5550 \text{ mol } C_6H_{12}O_6$$

$$? \text{ mol } H_2O = 50.0 \text{ g } H_2O \times \frac{1 \text{ mol } H_2O}{18.02 \text{ g } H_2O} = 2.77 \text{ mol } H_2O$$

$$? \text{ mol total} = 0.5550 \text{ mol } C_6H_{12}O_6 + 2.77 \text{ mol } H_2O = 3.32 \text{ mol total}$$

$$? X_{C_6H_{12}O_6} = \frac{0.5550 \text{ mol } C_6H_{12}O_6}{3.32 \text{ mol total}} = \boxed{0.167}$$

$$? X_{H_2O} = \frac{2.77 \text{ mol } H_2O}{3.32 \text{ mol total}} = \boxed{0.834}$$

14-4 Colligative Properties of Solutions

Physical properties of solutions that are affected by the total concentration and not the nature of solute particles are called **colligative properties**. These properties include: (1) vapor pressure lowering; (2) boiling point elevation; (3) freezing point depression; and, (4) osmotic pressure.

14-5 Vapor Pressure Lowering

The vapor pressure exerted by a solvent at a given temperature is always lowered by dissolving a nonvolatile solute in it. **Raoult's Law** states that the vapor pressure of a solvent in a solution with a nonvolatile solute decreases as the mole fraction of the solvent decreases. Mathematically, this may be expressed as:

$$P_{solvent} = X_{solvent}(P°_{solvent})$$

$P_{solvent}$ = the vapor pressure of solvent in the solution
$X_{solvent}$ = the mole fraction of solvent in the solution
$P°_{solvent}$ = the vapor pressure of the pure solvent

$P°$ is temperature dependent. Remember that a table of the vapor pressure of water at various temperatures can be found in Appendix E of the text.

Example 14-6 Vapor Pressure of a Solution of a Nonvolatile Solute

What is the vapor pressure of water in a solution at 85°C prepared as specified in Example 14-5? The glucose is nonvolatile; the vapor pressure of pure water at 85°C is 433.6 torr.

Plan

We found that the mole fraction, X_{H_2O}, of the solution was 0.832. We can find the vapor pressure in the solution by using Raoult's law directly.

Solution

$$?\ P_{H_2O} = X_{H_2O}(P°_{H_2O}) = 0.832(433.6\ \text{torr}) = \boxed{361\ \text{torr}}$$

If we have an ideal solution composed of two volatile components A and B, each contributes to the vapor pressure of the solution. The total vapor pressure of the solution is:

$$P_{Total} = P_A + P_B = X_A P°_A + X_B P°_B$$

Example 14-7 Ideal Solution of Two Volatile Liquids

At 40°C, P° for 2-butyne, C_4H_6 is 40 torr and P° for *cis*-2-bromo-2-butene, C_4H_7Br is 400 torr. Assume that these two substances make an ideal solution and compute the vapor pressure of a mixture of 2.6 moles of C_4H_6 and 4.9 moles of C_4H_7Br.

Plan
Because the number of moles is given, we can readily determine the mole fraction of each substance. We then apply the equation for the vapor pressure of a solution containing two volatile substances.

Solution
The total number of moles is: 2.6 mol C_4H_6 + 4.9 mol C_4H_7Br = 7.5 mol.

? mole fraction $C_4H_6 = \dfrac{2.6}{7.5} = 0.35 = X_A$

? mole fraction $C_4H_7Br = \dfrac{4.9}{7.5} = 0.65 = X_B$

? $P_{Total} = P_A + P_B = X_A P°_A + X_B P°_B = (0.35)(40\ torr) + (0.65)(400\ torr) =$ $\boxed{274\ torr}$

14-6 Boiling Point Elevation

The boiling point of a solution that contains a nonvolatile solute is always higher than that of the pure solvent because the vapor pressure has been decreased at all temperatures in proportion to the mole fraction of solute. The elevation of the boiling point of a solution is directly proportional to the molality of the solute:

$$\Delta T_b = K_b m$$

ΔT_b = the increase in the boiling point
K_b = the boiling point elevation constant, which is a characteristic of the **solvent**
m = the molality of the solute in the solution

Normal boiling points of a few pure solvents and their boiling point elevation constants appear in Table 14-2 in the text.

Example 14-8 Boiling Point Elevation

Calculate the boiling point of a solution that contains 12.2 grams of benzoic acid, $C_6H_5CO_2H$, dissolved in 250.0 grams of pure nitrobenzene, $C_6H_5NO_2$. The boiling point of pure nitrobenzene is 210.88°C and its boiling point elevation constant, K_b, is 5.24°C/m.

Plan

We first must find the molality of the benzoic acid solution. We can then apply the relationship $\Delta T_b = K_b m$ to find the increase in the boiling point.

Solution

$$? \text{ mol } C_6H_5CO_2H = \frac{12.2 \text{ g } C_6H_5CO_2H}{250.0 \text{ g } C_6H_5NO_2} \times \frac{1 \text{ mol } C_6H_5CO_2H}{122.1 \text{ g } C_6H_5CO_2H} \times \frac{1000 \text{ g}}{1 \text{ kg}}$$

$$= 0.400 \text{ m } C_6H_5CO_2H$$

$$? \Delta T_b = K_b m = (5.24°C/m)(0.400 \text{ m}) = 2.10°C$$

The boiling point of the solution is: 210.88°C + 2.10°C = $\boxed{212.98°C}$

14-7 Freezing Point Depression

The freezing point of a solution is always lower than that of the pure solvent if only pure solvent is frozen out. The freezing point depression, ΔT_f, is proportional to the freezing point depression constant, K_f, which is a characteristic of the solvent only, and the molality of the solute in the solution. Mathematically:

$$\Delta T_f = K_f m$$

ΔT_f = the decrease in the freezing point
K_f = the freezing point elevation constant, which is a characteristic of the **solvent**
m = the molality of the solute in the solution

Normal freezing points and freezing point depression constants for several solvents appear in Table 14-2 of the text.

Example 14-9 Freezing Point Depression

Calculate the freezing point of the solution described in Example 14-8. The freezing point of pure nitrobenzene is 5.7°C and its freezing point depression constant, K_f, is 7.00°C/m.

Plan
We know that the molality of the solution is 0.400 m. We use the equation above to find the amount by which the freezing point is lowered.

Solution
$$?\ \Delta T_f = K_f m = (7.00°C/m)(0.400\ m) = 2.80°C$$

The freezing point is: 5.7°C - 2.80°C = $\boxed{2.9°C}$

Example 14-10 Freezing Point Depression Constant

Camphor, used in embalming fluid, has a freezing point of 178.4°C. If 0.711 grams of nonionizing glucose, $C_6H_{12}O_6$, are dissolved in 12.4 g of camphor, the freezing point is 165.6°C. What is the freezing point depression constant, K_f, of camphor?

Plan
We first must find the molality of the solution. We can determine ΔT_f by subtracting the freezing point of the solution from that of the pure solvent. We finally find K_f by algebraically rearranging the freezing point depression formula to see that $K_f = \Delta T_f/m$.

Solution
$$?\ m\ C_6H_{12}O_6 = \frac{0.711\ g\ C_6H_{12}O_6}{12.4\ g\ camphor} \times \frac{1\ mol\ C_6H_{12}O_6}{180.2\ g\ C_6H_{12}O_6} \times \frac{1000\ g}{kg} = 0.318\ m\ C_6H_{12}O_6$$

$$?\ K_f = \frac{\Delta T_f}{m} = \frac{[178.4\ °C - 165.6°C]}{0.318\ m} = \frac{12.8°C}{0.318\ m} = \boxed{40.3\ °C/m}$$

14-8 Determination of Molecular Weights

The molecular weight of an unknown solid or liquid nonelectrolyte can be determined by dissolving a known mass of the substance in a known mass of solvent and measuring the boiling point elevation or the freezing point depression. In practice, it is generally easier to work with the latter, since freezing points are less susceptible to atmospheric pressure. Additionally, as a solution is boiled, its molality changes because of the removal of the solvent. The boiling point continues to rise as boiling occurs. A sample can be frozen and thawed repeatedly without the molality changing to any appreciable extent. The following example illustrates this technique for finding molecular weights.

Example 14-11 Molecular Weight from Colligative Properties

Elemental analysis indicates the simplest formula of an unknown nonelectrolyte is C_2H_4O. When 0.735 g of the compound is dissolved in 50.0 g of benzene, the solution freezes at 5.06°C. What are the molecular weight and the molecular formula of the compound? The freezing point of pure benzene is 5.48°C and its $K_f = 5.12°C/m$.

Plan

To find the molecular weight of the substance, we must find the number of moles represented by 0.735 g of the substance. We first use the freezing point to determine the molality of the solution with $m = \Delta T_f/K_f$. Then, because we know the mass of solvent in the solution, we can find the number of moles of solute. Dividing the mass by the number of moles gives the molecular weight. We divide the molecular weight by the empirical weight to find the number of empirical units in one molecule.

Solution

$$? m = \frac{\Delta T_f}{K_f} = \frac{[5.48°C - 5.06°C]}{5.12°C/m} = \frac{0.42°C}{5.12°C/m} = 0.082\ m = \frac{0.082\ \text{mol solute}}{1\ \text{kg solvent}}$$

$$? \text{mol solute} = \frac{0.082\ \text{mol solute}}{1\ \text{kg solvent}} \times 50.0\ \text{g benzene} \times \frac{1\ \text{kg}}{1000\ \text{g}} = 0.0041\ \text{mol solute}$$

$$? \text{g/mol} = \frac{0.735\ \text{g}}{0.0041\ \text{mol}} = 1.8 \times 10^2\ \text{g/mol}$$

C_2H_4O has an empirical weight of: $2(12.01) + 4(1.01) + 1(16.00) = 44.06$ g/mol

The approximate molecular weight divided by the simplest formula weight yields:

$$\frac{1.80 \times 10^2\ \text{g/mol}}{44.06\ \text{g/mol}} = 4.09$$

Since the molecular formula must be a whole number multiple of the simplest formula, the true molecular formula must be $\boxed{C_8H_{16}O_4}$.

The molecular weight is really 4 x (44.06) = $\boxed{176.24 \text{ g/mol}}$.

In either boiling point elevation or freezing point depression, the magnitude of the observed temperature change is often small. This will introduce a source of experimental error as above when this method is employed.

14-9 Dissociation of Electrolytes

Since colligative properties depend only on the concentration of solute particles in solution, greater effects are observed when ionized or partially ionized substances are solutes. When ions are produced, the total number of particles present increases, consequently increasing the concentration. The resulting greater magnitude of freezing point depression can be used to determine the percent ionization in solutions of weak electrolytes that ionize to a small extent.

The percentage ionization is given mathematically as:

$$\% \text{ ionization} = \frac{m_{ionized}}{m_{original}} \times 100$$

In Chapter Four we stated that strong electrolytes are 100% dissociated in dilute aqueous solution. This is true; however, solutions of strong electrolytes sometimes behave as if they are less than completely dissociated. This happens because, at any instant in a solution of a strong electrolyte, a certain percentage of the solvated anions and cations that are in constant motion will collide with each other and remain associated. While these "ion pairs" associate with each other, they behave as an "unionized" pair thereby decreasing the effective number of particles in the solution. (This does **not** mean that molecules of ionic substances exist in aqueous solution.) As the concentration of a solution increases, more and more ion pairs are formed and the "effective" percentage ionization or dissociation decreases. Additional interactions between the solvated ions with charge greater than 1 or less than -1 also cause nonideal behavior.

One measure of the extent of ionization or dissociation is the van't Hoff i factor. This is a ratio of the **effective** molality of all particles in solution to the **stated** molality, where the stated molality would be the value if **no** dissociation or ionization occurred at all.

$$i = \frac{\Delta T_f \text{ (observed)}}{\Delta T_f \text{(if nonelectrolyte)}} = \frac{K_f m_{effective}}{K_f m_{stated}} = \frac{m_{effective}}{m_{stated}}$$

A nonelectrolyte has an i value equal to 1. No ions are formed, therefore, the total number of particles equals the original or "stated" molality. For an ionic compound, or an ionizing covalent compound, the ideal value of i would be that found from 100% dissociation of the substance into anions and cations. For example, if "Y" m Na_2SO_4 were to totally dissociate in solution:

$$Na_2SO_4 \rightarrow 2Na^+ + SO_4^{2-}$$

	Na_2SO_4	$2Na^+$	SO_4^{2-}
Initial	Y m	0 m	0 m
Change	-Y m	+2 Y m	+Y m
Final	0 m	2Y m	Y m

The total molality of particles in solution would be $2Y + Y = 3Y$.

The ideal van't Hoff i factor would have a value of: $\quad i = \dfrac{m_{effective}}{m_{stated}} = \dfrac{3Y}{Y} = 3$

Table 14-3 in the text has some actual and ideal i factors for several aqueous systems. The concepts of the i value and extent of ionization can also be applied to boiling point elevation.

Example 14-12 Percent Ionization and i Value

A 0.1096 m aqueous solution of formic acid, HCOOH, freezes at -0.210°C. Determine the i value and the percentage ionization in this solution.

Plan

(a) We can calculate the effective molality by using the observed freezing point and the freezing point depression constant. We can then find the i value by comparing the effective molality to the stated molality. The percent dissociation depends on the molality ionized and the stated molality. (b) The freezing point depression is caused by the effective molality, which is the sum of all dissolved species, in this case, HCOOH, H$^+$ and HCOO$^-$. We need to construct an expression for the effective molality in terms of the amount of ions. We represent the amount of HCOOH that ionizes by x and write the concentrations of all species in terms of this unknown.

Solution

(a) $\quad m_{effective} = \dfrac{\Delta T_f}{K_f} = \dfrac{0°C - (-0.210°C)}{1.86°C/m} = 0.113 m$

$? \; i = \dfrac{0.113 \; m}{0.1096 \; m} = \boxed{1.03}$

(b) This compound ionizes as shown below.

	HCOOH	→	H$^+$	+	HCOO$^-$
Initial	0.1096 m		0 m		0 m
Change	-x m		+x m		+x m
Final	(0.1096-x) m		x m		x m

The total molality of all particles in the solution would be:

$$(0.1096 - x)\,m + x\,m + x\,m = (0.1096 + x)\,m = m_{effective} = 0.113\,m$$

$$x = (0.113 - 0.1096)m = 0.003\,m$$

$$?\,\%\text{ionization} = \frac{m_{ionized}}{m_{stated}} \times 100 = \frac{0.003\,m}{0.1096\,m} \times 100 = \boxed{3\%}$$

Example 14-13 Determination of i Value

Determine the i value in a 0.0531m aqueous solution of MgCl$_2$. The solution freezes at -0.255°C. The K_f for water is 1.86°C/m.

Plan
We can solve for i as in the previous example.

Solution

$$m_{effective} = \frac{\Delta T_f}{K_f} = \frac{0°C - (-0.255°C)}{1.86°C/m} = 0.137m$$

$$?\,i = \frac{m_{effective}}{m_{stated}} = \frac{0.137\,m}{0.0531\,m} = \boxed{2.58}$$

If the i value of a solute at a certain concentration is available, we can use it to determine the freezing point of a solution of an electrolyte.

Example 14-14 Freezing Point of an Electrolyte Solution

What is the freezing point of a 0.100 m aqueous solution of K_2CrO_4? From Table 14-3 in the text, the i value of this solution is 2.39.

Plan

We know that the freezing point depends on the effective molality. We can find the effective molality by using the known i value and the stated molality. We can then use the equation $\Delta T_f = K_f(m_{effective})$ to find the amount by which the freezing point is lowered.

Solution

$$i = \frac{m_{effective}}{m_{stated}} \text{ therefore } m_{effective} = i \, (m_{stated}) = 2.39(0.100 \, m) = 0.239 \, m$$

$$? \, \Delta T_f = K_f(m_{effective}) = (1.86°C/m)(0.239 \, m) = 0.445°C$$

The freezing point of the solution is: $0.000°C - 0.445°C = \boxed{-0.445°C}$

14-10 Osmotic Pressure

If a solution is separated from pure solvent (or a more dilute solution) by a membrane that is permeable only to solvent molecules, the natural tendency is for solvent molecules to migrate more rapidly from the pure solvent (or more dilute solution) into the solution than in the reverse direction. This process, which dilutes the more concentrated solution, is called **osmosis**. Osmosis continues until the concentration on each side of the membrane is the same; however it can be counterbalanced by an opposing pressure called the **osmotic pressure**, Π. The osmotic pressure for a given solution is:

$$\Pi = iMRT = \frac{inRT}{V}$$

i = the van't Hoff i factor
M = the molarity of the solute in solution
R = the ideal gas constant, $\left[0.0821 \, \frac{L \cdot atm}{mol \cdot K}\right]$
T = the temperature in Kelvin
n = number of moles of solute
V = volume of solution

Example 14-15 Osmotic Pressure

What is the osmotic pressure of a solution made by dissolving 3.78 g CCl_4 in enough C_5H_{12} to make 350.0 mL of solution at 18°C?

Plan
We convert the mass of solute to a number of moles, the volume of solution to liters, and the temperature to kelvins. We can then evaluate the expression for the osmotic pressure.

Solution

$$R = \left[0.0821 \frac{L\ atm}{mol\ K}\right]; \quad V = 350.0\ mL = 0.350\ L; \quad T = 18°C + 273 = 291\ K$$

$$n = 3.78\ g\ CCl_4 \times \frac{1\ mol\ CCl_4}{153.81\ g\ CCl_4} = 0.0246\ mol\ CCl_4$$

$$\Pi = \frac{n}{V}RT = \frac{(0.0246\ mol\ CCl_4)\left[0.0821 \frac{L\ atm}{mol\ K}\right](291K)}{0.3500\ L} = \boxed{1.68\ atm}$$

Measurements of osmotic pressures of dilute solutions of large molecules, such as those in biological systems, are often used to determine molecular weights of solutes because only a small number of moles of solute gives rise to an easily observable osmotic pressure. Conversely, very low concentrations often give a change in magnitude that is too slight to be easily detected in boiling or freezing points.

Example 14-16 Molecular Weight

When 4.00 grams of biological compound are dissolved in enough benzene to give 2.00 liters of solution, the solution has an osmotic pressure of 3.20 torr at 25°C? What is the molecular weight of the sample?

Plan
To find the molecular weight of the substance, we need to find the number of moles represented by 4.00 g of the substance. We first use the osmotic pressure to determine the molarity of the solution. Then, because we know the volume of the solution, we can find the number of moles of solute.

Solution

$T = 273 + 25°C = 298K$, $\Pi = 3.20 \text{ torr} \times \dfrac{1 \text{ atm}}{760 \text{ torr}} = 0.00421 \text{ atm}$

$\Pi = MRT$ so $M = \dfrac{\Pi}{RT} = \dfrac{0.00421 \text{ atm}}{\left[0.0821 \dfrac{L \cdot atm}{mol \cdot K}\right](298K)} = 1.72 \times 10^{-4} \text{ M}$

? mol solute $= 1.72 \times 10^{-4} \dfrac{\text{mol}}{L} \times (2.00 \text{ L}) = 3.44 \times 10^{-4} \text{ mol}$

? g/mol $= \dfrac{4.00 \text{ g}}{3.44 \times 10^{-4} \text{ mol}} = \boxed{1.16 \times 10^{4} \text{ g/mol}}$

EXERCISES

Concentration of Solutions

1. Calculate the molality of a solution made when 30.0 grams of NaOH are dissolved in 125.0 grams of water.

2. What is the mole fraction of NaOH in the solution made in Exercise 1?

3. Urea, $CO(NH_2)_2$, is a nonvolatile, nonelectrolyte that is a component of urine. How many grams of urea are needed to prepare a 0.388 m solution, using 100.0 grams of water as the solvent? What is the mass percent urea in this solution?

4. How many milliliters of H_2O (density = 1.00 g/mL) must be added to 50.0 grams of KCl to make a 3.00 m solution of KCl?

5. What are the molality and the mole fraction of solute in a 20.0% by mass aqueous solution of KCN?

6. What are the molality and the mole fraction of solute in a 3.363 M aqueous solution of KI? The specific gravity of the solution is 1.398.

7. How many grams of ethanol, C_2H_5OH, must be added to 65.0 grams of benzene, C_6H_6 to give a mole fraction of ethanol of 0.333?

8. Formic acid, HCOOH, is responsible for the sting of an ant bite. An aqueous solution of formic acid has a molality of 11.19 m and a solution density of 1.0842 g/mL. What are the molarity, the mole fraction, and the molality of this solution

Raoult's Law

9. The vapor pressure of pure toluene, C_7H_8, is 137 torr at 60°C. What is the vapor pressure of toluene in a solution made by dissolving 25.0 grams of the nonelectrolyte glucose, $C_6H_{12}O_6$, in 50.0 grams of toluene at 60°C?

10. The vapor pressure of water is 118 torr at 55°C. How many grams of urea, $CO(NH_2)_2$ must be added to 67.4 grams of water to lower the vapor pressure to 109 torr?

11. How much is the vapor pressure lowered when a 12.0 gram sample of nonvolatile fructose, $C_6H_{12}O_6$, is dissolved in 40.0 grams of ethanol, C_2H_5OH, (also known as "grain" or "drinking" alcohol) at 78.4°C? For pure ethanol, the vapor pressure is equal to 760 torr at 78.4°C.

12. What is the vapor pressure of a solution made by mixing 4.3 moles of C_4H_6 (P° = 40 torr at 40°C) and 1.8 moles C_4H_7Br (P° = 400 torr at 40°C) at 40°C?

Freezing Point Depression and Boiling Point Elevation

13. What are the freezing point and the boiling point of the solution made in Exercise 11? For ethanol, K_f is $1.99°C/m$, the normal freezing point is $-114.6°C$; K_b is $1.22°C/m$ and the normal boiling point is $78.4°C$.

14. A solution of 5.46 grams of an unknown dissolved in 30.0 grams of ethanol gives a freezing point of $-117.1°C$. What is the molecular weight of the unknown? See the constants in Exercise 13.

15. What will be the freezing point and the boiling point of a solution prepared by dissolving 25.0 grams of the nonvolatile, nonelectrolyte $CO(NH_2)_2$ in 75.0 grams of nitrobenzene, $C_6H_5NO_2$? For nitrobenzene, $K_b = 5.24°C/m$ and $K_f = 7.00°C/m$. The normal boiling point is $210.88°C$ and the normal freezing point is $5.70°C$.

16. Camphor, $C_{10}H_{16}O$, with a fragrant and penetrating odor, is used in cosmetics as a preservative. Camphor has a normal freezing point of $178.8°C$ and a K_f of $37.3°C/m$. When a 2.96 gram sample of a nonelectrolyte with an empirical formula of C_2H_4Cl is dissolved in 47.21 grams of camphor, the freezing point of the solution is $160.4°C$. What are the molecular weight and the molecular formula of the solute?

17. Fatty acids, such as stearic acid, $C_{18}H_{38}O_2$, are nonpolar because of their long hydrocarbon chains. Because they are nonpolar, they have relatively low freezing points. How many grams of iodine, I_2, must be dissolved in 22.3 grams of stearic acid to lower its freezing point to $58.5°C$? The normal freezing point $= 70.1°C$ and $K_f = 7.40°C/m$.

18. The normal boiling point of toluene, C_7H_8, a common industrial solvent, is $110.61°C$. When a 6.95 gram sample of the nonelectrolyte glucose, $C_6H_{12}O_6$, is dissolved in 22.4 grams of toluene, the boiling point of the solution is $114.87°C$. What is the boiling point elevation constant, K_b, of toluene?

Dissociation or Ionization of Electrolytes

19. A $0.20\ m$ aqueous solution of formic acid, HCOOH, freezes at $-0.383°C$. Formic acid is a monoprotic acid. Calculate the percentage ionization and the van't Hoff i factor of the formic acid in this solution.

20. Methylamine, CH_3NH_2 is a weak base that can accept one proton from water in a Brønsted-Lowry acid-base reaction. A $0.0100\ m$ aqueous solution of methylamine is 18.9% ionized. What is the freezing point of the solution?

21. A $0.0500\ m$ aqueous solution of KCl freezes at $-0.175°C$. What are (a) the ideal i value and (b) the actual van't Hoff i value in this solution?

22. A 0.0620 m aqueous $ZnSO_4$ solution freezes at -0.150°C. What are (a) the ideal i value and (b) the actual van't Hoff i value in this solution?

23. A 0.100 m aqueous solution of K_2SO_4 freezes at -0.458°C. What are (a) the ideal i value and (b) the actual van't Hoff i value in this solution?

24. What is the freezing point of a 0.00600 m aqueous solution of $K_3Fe(CN)_6$? The van't Hoff i value for this solution is 3.665. What is the ideal i value for this electrolyte if it dissociates according to the reaction below?

$$K_3Fe(CN)_6(aq) \rightarrow 3K^+(aq) + Fe(CN)_6^{3-}(aq)$$

25. What is the freezing point of a 0.00500 m aqueous solution of $MgSO_4$ if the van't Hoff i value is 1.69?

Osmotic Pressure

26. A 0.244 gram sample of a high molecular weight compound dissolves in benzene to give 36.75 mL of a solution that exhibits an osmotic pressure of 11.6 torr at 23°C. Estimate the molecular weight of the compound.

27. Calculate the osmotic pressure in torr associated with 0.0010 M sucrose in water at 25°C and compare its magnitude with the boiling point elevation and freezing point depression of the solution. Assume that for this dilute solution molality and molarity are approximately the same.

28. At what temperature will a solution made from 0.172 grams of sucrose ($C_{12}H_{22}O_{11}$, also known as "table sugar") dissolved in enough water to make 375.0 mL of solution have an osmotic pressure be 25.0 torr?

29. Hemoglobin is a protein responsible for oxygen transport in the body. A solution is made from 0.138 gram of hemoglobin dissolved to make 5.00 mL of solution at 22°C. The osmotic pressure of the solution is 7.94 torr. Estimate the molecular weight of the hemoglobin.

Miscellaneous Exercises

30. What are the molarity, the molality, and the mole fraction of solute in an aqueous solution that is 9.97% by mass K_2SO_4? The specific gravity of the solution is 1.023.

31. A 0.200 molal aqueous solution of weak monoprotic nitrous acid, HNO_2, has a freezing point of -0.390°C. What is the percent ionization of nitrous acid in this solution?

32. Nonpolar carbon tetrachloride, CCl_4, is often used as a solvent for fatty acids, such as stearic acid (Exercise 17) or pentadecanoic acid, $C_{15}H_{30}O_2$. How many grams of pentadecanoic acid must be added to 125.0 grams of carbon tetrachloride at 57.8°C to lower its vapor pressure to 371 torr? The vapor pressure of pure CCl_4 at 57.8°C is 400.0 torr.

33. The vapor pressure of pure water is 17.5 torr at 20°C. What is the vapor pressure of water when 2.50 grams of the strong electrolyte NaCl is dissolved in 20.0 grams of water at 20°C? Assume an ideal value of i.

34. Vitamin C, also known as ascorbic acid, is water soluble. Ascorbic acid is 40.9 % carbon, 4.58% hydrogen, and 54.5% oxygen. When 7.32 grams of ascorbic acid are dissolved in 33.3 grams of water, the freezing point of the solution is -2.33°C. What is the molecular formula of ascorbic acid?

35. When a 1.90 gram sample of $MgCl_2$ is dissolved in 400.0 grams of water, the freezing point of the solution is -0.251°C. What is the van't Hoff i factor for magnesium chloride at this concentration?

36. Ethylene glycol, $C_2H_4(OH)_2$, is a sweet tasting, but toxic, substance used in many antifreeze preparations. How many kilograms of ethylene glycol must be added to 4.00 kilograms of water to lower the freezing point to -12.0°C?

37. When a 41.9 gram sample of the nonelectrolyte urea, $CO(NH_2)_2$, is dissolved in 125.0 grams of formic acid, HCOOH, the solution has a freezing point of 3.25°C. The normal freezing point of formic acid is 8.40°C. What is the freezing point depression constant of formic acid?

38. How many grams of iodine, I_2, must be dissolved to make 225.0 mL of carbon tetrachloride solution with an osmotic pressure of 68.5 torr at 26°C?

39. What are the molality and the molarity of KBr in an aqueous solution that is 38.00% by mass KBr. The solution density is 1.3525 g/mL.

ANSWERS TO EXERCISES

1. 6.00 m NaOH
2. X = 0.0975
3. 23.3 g $CO(NH_2)_2$, 18.9%
4. 224 mL H_2O
5. 3.84 m KCN; X = 0.0646
6. X = 0.06728; 4.003 m KI
7. 19.1 g C_2H_5OH
8. 8.008 M; 34.00%; X = 0.1677
9. 743 torr
10. 18.6 g $CO(NH_2)_2$
11. 54.2 torr
12. 146 torr
13. T_b = 80.4°C; T_f = -117.9°C
14. 1.4×10^2 g/mol
15. T_b = 240.0°C; T_f = -33.2°C
16. 127 g/mol; $C_4H_8Cl_2$
17. 8.88 g I_2
18. 3.33°C/m
19. i = 1.03; 3%
20. -0.0221°C
21. i = 1.88; ideal i = 2
22. i = 1.30; ideal = 2
23. i = 2.46; ideal i = 3
24. -4.45°C; ideal i = 4
25. -0.0157°C
26. 1.06×10^4 g/mol
27. Π = 18.6 torr; ΔT_b = 0.00512°C; ΔT_f = 0.00186°C
The temperature differences are probably too small to be easily monitored using freezing point depression or boiling point elevation. The osmotic pressure is, therefore, more useful for the dilute solutions often formed from high molecular weight solutes.

28. 299K = 26°C
29. 6.40×10^4 g/mol
30. X = 0.103; 0.637 m K_2SO_4; 0.587 M K_2SO_4
31. 5.0%
32. 15.6 g $C_{15}H_{30}O_2$
33. 16.2 torr
34. $C_6H_8O_6$ (MW = 175 g/mol, calculated; 176 g/mol, true)
35. i = 2.70
36. 1.60 kg
37. 2.77°C/m
38. 0.210 g I_2
39. 5.150 m KBr; 4.319 M KBr

Chapter 15: Chemical Thermodynamics

15-1 Some Thermodynamic Terms

Energy is the capacity to do work or transfer heat. The **First Law of Thermodynamics**, better known as the Law of Conservation of Energy, states that the total amount of energy in the universe is a constant. Energy is neither created nor destroyed in a chemical process or a physical change. An **exothermic** process releases energy in the form of heat, while an **endothermic** process absorbs energy in the form of heat.

Thermodynamics is the study of energy transfers or changes that accompany physical or chemical processes. Thermodynamics is used to determine the spontaneity of a process. A **spontaneous** process is one that, once initiated under a given set of conditions, tends to create products without further input of energy. A **nonspontaneous** process will not occur without continuous energy input. The spontaneity of a process is independent of the rate at which the process occurs.

The substances of interest in a process are called the **system** and everything else is the **surroundings**. The **universe** is the system plus the surroundings. The **thermodynamic state of a system** is defined by a set of conditions specifying all properties of the system: composition (number and kind of moles), pressure, temperature and physical state. Once the state is specified, all other properties are fixed. **State functions** are properties whose value depends only on the state of the system not on the way in which that state was achieved (pathway independent). We can describe changes in the state function in terms of differences between the initial and the final states. For any state function, X:

$$\Delta X = X_{final} - X_{initial} = X_{products} - X_{reactants}$$

15-2 Enthalpy Changes and Calorimetry

Enthalpy, given the symbol **H**, is defined as the heat lost or absorbed by a system at constant pressure, when the only work involved is compression or expansion work. Enthalpy is a state function, so $\Delta H = H_{final} - H_{initial}$. Neither $H_{initial}$ nor H_{final} can be absolutely determined; however, changes in enthalpy can be measured by monitoring the heat absorbed or released by a system at constant pressure.

Calorimetry is a technique for determining the energy change associated with a process. To determine enthalpy changes, we use a "coffee-cup" calorimeter, which is open to the atmosphere. By using this type of calorimeter, we can determine q_p, the heat absorbed or released by the system at constant pressure.

$$\begin{pmatrix} \text{amount of heat} \\ \text{released by reaction} \end{pmatrix} = \begin{pmatrix} \text{amount of heat} \\ \text{gained by calorimeter} \end{pmatrix} + \begin{pmatrix} \text{amount of heat} \\ \text{gained by solution} \end{pmatrix}$$

Example 15-1 Heat Capacity of a Calorimeter

A student mixes 75.0 grams of water at 62.31°C in a coffee-cup calorimeter with 75.0 grams of water at 25.84°C. The final temperature of the mixture is 41.71°C. What is the heat capacity of the calorimeter? The specific heat of water is 4.184 J/g•°C?

Plan

We can use the relationship heat = mass x specific heat x Δt to determine the heat released by the "hot" water and the heat gained by the "cold" water. We then rearrange the relationship above to determine the heat capacity of the calorimeter in J/°C.

Solution

$$? \text{ J lost by "hot" water} = 75.0 \text{ g} \times \frac{4.184 \text{ J}}{\text{g} \cdot °C} \times (62.31°C - 41.71°C) = 6.46 \times 10^3 \text{ J}$$

$$? \text{ J gained by "cold" water} = 75.0 \text{ g} \times \frac{4.184 \text{ J}}{\text{g} \cdot °C} \times (41.71°C - 25.84°C) = 4.99 \times 10^3 \text{ J}$$

$$? \text{ J gained by calorimeter} = 6.46 \times 10^3 \text{ J} - 4.99 \times 10^3 \text{ J} = 1.47 \times 10^3 \text{ J}$$

$$? \frac{\text{J}}{°C} = \frac{1.47 \times 10^3 \text{ J}}{41.71°C - 25.84°C} = \boxed{92.6 \frac{\text{J}}{°C}}$$

Example 15-2 Heat Measurements Using a Calorimeter

When a 1.45 gram sample of LiCl is dissolved in 75.00 g of water in the calorimeter in Example 15-1, the temperature of the solution changes from 25.00°C to 28.95°C. Assume that the specific heat of the solution is the same as for water, 4.184 J/g•°C. What is the heat of solution for this process?

Plan

The amount of heat released by the solution process is equal to the sum of the heat gained by the calorimeter and by the solution. The mass of the solution is the sum of the masses of the solute and the solvent.

Solution

Mass solution = mass of LiCl + mass of water = 1.45 g + 75.00 g = 76.45 g.

$$? \text{ J gained by solution} = 76.45 \text{ g} \times 4.184 \frac{J}{g \, °C} \times (28.05°C - 25.00°C) = 976 \text{ J}$$

$$? \text{ J gained by calorimeter} = \frac{92.6 \text{ J}}{°C} \times (28.05 - 25.00°C) =$$

$$? \text{ J} = 976 \text{ J} + 282 \text{ J} = 1258 \text{ J}$$

The solution process liberated $\boxed{1258 \text{ J}}$ of heat. We could also report this as: $q_p = \boxed{-1.26 \times 10^3 \text{ J}}$ or $\boxed{-1.26 \text{ kJ}}$, with the negative sign indicating that heat was released.

15-3 Thermochemical Equations

A **thermochemical equation** is a balanced equation that also includes a designation of its value of ΔH. The change in enthalpy is an extensive quantity that is related to the number of moles of each component in the equation with units expressed in kJ/mol rxn. For example, in the reaction below:

$$2CO_2(g) \rightarrow 2\,CO(g) + O_2(g) \quad \Delta H = +566 \text{ kJ/mol rxn}$$

This amount of heat is required when **two moles** of CO_2 react to form **two moles** of CO and **one mole** of O_2. We can refer to this amount of reaction as one **mole of reaction**, which will be abbreviated as "mol rxn". This interpretation of the balance equation allows us to write various unit factors such as:

$$\frac{2 \text{ mol CO}_2}{1 \text{ mol rxn}}, \frac{2 \text{ mol CO}}{1 \text{ mol rxn}}, \frac{1 \text{ mol O}_2}{1 \text{ mol rxn}}$$

and so on. If a different amount of a material is used, ΔH must be scaled accordingly. For brevity, the units of ΔH are sometimes written as kJ/mol or even just as kJ. Regardless of the units, be sure to interpret the thermodynamic change per mole of reaction for the balanced equation to which it refers. The coefficients **must** be interpreted as number of moles.

Example 15-3 Amount of Heat versus Extent of Reaction

Use the thermochemical equation below to determine the number of kilojoules released by the formation of 72.0 grams of BrF_3.

$$Br_2(\ell) + 3F_2(g) \rightarrow 2BrF_3 \quad \Delta H = -511.2 \text{ kJ/mol rxn}$$

Plan

We see that production of 2 moles of BrF_3 releases 511.2 kJ. We need to find the number of moles of BrF_3 involved and then use dimensional analysis.

Solution

$$? \text{ kJ} = 72.0 \text{ g BrF}_3 \times \frac{1 \text{ mol BrF}_3}{136.91 \text{ g BrF}_3} \times \frac{-511.2 \text{ kJ}}{1 \text{ mol rxn}} \times \frac{1 \text{ mol rxn}}{2 \text{ mol BrF}_3} = \boxed{-134 \text{ kJ}}$$

This tells us that 134 kJ of heat is released when this amount of BrF_3 is produced.

15-4 Standard Molar Enthalpies of Formation, $\Delta H°_f$

The **thermochemical standard state** of a substance is its most stable state at one atmosphere pressure and a specified constant temperature, usually 25°C, unless otherwise noted. A superscript zero follows the thermodynamic symbol to refer to standard state. A following subscript denotes the temperature. If dissolved, the substance should be one-molar (1 M, more strictly unit activity), while gases should be at 1 atmosphere pressure. All solids and liquids should be in the pure state. The heat change accompanying the formation of one mole of a substance from its constituent elements in their standard states is called the **standard enthalpy of formation** of the substance, $\Delta H°_f$. Consequently, the standard enthalpy of formation of any element in its standard state is zero. The symbol $\Delta H°_{rxn}$ stands for the enthalpy of **any** reaction at standard conditions.

Some $\Delta H°_f$ values are tabulated in Appendix K in the text. When using this table, be sure to che[ck] for the proper state, allotrope, etc., of the substance. The following are examples of a few formation reactions and their standard enthalpies of formation.

$$Hg(\ell) + Cl_2(g) \rightarrow HgCl_2(s)$$
$$6C(s, graphite) + 3H_2(g) \rightarrow C_6H_6(\ell)$$
$$H_2(g) + 1/2 O_2(g) \rightarrow H_2O(\ell)$$
$$H_2(g) + 1/2 O_2(g) \rightarrow H_2O(g)$$

$\Delta H°_f\, HgCl_2(s) = -224$ kJ/mol rxn
$\Delta H°_f\, C_6H_6(\ell) = +49$ kJ/mol rxn
$\Delta H°_f\, H_2O(\ell) = -285.8$ kJ/mol rxn
$\Delta H°_f\, H_2O(g) = -241.8$ kJ/mol rxn

15-5 Hess' Law of Heat Summation

Hess' law of heat summation states that the molar enthalpy change for the conversion of a given set of reactants into a given set of products is always the same whether the conversion occurs in one step or a series of steps. For example, the formation reaction of tin and chlorine to form tin(IV) chloride is:

$$Sn(s, white) + 2Cl_2(g) \rightarrow SnCl_4(\ell) \qquad \Delta H°_f\, SnCl_4(\ell) = -511 \text{ kJ/mol rxn}$$

The reaction could also occur in two or more steps. The sum of the enthalpies for those steps is equal to the enthalpy of the one-step process. One possibility for the previous reaction is:

(1) $\quad Sn(s, white) + Cl_2(g) \rightarrow SnCl_2(s) \qquad \Delta H°_f\, SnCl_2(s) = -350$ kJ/mol rxn
(2) $\quad SnCl_2(s, white) + Cl_2(g) \rightarrow SnCl_4(\ell) \qquad \Delta H°_{rxn} = -161$ kJ/mol rxn

(1)+(2) $\quad Sn(s, white) + 2Cl_2(g) \rightarrow SnCl_4(\ell) \qquad \Delta H°_f\, SnCl_4(\ell) = -511$ kJ/mol rxn

In general, for an n-step process:

$$\Delta H°_{rxn} = \Delta H°_{rxn(1)} + \Delta H°_{rxn(2)} + \Delta H°_{rxn(3)} + \cdots + \Delta H°_{rxn(n)}$$

When a needed reaction is multiplied to give the step desired, its $\Delta H°_{rxn}$ is multiplied by the same number. When a reaction is reversed, its $\Delta H°_{rxn}$ is multiplied by -1.

A corollary of Hess' Law is summarized by the equation below. For any reaction:

$$\Delta H°_{rxn} = \sum n\Delta H°_{f\text{ products}} - \sum n\Delta H°_{f\text{ (reactants)}}$$

The second reaction step above illustrates this relationship:

$$\begin{array}{cccc}
& SnCl_2(s) & + \quad Cl_2(g) \quad \rightarrow & SnCl_4(\ell) \\
\Delta H°_f & -350 \text{ kJ/mol } SnCl_2 & 0 \text{ kJ/mol } Cl_2 & -511 \text{ kJ/mol } SnCl_4
\end{array}$$

$\Delta H°_{rxn} = 1 \text{ mol } SnCl_4(\ell) \Delta H°_f \; SnCl_4(\ell) - [1 \text{ mol } SnCl_2(s) \Delta H°_f \; SnCl_2(s)) + 1 \text{ mol } Cl_2 \Delta H°_f \; Cl_2(g)]$

$$= \frac{1 \text{ mol } SnCl_4(\ell)}{\text{mol rxn}} \times \frac{-511 \text{ kJ}}{\text{mol } SnCl_4(\ell)}$$

$$- \left[\frac{1 \text{ mol } SnCl_2(s)}{\text{mol rxn}} \times \frac{-350 \text{ kJ}}{\text{mol } SnCl_2(s)} + \frac{1 \text{ mol } Cl_2(g)}{\text{mol rxn}} \times \frac{0 \text{ kJ}}{\text{mol } Cl_2(g)} \right]$$

$= -511 \text{ kJ/mol rxn} + 350 \text{ kJ/mol rxn} = -161 \text{ kJ/mol rxn}$

This result agrees with the $\Delta H°_{rxn}$ seen for the reaction in step (2).

For brevity we shall omit units in the intermediate steps of subsequent calculations and simply assign proper units to the answer.

Example 15-4 Combining Thermochemical Equations–Hess' Law

Find the $\Delta H°_{rxn}$ for following reaction, given the three numbered reactions.

$CaCl_2(s) + H_2O(\ell) + CO_2(g) \rightarrow CaCO_3(s) + 2HCl(aq)$

Given: $\Delta H°_{rxn}$ (kJ/mol rxn)

(1) $CaH_2(s) + 2HCl(aq) \rightarrow CaCl_2(aq) + 2H_2(g)$ -271
(2) $CaH_2(s) + 2H_2O(\ell) \rightarrow Ca(OH)_2(aq) + 2H_2(g)$ -242
(3) $Ca(OH)_2(aq) + CO_2(g) \rightarrow CaCO_3(s) + H_2O(\ell)$ -96

Plan
By inspection, we see that we need to reverse equation (1) to put the $CaCl_2$ on the reactant side. When we do so, we must reverse the sign of ΔH. By adding reactions (2) and (3) to the reverse of (1), we obtain the equation of interest. We add the ΔH values to obtain $\Delta H°_{rxn}$.

Solution

	$\Delta H°_{rxn}$ (kJ/mol rxn)
Reverse (1) $CaCl_2(aq) + 2H_2(g) \rightarrow CaH_2(s) + 2HCl(aq)$	-(-271)
(2) $CaH_2(s) + 2H_2O(\ell) \rightarrow Ca(OH)_2(aq) + 2H_2(g)$	-242
(3) $Ca(OH)_2(aq) + CO_2(g) \rightarrow CaCO_3(s) + H_2O(\ell)$	-96
$CaCl_2(s) + H_2O(\ell) + CO_2(g) \rightarrow CaCO_3(s) + 2HCl(aq)$	$\boxed{-67 \text{ kJ/mol rxn}}$

Example 15-5 Using $\Delta H°_f$ Values–Hess' Law

Use tabulated values for $\Delta H°_f$ to calculate $\Delta H°_{rxn}$, in kilojoules, for the reaction below:

$$2SOCl_2(\ell) \quad + \quad O_2(g) \quad \rightarrow \quad 2SO_2Cl_2(\ell)$$

$\Delta H°_f$ -206 kJ/mol $SOCl_2$ 0 kJ/mol O_2 -389 kJ/mol Cl_2

Plan

We apply Hess' Law in the form $\Delta H°_{rxn} = \sum n\Delta H°_{f\,(products)} - \sum n\Delta H°_{f(reactants)}$.

Solution

$\Delta H°_{rxn} = \sum n\Delta H°_{f\,(products)} - \sum n\Delta H°_{f\,(reactants)}$

$= 2(\Delta H°_f\, SO_2Cl_2(\ell)) - [2(\Delta H°_f\, SOCl_2(\ell)) + 1(\Delta H°_f\, O_2(g))]$

$= 2(-389) - [2(-206) + 1(0)] = -778 + 412 = \boxed{-366 \text{ kJ/mol rxn}}$

Example 15-6 Using $\Delta H°_f$ Values–Hess' Law

$\Delta H°_{rxn} = 29.4$ kJ for the reaction below. Given the following standard molar enthalpies of formation, calculate $\Delta H°_f$ for $H_3AsO_4(aq)$.

$$3NH_4Cl(aq) + H_3AsO_4(aq) \rightarrow (NH_4)_3AsO_4(aq) + 3HCl(aq)$$

$\Delta H°_f$ -300.2 kJ/mol   ?   -1268 kJ/mol   -167.4 kJ/mol

Plan

We apply Hess' Law as in Example 15-5 and then rearrange the equation to solve for the missing $\Delta H°_f$ value.

Solution

$$\Delta H°_{rxn} = \sum n\Delta H°_f \text{ (products)} - \sum n\Delta H°_f \text{ (reactants)}$$

$$= 1(\Delta H°_f (NH_4)_3AsO_4(aq)) + 3(\Delta H°_f HCl(g)) -$$

$$[3(\Delta H°_f NH_4Cl(aq)) + 1(\Delta H°_f H_3AsO_4(aq)]$$

$$29.4 \text{ kJ} = 1(-1268 + 3(-167.4) - 3(-300.4) - 1(\Delta H°_f H_3AsO_4(aq))$$

Rearranging yields:

$$\Delta H°_f H_3AsO_4(aq) = -29.4 - 1268 - 502.2 + 901.2 = \boxed{-898 \text{ kJ/mol of } H_3AsO_4}$$

15-6 Bond Energies

Strengths of chemical bonds are measured in terms of bond energies. Higher bond energies correspond to stronger bonds. The **bond energy (B.E.)** of a particular bond is defined as the amount of energy that must be added to break one mole of the bonds in gaseous molecules to form gaseous products at constant temperature and pressure. The strengths of bonds between particular atoms vary from compound to compound. Tables 15-2 (single bonds) and 15-3 (multiple bonds) in the text contain average values for bond energies. If bond energies of reactants and products are known, the enthalpy of reaction for a gas phase reaction may be estimated using the formula:

$$\Delta H°_{rxn} = \sum B.E._{(reactants)} - \sum B.E._{(products)}$$

Example 15-7 Bond Energies

Use bond energies to estimate the enthalpy change for the reaction below. Compare the result with that obtained from heat of formation data.

$$N_2(g) + 3H_2(g) \rightarrow 2NH_3(g)$$

The bond energies of interest are:

N≡N = 946 kJ/mol N≡N bonds, H-H = 435 kJ/mol H-H bonds, N-H = 389 kJ/mol NH bonds

Plan

We break 1 mole of N≡N bonds and 3 moles of H-H bonds, while making 2x3 or 6 moles of N-H bonds. We add and subtract the appropriate bond energies. We use Hess' Law to determine $\Delta H°_{rxn}$ as in previous examples.

Solution

$\Delta H°_{rxn} = \sum B.E._{(reactants)} - \sum B.E._{(products)}$

$= [1(\Delta H_{N≡N}) + 3(\Delta H_{H-H})] - 6(\Delta H_{N-H})$

$= [1(946) + 3(435)] - 6(389)$

$= 946 + 1305 - 2334 = \boxed{-83 \text{ kJ/mol rxn}}$

This estimated value can be compared with $\Delta H°_{rxn}$ calculated from $\Delta H°_f$ values.

$\Delta H°_{rxn} = 2(\Delta H°_{f\ NH_3(g)}) - 1(\Delta H°_{f\ N_2(g)}) - 3H_2(\Delta H°_{f\ H_2(g)})$

$= 2(-46.11) - 1(0) - 3(0)$

$= -92.22 - 0 - 0 = \boxed{-92.22 \text{ kJ/mol rxn}}$

The two values agree to within about 10%. This illustrates the fact that bond energies only give approximate $\Delta H°_{rxn}$ values.

Example 15-8 Bond Energies

Oxygen and fluorine react to produce the covalent compound oxygen difluoride. Calculate the bond energy of an O-F bond from the following data.

$$2F_2(g) + O_2(g) \rightarrow 2OF_2(g) \quad \Delta H°_{rxn} = +46 \text{ kJ}$$

The bond energies are: F-F = 158.0 kJ/mol, O=O = 498.4 kJ/mol

Plan

There are 2 moles of F-F bonds and 1 mole of O=O bonds broken, but 2x2 = 4 moles of O-F bonds are made. We add and subtract the appropriate bond energies and rearrange to find the unknown O-F bond energy.

Solution

$$\Delta H°_{rxn} = \Sigma \text{ bond energies}_{(reactants)} - \Sigma \text{ bond energies}_{(products)}$$

$$= [2(\Delta H_{F-F}) + 1(\Delta H_{O=O}) - 4(\Delta H_{O-F})$$

$$46 = 2(158.0) + 1(498.4) - 4(\Delta H_{O-F})$$

$$46 = 316.0 + 498.4 - 4(\Delta H_{O-F})O-F) = 814.4 - 4(\Delta H_{O-F})$$

$$768 = 4(\Delta H_{O-F})$$

The bond energy of O-F = $\dfrac{768 \text{ kJ}}{4}$ = $\boxed{192 \text{ kJ/mol O-F bonds}}$.

15-7 Changes in Internal Energy

The **internal energy**, **E**, of the reactants of a system is the sum of all the energies of the individual reacting species. Similarly, the internal energy of the products is the sum of all the energies of the individual products. Although the internal energy of a species can not be precisely determined, it is a state function and the changes can be determined by measurements of the **heat (q)** absorbed by the system and **work (w)** done by the system.

$$\Delta E = E_{final} - E_{initial} = E_{products} - E_{reactants} = q + w$$

The conventions regarding the signs of q and w are:

q is positive if heat is **absorbed** by the system from the surroundings.
q is negative if heat is **released** by the system to the surroundings.
w is positive if work is done **on** the system by the surroundings.
w is negative if work is done **by** the system on the surroundings.

A positive sign of ΔE indicates that the internal energy increases for a specified process, while a negative sign indicates that it decreases. The only important type of work involved in most chemical and physical processes is pressure-volume work. **Work = w = -PΔV**, where **P** is the external pressure and $\Delta V = V_{final} - V_{initial}$ (the change in the volume occupied by the system).

Example 15-9 Internal Energy

A gas is heated with 165 joules while it expands from 0.120 liters to 0.180 liters under a constant opposing pressure of 7.50 atm. What are q, w, and ΔE for this process?

Plan

We consult the sign conventions listed above and use $w = -P\Delta V$. We also must convert liter·atmospheres to joules. The conversion given in the Appendix of the text is that 1 L·atm = 101.32 J.

Solution

$q = \boxed{+165 \text{ J}}$ since the sample absorbed heat

$w = -P\Delta V = -7.50 \text{ atm}(0.180 \text{ L} - 0.120 \text{ L}) \times \dfrac{101.32 \text{ J}}{1 \text{ L·atm}} = \boxed{-45.6 \text{ J}}$

$\Delta E = q + w = +165 \text{ J} + (-45.6 \text{ J}) = \boxed{+119 \text{ J}}$

For systems in which only liquids and solids are involved, volume changes are usually small and a useful approximation is that $\Delta V \approx 0$. A given number of moles of a gas occupies a much larger volume than does the same number of moles of a liquid or solid. If pressure and temperature are held constant, gases that are produced by a reaction do work on the atmosphere. Conversely, as a gas is consumed in a reaction, the atmosphere can do work on a system by expanding into its space. For those processes in which the number of moles of product gas equals the number of moles of reactant gas, $\Delta V = 0$, and no work is done.

For an ideal gas, $PV = nRT$, so $V = \dfrac{nRT}{P}$. At constant pressure and temperature:

$$V_{final} = \dfrac{n_{final} RT}{P} \text{ and } V_{initial} = \dfrac{n_{initial}RT}{P}$$

$$P\Delta V = P(V_{final} - V_{initial}) = P\left[\dfrac{n_{final} RT}{P} - \dfrac{n_{initial}RT}{P}\right] = \Delta nRT$$

With the assumption that solids and liquids have volume changes negligible to those of gases as they react, we can approximate the work involved in any system by $w = -P\Delta V \approx -\Delta n_{gas}RT$. In this relationship, the change in number of moles of gas is $\Delta n_{gas} = n_{product(g)} - n_{reactant(g)}$. The number of moles of liquid or solid **must** be omitted. The ideal gas constant, R, has appropriate units (see Appendix D). Temperature must be expressed in kelvins.

Example 15-10 Determining the Amount of Work

Estimate the work involved (in joules) when 2.00 mol of CO_2 is decomposed to CO and O_2 at constant pressure and 25°C. Is work done by or on the system?

Plan

We need to write the balanced decomposition reaction for this process. Because only gases are involved, we know that work = $-\Delta n_{gas}RT$ will apply exactly. We need to find the change in number of moles of gas. The R value that involves joules is 8.314 J/mol·K.

Solution

The balanced equation is: $2CO_2(g) \rightarrow 2CO(g) + O_2(g)$

Δn_{gas} = (2.00 mol CO + 1.00 mol O_2) - 2.00 mol CO_2 = +1.00 mol gas

$w = -(+1.00 \text{ mol})(8.314 \text{ J/mol·K})(298K) = \boxed{-2.48 \times 10^3 \text{ J}}$

Work is done $\boxed{\text{by the system}}$ since the system expands.

Work is an extensive function. If we had not specified a number of moles that was in the balanced equation, we would need to make the appropriate stoichiometric conversions.

In a **bomb calorimeter**, the reaction is carried out in a strong steel container, which is immersed in a large vat of water. The calorimeter has a known heat capacity, which is determined by performing a standard reaction. Heat is transferred from the system to the calorimeter-water surroundings. When a bomb calorimeter is used to measure heat transfer, the volume is held constant, and $\Delta V = 0$. That means that no work can be done; w = 0. Therefore, $\Delta E = q_v$, the heat absorbed at constant volume.

Example 15-11 Bomb Calorimeter

A 1.00 mol sample of ICl for the following reaction is added to a bomb calorimeter with water equivalent of 1.14 kJ/°C and immersed in 5000.0 g H_2O. The reaction is initiated and the temperature of the water and calorimeter increases from 25.00°C to 25.81°C. The specific heat of water is 4.184 J/g·°C. What is ΔE of the reaction in kJ/mol rxn?

$ICl(g) \rightarrow 1/2 I_2(s) + 1/2 Cl_2(g)$

Plan

We know that the heat given off by the reaction is absorbed by the water and the calorimeter. We can find the heat absorbed by using the specific heat of water, the mass of water, and the calorimeter constant and the temperature change. The sum of these will be the total heat absorbed by the calorimeter. ΔE of the reaction will be $-\Delta E$ of the calorimeter, because the calorimeter absorbs the heat released by the system. Finally, we must convert the heat to heat/mol using the mass and molar mass of the reactant.

Solution

$$? \,°C = (25.81°C - 25.00°C) = 0.81\,°C$$

$$? \text{ kJ to warm water} = (5000.0 \text{ g}) \times 4.184 \,\frac{J}{g \cdot °C} \times 0.81°C \times \frac{1 \text{ kJ}}{1000 \text{ J}} = +17 \text{ kJ}$$

$$? \text{ kJ to warm calorimeter} = \frac{1.14 \text{ kJ}}{°C} \times 0.81\,°C = +0.92 \text{ kJ}$$

$$? \,\Delta E_{system} = -\Delta E_{calorimeter} = -(+17 \text{ kJ} + 0.92 \text{ kJ}) = (-18 \text{ kJ})$$

This internal energy change is for one mole of ICl, so ΔE in kJ/mol rxn is:

$$\Delta E = \frac{-18 \text{ kJ}}{1.00 \text{ mol ICl}} \times \frac{1 \text{ mol rxn}}{1 \text{ mol ICl}} = \boxed{-18 \text{ kJ/mol rxn}}$$

15-8 Relationship Between ΔH and ΔE

The definition of enthalpy is $H = E + PV$. Therefore, for a change at constant temperature and pressure:

$$\Delta H = \Delta E + P\Delta V \approx \Delta E + \Delta n_{gas}RT$$

The difference between ΔH and ΔE is the amount of expansion work involved. Unless there is a change in the number of moles of gas, the $P\Delta V$ term is usually very small, and $\Delta H \approx \Delta E$. Additionally, at constant pressure: $\Delta H = \Delta E + P\Delta V = q_p + w + P\Delta V$, but $w = -P\Delta V$, so $\Delta H = q_p + w + P\Delta V = q_p - P\Delta V + P\Delta V$. Therefore, as we stated earlier:

$$\Delta H = q_p$$

Example 15-12 Relationship Between ΔH and ΔE

What is the change in enthalpy, ΔH, of the reaction in Example 15-11 at 25°C?

Plan

We know that ΔE of the process is -18 kJ/mol rxn. We can find the change in the number of moles of gas. We then apply the equation above. (We note that I_2 is a solid and not used in this calculation because the ideal gas equation applies to **gases**.)

Solution

$? \, \Delta n_{gas} = (1/2 \text{ mol } Cl_2) - 1 \text{ mol } ICl = -1/2 \text{ mol gas/mol rxn}$

$? \, \Delta H = \Delta E + \Delta n_{gas} RT$

$= -18 \text{ kJ/mol rxn} + \left[\left(\dfrac{-1/2 \text{ mol gas}}{\text{mol rxn}} \right) \times 8.314 \times 10^{-3} \dfrac{\text{kJ}}{\text{mol·K}} \times 298 \text{K} \right]$

$= -18 \text{ kJ/mol rxn} - 1.24 \text{ kJ/mol rxn} = \boxed{-19 \text{ kJ/mol rxn}}$

15-9 The Two Parts of Spontaneity

Most processes that happen spontaneously, such as combustion reactions, are exothermic, giving off heat ($\Delta H°_{rxn} < 0$). Some endothermic ($\Delta H°_{rxn} > 0$) processes, such as the dissolving of sodium chloride in water, happen spontaneously. This is possible because there is a sufficient increase in disorder or randomness (entropy). Therefore, when we consider the spontaneity or nonspontaneity of a process, we must consider both the enthalpy or the process and the amount of disorder. Processes are more likely to be spontaneous if the products have lower heat content and greater disorder than the reactants.

15-10 The Second Law of Thermodynamics

The **Second Law of Thermodynamics** states that in any spontaneous change, the **universe** tends toward greater disorder. The measure of disorder is called **entropy**, a state function given the symbol S. Entropy and entropy changes of the universe cannot be directly measured in the laboratory, however, entropy changes in a system, ΔS, can be evaluated by relating them to other quantities (such as equilibrium constants and free energy) as will be discussed later.

15-11 Entropy, S

The **Third Law of Thermodynamics** states that the entropy of a pure, perfectly ordered crystalline substance is zero at absolute zero (0 K). As the temperature increases, disorder increases; therefore even pure elements have a non-zero entropy at any temperature above absolute zero. The entropy of a substance at any condition is called the **absolute entropy** and is given the symbol $S°$. Absolute entropy values for some substances at 25°C are given in Appendix K of the text. In general, $S°$ increases in the order: solids < liquids < gases, since the disorder of particles increases in that order. Because entropy is also a state function, a relationship similar to Hess' Law also applies for determining changes in entropy:

$$\Delta S°_{rxn} = \Sigma\, nS°_{(products)} - \Sigma\, nS°_{(reactants)}$$

As we saw earlier, the "mol" term in the units for a substance refers to a mole of the substance, while for a reaction, it refers to a mole of reaction. Each term will have the units:

$$\frac{\text{mol substance}}{\text{mol rxn}} \times \frac{J}{(\text{mol substance})\cdot K} = \frac{J}{(\text{mol rxn})\cdot K}$$

The result is usually abbreviated as J/mol·K or even as J/K. We will omit units in the intermediate steps, but show the appropriate units in the result.

Example 15-13 Calculation of $\Delta S°_{rxn}$

Using $S°$ values, determine the standard entropy change for the reaction below at 25°C.

	NH$_3$(g) +	HNO$_3$(ℓ)	→	N$_2$O(g) +	2H$_2$O(ℓ)
$S°$(J/mol·K)	192.3	155.6		219.7	69.91

Plan

We will use the equation above to calculate $\Delta S°_{rxn}$ for the process.

Solution

$\Delta S°_{rxn} = \Sigma\, nS°_{(products)} - \Sigma\, nS°_{(reactants)}$

$= 1(S°_{N_2O(g)}) + 2(S°_{H_2O(\ell)}) - [1(S°_{NH_3(g)}) + 1(S°_{HNO_3(\ell)})]$

$= 1(219.7) + 2(69.91) - [1(192.3) + 1(155.6)]$

$= 219.7 + 139.2 - 192.3 - 155.6 = \boxed{11.6 \text{ J/mol rxn·K}}$

Because $\Delta S°$ for the reaction in Example 15-13 is positive, disorder is increased, a situation favorable for, but not assuring, the spontaneity of the reaction at these conditions.

15-12 Free Energy Changes, ΔG, and Spontaneity

A reaction is thermodynamically spontaneous if, when stoichiometric amounts of the products and reactants are mixed, the forward reaction proceeds. The spontaneity of a standard reaction at constant temperature and pressure can be predicted by the **Gibbs free energy change, ΔG**. A **standard reaction** is one in which the number of moles of reactants shown in the balanced equation, all in their standard states, are completely converted to the numbers of moles of products shown in the balanced equation, also in their standard states. ΔG is a state function that predicts the maximum useful energy obtained in the form of work from a process. The Gibbs free energy is related to both ΔH and ΔS by the following Gibbs-Helmholtz equation:

$$\Delta G = \Delta H - T\Delta S \quad (T, P \text{ constant}, T \text{ in kelvins})$$

If ΔG is **negative**, free or useful energy is released by the system and the process is spontaneous under the given conditions. The amount by which the Gibbs free energy decreases is a measure of the maximum amount of useful work that could be done by a system. Conversely, a **positive ΔG** corresponds to a nonspontaneous process. The reverse of any spontaneous process is nonspontaneous because $\Delta G_{forward} = -\Delta G_{reverse}$. When a system is at equilibrium, neither the forward nor reverse reaction is spontaneous and $\Delta G = 0$.

The spontaneity of a process is **favored**, but not required, if either $\Delta H < 0$ or $\Delta S > 0$. Since each may be either positive or negative, the spontaneity with respect to their signs is summarized into four classes in shown in Table 15-1.

Table 15.1 Relationship of Thermodynamic Quantities to Spontaneity

ΔH	ΔS	ΔG	Temperature Range of Spontaneity
−	+	−	all temperatures
+	−	+ or −	below some calculable temperature
−	−	+ or −	above some calculable temperature
+	+	+	no temperatures
+	−		

If a process is at standard conditions: $\Delta G°_{rxn} = \Delta H°_{rxn} - T\Delta S°_{rxn}$. We can determine $\Delta G°_{rxn}$ values for a special type of reaction, called a **standard reaction**. In a standard reaction, exactly the numbers of moles of reactants as are given in the balanced equation are **completely** converted to stoichiometric numbers of moles of products. All reactants and products must be in standard states, which means that dissolved species must be one-molar (1M) and the partial pressure of a gas must be one atmosphere (1 atm).

The value of $\Delta G°$ can be calculated by either:

1. evaluation of $\Delta H°$ and $\Delta S°$ from $\Delta H°_f$ and $S°$ values of the species involved, or by
2. utilization of the relationship, similar to Hess' Law, that:

$$\Delta G°_{rxn} = \sum n\Delta G°_f \text{ (products)} - \sum n\Delta G°_f \text{ (reactants)}$$

Standard free energy of formation values, $\Delta G°_f$ are tabulated in Appendix K in the text. Elements in their standard states have values of $\Delta G°_f = 0$. Treatment of the units for the calculation of $\Delta G°$ is the same as that for $\Delta H°$.

Example 15-14 Spontaneity of a Standard Reaction

Use tabulated $\Delta G°_f$ values to calculate the standard free energy change in kilojoules for the reaction below at 298K. Is the reaction spontaneous at 298K?

$$2NH_3(g) + 3CuO(s) \rightarrow 3H_2O(\ell) + N_2(g) + 3Cu(s)$$

$\Delta G°_f$ (kJ/mol):   −16.15   −130   −237.2   0   0

Plan

We can use the second method of evaluating $\Delta G°_{rxn}$ because $\Delta G°_f$ values are available.

Solution

$\Delta G°_{rxn} = \sum n\Delta G°_f \text{ (products)} - \sum n\Delta G°_f \text{ (reactants)}$

$= 3(\Delta G°_f \text{ } H_2O(\ell)) + 1(\Delta G°_f \text{ } N_2(g)) + 3(\Delta G°_f \text{ } Cu(s)) - [2(\Delta G°_f \text{ } NH_3(g)) + 3\Delta G°_f \text{ } CuO(s)]$

$= 3(-237.2) + 1(0) + 3(0) - 2(-16.15) - 3(-130)$

$= -711.6 + 0 + 0 + 32.30 + 390 = \boxed{-289 \text{ kJ/mol rxn}}$

Since $\Delta G°_{rxn}$ is negative, the forward reaction is $\boxed{\text{spontaneous}}$ at 298 K. This means that the reaction can, but does not necessarily, occur at an observable rate under these conditions.

Example 15-15 Spontaneity of a Standard Reaction

Evaluate $\Delta G°_{rxn}$ for the reaction in Example 15-14 using $\Delta H°_f$ and $S°$ values.

$$2NH_3(g) + 3CuO(s) \rightarrow 3H_2O(\ell) + N_2(g) + 3Cu(s)$$

	$2NH_3(g)$	$3CuO(s)$	$3H_2O(\ell)$	$N_2(g)$	$3Cu(s)$
$\Delta H°_f$ (kJ/mol)	-46.11	-157	-285.8	0	0
$S°$ (J/mol·K)	192.3	42.63	69.91	191.5	33.15

Plan
We will use Hess' Law to determine $\Delta H°_{rxn}$ and $\Delta S°_{rxn}$. We will then use the Gibbs-Helmholtz equation to determine $\Delta G°_{rxn}$, remembering to convert the units of $S°$ from J/mol·K to kJ/mol·K.

Solution

$\Delta H°_{rxn} = 3(\Delta H°_f\, H_2O(\ell)) + 1(\Delta H°_f\, N_2(g)) + 3(\Delta H°_f\, Cu(s))$
$\qquad\qquad - [2(\Delta H°_f\, NH_3(g)) + 3(\Delta H°_f\, CuO(s))]$

$\qquad = 3(-285.8) + 1(0) + 3(0) - 2(-46.11) - 3(-157)$

$\qquad = -857.4 + 0 + 0 + 92.22 + 471 = -294$ kJ/mol rxn

$\Delta S°_{rxn} = 3(S°_{H_2O(\ell)}) + 1(S°_{N_2(g)}) + 3(S°_{Cu(s)}) - [2(S°_{NH_3(g)}) + 3(S°_{CuO(s)})]$

$\qquad = 3(69.91) + 1(191.5) + 3(33.15) - 2(192.3) - 3(42.63)$

$\qquad = 209.73 + 191.5 + 99.45 - 384.6 - 127.89$

$\qquad = -11.81$ J/mol rxn·K = -0.01181 kJ/mol rxn·K

$\Delta G°_{rxn} = \Delta H°_{rxn} - T\Delta S°_{rxn} = -294$ kJ/mol rxn $- (298$ K$)(-0.01181$ kJ/mol rxn·K$)$

$\qquad = -294$ kJ/mol rxn $+ 3.52$ kJ/mol rxn

$\qquad = \boxed{-290 \text{ kJ/mol rxn}}$ (Rounding causes the slight difference observed)

15-13 The Temperature Dependence of Spontaneity

For any system at equilibrium:

$$\Delta G = 0 = \Delta H - T\Delta S \text{ then: } \Delta H = T\Delta S \text{ or } T = \frac{\Delta H}{\Delta S}$$

The temperature of equilibrium may be **estimated** using tabulated $\Delta H°_{298}$ and $\Delta S°_{298}$ values. The result will only be an estimate because of the dependence of ΔH and, especially, ΔS on temperature. Additionally, sometimes the conditions for the calculations are not standard conditions. As temperature is changed, a process may change from spontaneous to nonspontaneous, or vice-versa, if an equilibrium temperature exists and is passed.

Example 15-16 Temperature Range of Spontaneity

Estimate the temperature range over which the reaction in Example 15-15 is spontaneous.

Plan

For this process, $\Delta H° < 0$ and $\Delta S° < 0$. This means that the reaction will be spontaneous below some calculable temperature. We can estimate the equilibrium temperature at which $\Delta G = 0$, by assuming that ΔH and ΔS at the equilibrium temperature are equal to that at 25°C. The reaction will be spontaneous below this calculated temperature.

Solution

At equilibrium $\Delta G_{rxn} = 0 = \Delta H_{rxn} - T\Delta S_{rxn}$, therefore,

$$T = \frac{\Delta H_{rxn}}{\Delta S_{rxn}} = \frac{-294 \text{ kJ/mol rxn}}{0.01181 \text{ kJ/mol rxn} \cdot \text{K}} = 2.49 \times 10^4 \text{ K}$$

At temperatures over 2.49×10^4 K, the positive value of $T\Delta S$ becomes larger than the value of ΔH. Therefore, the reaction will be spontaneous at temperatures $\boxed{\text{below } 2.49 \times 10^4 \text{ K}}$, a very high temperature.

Exercise 15-17 Estimation of a Boiling Point

Use thermodynamic data below to estimate the boiling point of mercury.

$$Hg(\ell) \rightarrow Hg(g)$$

	Hg(ℓ)	Hg(g)
$\Delta H°_f$ (kJ/mol)	0	61.3
$S°$ (J/mol·K)	76.0	174.8

Plan

We know that at the boiling point the gas will be in equilibrium with the liquid. We have the appropriate date to determine $\Delta H°_{rxn}$ and $\Delta S°_{rxn}$. Then, we need to solve for the temperature at which $\Delta G = 0$ as in Example 15-16.

Solution

$$\Delta H°_{rxn} = 1(\Delta H°_f\ Hg(g)) - 1(\Delta H°_f\ Hg(\ell)) = 1(61.3.0) - 1(0) = 61.3 \text{ kJ/mol rxn}$$

$$\Delta S°_{rxn} = 1(S°_{Hg(g)}) - 1(S°_{Hg(\ell)}) = 1(174.8) - 1(76.0) = 98.8 \text{ J/mol rxn·K}$$

$$T = \frac{\Delta H}{\Delta S} = \frac{61.3 \text{ kJ/mol}}{98.8 \frac{J}{mol\cdot K} \times \frac{1 \text{ kJ}}{1000 \text{ J}}} = \boxed{620K} = \boxed{34\ °C}$$

The handbook value of the boiling point of mercury is 356.58°C. This indicates that using thermodynamic standard state values at 298K gives only an approximation of the values at other temperatures. The magnitude of the deviation of the true boiling or melting point from the value estimated using standard state enthalpies and entropies tends to be greater as temperatures vary more significantly from 298K.

EXERCISES

Refer to Appendix K in the text for thermodynamic data as needed.

Calorimetry

1. The reaction below is performed in a bomb calorimeter with a heat capacity of 3.138 kJ/°C and immersed in 7000.0 grams of H_2O. When 0.200 mol Br_2 and a stoichiometric amount of F_2 react, the temperature of the water and calorimeter changes from 25.00°C to 28.12°C. What is ΔE in kJ/mol of BrF_3?

$$Br_2(\ell) + 3F_2(g) \rightarrow 2BrF_3(g)$$

2. A bomb calorimeter, immersed in 650.0 grams of water, has a heat capacity of 628 J/°C. When a 0.750 gram sample of C_2H_4 is burned, the temperature of the calorimeter and water changes from 25.000°C to 34.755°C. What is ΔE in kJ/mol of C_2H_4 for this process?

3. A student places a piece of copper with a mass of 125.0 grams and an initial temperature of 99.00°C into 100.0 mL of water at 25.00°C in a coffee-cup calorimeter. The final temperature of the mixture is 31.37°C. The specific heat of copper is 0.385 J/g·°C and that of water is 4.18 J/g·°C. What is the heat capacity of the calorimeter?

4. A student measures an initial temperature of 25.00°C for a 100.0 gram sample of water in the coffee-cup calorimeter from Exercise 3. When the student adds a 35.5 gram sample of NH_4I to the water, the final temperature of the solution is 19.92°C. The solution specific heat is 4.18 J/g·°C. What is $\Delta H°_{rxn}$ for the heat of solution of NH_4I in J/g of NH_4I and kJ/mol of NH_4I?

5. A 3.41 g sample of solid NaOH is dissolved in 100.0 g H_2O at 25.00°C in the coffee-cup calorimeter from Exercise 3. The temperature of the mixture rises to 32.22°C. Calculate q_p in joules and the enthalpy, ΔH, for the solution process in J/g_{NaOH} and in kJ/mol_{NaOH}? Assume that the specific heat of the solution is the same as that of H_2O.

Thermochemical Equations

6. Nitroglycerine, $C_3H_5(NO_3)_3$, is the major component in dynamite. It explodes according to the reaction below. How many kilojoules are released when 4.50 grams of nitroglycerine react? For this process, $\Delta H°_{rxn}$ = -5692 kJ/mol rxn.

$$4C_3H_5(NO_3)_3(\ell) \rightarrow 6N_2(g) + O_2(g) + 10H_2O(g) + 12CO_2(g)$$

7. Benzene, C_6H_6, is a widely used solvent in spite of its acute toxicity. As do most hydrocarbons, it readily burns in oxygen in the highly exothermic process below. How many kilojoules are released in the combustion of 125.0 grams of benzene?

$$2C_6H_6(\ell) + 9O_2(g) \rightarrow 6CO_2(g) + 6H_2O(\ell)$$

8. When 9.50 grams of NO_2 react to form NO and O_2, 11.7 kilojoules are absorbed. What is $\Delta H°_{rxn}$ for this process? In the process, lowest whole number coefficients are used to balance the equation.

Enthalpy

9. The heat of formation of copper(I) chloride is -137 kJ/mol rxn. How many kilojoules are released when 4.46 grams of copper are reacted with a stoichiometric amount of chlorine to produce copper(I) chloride.

10. Calculate the standard molar enthalpy change, $\Delta H°_{rxn}$, at 25°C for each of the following reactions.

    (a) $BrF_3(g) + 2H_2(g) \rightarrow HBr(g) + 3HF(g)$

    (b) $2PCl_5(g) + 5I_2(s) \rightarrow 2P(g) + 10ICl(g)$

    (c) $2Fe_3O_4(s) + H_2O_2(\ell) \rightarrow 3Fe_2O_3(s) + H_2O(\ell)$

    (d) $2NH_4NO_3(s) \rightarrow 4H_2O(g) + 2N_2(g) + O_2(g)$

    (e) $SiF_4(g) + 2 H_2O(\ell) \rightarrow SiO_2(s) + 4HF(g)$

    (f) $2C_8H_{18}(\ell) + 25O_2(g) \rightarrow 16CO_2(g) + 18H_2O(g)$

11. $\Delta H°_{rxn} = -524.0$ kJ/mol rxn for the following reaction. Use the $\Delta H°_f$ data for the other substances to determine $\Delta H°_f$ $CO(NH_2)_2(s)$ at 25°C:

$$C(s) + O_2(g) + 2NH_3(g) \rightarrow CO(NH_2)_2(s) + H_2O(\ell)$$

12. Given the following at 25°C:

	$\Delta H°_{rxn}$ (kJ/mol rxn)
$Fe_2O_3(s) + 3CO(g) \rightarrow 2Fe(s) + 3CO_2(g)$	-24.8
$3Fe_2O_3(s) + CO(g) \rightarrow 2Fe_3O_4(s) + CO_2(g)$	-46.4
$Fe_3O_4(s) + CO(g) \rightarrow 3FeO(s) + CO_2(g)$	+19

Calculate the $\Delta H°_{rxn}$ for the reaction below $FeO(s) + CO(g) \rightarrow Fe(s) + CO_2(g)$

13. Calculate $\Delta H°_f$ for $CuCl_2(s)$ using the reactions and $\Delta H°_{rxn}$ values below. (4)

$2Cu(s) + Cl_2(g) \rightarrow 2CuCl(s)$ $\quad \Delta H°_{rxn} = -274.4$ kJ/mol

$2CuCl(s) + Cl_2(g) \rightarrow 2CuCl_2(s)$ $\quad \Delta H°_{rxn} = -165.8$ kJ/mol

14. Given the information below, find $\Delta H°_{rxn}$ for $3CH_4(g) + 4O_3(g) \rightarrow 3CO_2(g) + 6H_2O(\ell)$.

$CH_4(g) + 2O_2(g) \rightarrow CO_2(g) + 2H_2O(g)$ $\quad \Delta H°_{rxn} = -802$ kJ

$3O_2(g) \rightarrow 2O_3(g)$ $\quad \Delta H°_{rxn} = +286$ kJ

$H_2O(\ell) \rightarrow H_2O(g)$ $\quad \Delta H°_{rxn} = +44$ kJ

15. $\Delta H°_{rxn} = -845.0$ kJ/mol rxn for the following reaction. Use the $\Delta H°_f$ data for the other substances to determine $\Delta H°_f\, P_4(g)$ at 25°C:

$$5CCl_4(\ell) + P_4(g) \rightarrow 4PCl_5(g) + 5C(s,graphite)$$

16. Given the following at 25°C:

	$\Delta H°_{rxn}$(kJ/mol rxn)
$2C_6H_6(\ell) + 15O_2(g) \rightarrow 12CO_2(g) + 6H_2O(\ell)$	-6535
$CO(g) + 1/2O_2(g) \rightarrow CO_2(g)$	-283.0
$CO(g) + Cl_2(g) \rightarrow COCl_2(g)$	-113

Calculate $\Delta H°_{rxn}$ for the reaction:

$$12Cl_2(g) + 2C_6H_6(\ell) + 9O_2(g) \rightarrow 12COCl_2(g) + 6H_2O(\ell)$$

Bond Energies

17. Acetylene, C_2H_2, is used in torches for high temperature welding. Use bond energies to estimate the energy released in the combustion of acetylene. Compare the result to that found using $\Delta H°_f$ data.

$$C_2H_2(g) + 5O_2(g) \rightarrow 4CO_2(g) + 2H_2O(g)$$

18. Use Cl-Cl, Cl-F, and F-F bond energies to estimate the $\Delta H°_{rxn}$ for the process.

$$3F_2(g) + Cl_2(g) \rightarrow 2ClF_3(g)$$

19. For the reaction below $\Delta H°_{rxn} = -56$ kJ/mol rxn. Use the other tabulated bond energies to compute the H-Cl bond energy and compare your result to the tabulated H-Cl value.

$$PCl_5(g) + 4H_2(g) \rightarrow PH_3(g) + 5HCl(g)$$

Chapter 15 Chemical Thermodynamics

20. Calculate the average bond energy in kJ/mol of bonds for the S-F bond from the following data:

    $$S(s) + 3F_2(g) \rightarrow SF_6(g) \quad \Delta H°_{rxn} = -1209 \text{ kJ}$$

    $\Delta H°_f$ for $S(g) = 278.8$ kJ/mol and $\Delta H°_f$ for $F(g) = 78.99$ kJ/mol

Internal Energy, Heat, and Work

21. Calculate the change in internal energy of a system if it:
    (a) absorbs 300 joules of heat and does 200 joules of work on its surroundings.
    (b) absorbs 150 joules of heat and the surroundings do 450 joules of work on it.
    (c) releases 64.4 joules of heat and expands from 0.255 liters to 0.618 liters against a constant opposing pressure of 1.89 atm.
    (d) does 100 joules of work on the surroundings.
    (e) does 1.50 kilojoules of work in the surroundings and releases 5.00 kilojoules of heat.

22. When metallic zinc is dropped into hydrochloric acid, hydrogen is produced.
    (a) How much work is done if the dry hydrogen, collected in a 4.00 liter container, exerts a pressure of 742 torr at 30.0°C?
    (b) How many moles and grams of hydrogen were produced?

23. A sample is cooled by 457 J as it expands from 3.88 liters to 9.46 liters against a constant opposing pressure of 0.405 atm. What are the heat, the work, and ΔE for this system?

24. Estimate the work in kJ done by the reaction of 75.0 grams of $(NH_4)_2Cr_2O_7$ reacted at 1.00 atm and 25°C. Assume that the solids have negligible volume.

    $$(NH_4)_2Cr_2O_7(s) \rightarrow Cr_2O_3(s) + N_2(g) + 4H_2O(g)$$

25. Estimate the work involved when 30.0 mol of nitroglycerine reacts at 25°C. See Exercise 6 for necessary data. What is ΔE in kJ/mol rxn at this temperature?

Entropy

26. Calculate the standard molar entropy change, $\Delta S°_{rxn}$, for each of the reactions below at 25°C.
    (a) $H_2S(g) + 4F_2(g) \rightarrow SF_6(g) + 2HF(g)$
    (b) $SiC(s) + 2H_2O(\ell) \rightarrow SiH_4(g) + CO_2(g)$

(c) $Fe_2O_3(s) + CO(g) \rightarrow 2Fe(s) + 3CO_2(g)$

(d) $N_2O_5(g) + H_2O(g) \rightarrow 2 HNO_3(g)$

Free Energy and Temperature Range of Spontaneity

27. Calculate $\Delta G°_{rxn}$ values at 25°C for each of the reactions in Exercise 27 using tabulated $\Delta G°_f$ values. Which reactions are spontaneous at 25°C?

28. Calculate $\Delta G°_{rxn}$ values at 25°C for each of the reactions in Exercise 27 using tabulated values of $\Delta H°_f$ and $S°$.

29. Calculate the temperature range of spontaneity (if any exists) for each reaction in Exercise 26 using the data from Exercise 28.

30. Estimate the normal boiling point (gas and liquid in equilibrium) of the following liquids using thermodynamic data.

    (a) $SnCl_4(\ell)$ (b) $C_2H_5OH(\ell)$ (c) $CCl_4(\ell)$ (d) $Fe(CO)_5(\ell)$

31. Estimate the temperature range over which iodine sublimes at one atmosphere pressure.

32. Calculate $\Delta S°$ at 25°C for the following reaction from the data given.

$$PbS(s) + 2HCl(g) \rightarrow PbCl_2(s) + H_2S(g)$$

	$PbS(s)$	$2HCl(g)$	$PbCl_2(s)$	$H_2S(g)$
ΔH_f (kJ/mol)	-100.4	-92.31	-359.4	-20.6
$\Delta G°_f$ (kJ/mol)	-98.7	-95.30	-314.1	-33.6

33. Calculate the temperature range in which the reaction in Exercise 32 is spontaneous.

Miscellaneous Exercises

34. The heat of formation of Sb_4O_6 is -1441 kJ/mol. How many kilojoules are released by the reaction of 62.55 grams of antimony with excess oxygen?

35. Given:

	$\Delta H°_{rxn}$ (kJ/mol rxn)
$HF(g) \rightarrow H(g) + F(g)$	+568
$H_2(g) \rightarrow 2H(g)$	+436
$H_2(g) + F_2(g) \rightarrow 2HF(g)$	-542

Find $\Delta H°_{rxn}$ for $F_2(g) \rightarrow 2F(g)$.

36. A bomb calorimeter contains 2250.0 grams of water and has heat capacity of 3.138 kJ/°C. When a 0.2500 gram sample of CS_2 is burned in the calorimeter, the temperature of the calorimeter and water changes from 25.000°C to 25.289°C. What is ΔE in kJ/mol CS_2?

37. Solid tin(IV) oxide can be dissolved by hydrochloric acid. One possible reaction for this process is shown below. Calculate $\Delta H°_{rxn}$, $\Delta S°_{rxn}$, and $\Delta G°_{rxn}$ for this process. Also compute the temperature range over which this process spontaneously occurs.

$$SnO_2(s) + 4HCl(aq) \rightarrow SnCl_4(\ell) + 2H_2O(\ell)$$

38. A gas is heated by 14.0 kJ as it expands from 2.85 liters to 19.60 liters against a constant opposing pressure of 1.48 atmospheres. Calculate q, w, and ΔE for the gas sample in kilojoules.

39. A 50.0 mL sample of 1.0 M HCl solution at 23.9°C is added to a 50.0 mL sample of 1.0 M NaOH solution at the same temperature in an insulated container. The final temperature of the mixture is 29.7°C. Assume that the calorimeter absorbs no heat and that the specific heat of the mixture is 4.01 J/g·°C. What is ΔH for the neutralization in kJ/mol of the limiting reactant?

40. Chloroform, $CHCl_3$, has been used as an anesthetic.

    (a) Use heats of formation data to determine $\Delta H°_{vap}$ for the chloroform at 25°C.

    (b) What is the heat involved in the vaporization of 12.5 grams of chloroform at 25°C?

    (c) The density of chloroform is 1.49 g/mL. What is the work involved (in joules) when a 12.5 gram sample of chloroform is vaporized at 25°C?

    (d) Repeat the computation in part (c) ignoring the original volume of the liquid. How do the two results compare?

    (e) Use S° data to calculate $\Delta S°_{vap}$ for chloroform at 25°C.

    (f) Estimate the normal boiling point of chloroform.

41. Ethene, C_2H_4, also known as ethylene, is the starting material for many polymers, such as polyethylene, which is used for a wide variety of consumer products. This gaseous material burns quite readily.

    (a) Write the combustion reaction for gaseous ethene, assuming that the water produced is in the vapor state.

    (b) Use bond energies to estimate the heat of combustion for this reaction.

    (c) Repeat the calculation in (c) using $\Delta H°_f$ data.

ANSWERS TO EXERCISES

1. -253 kJ/mol of BrF_3
2. -1221 kJ/mol of C_2H_4
3. 92.6 J/°C
4. 94.3 J/g; $+13.7$ kJ/mol of NH_4I
5. 3.79×10^3 J; 1.11×10^3 J/g; 44.4 kJ/mol NaOH
6. -28.2 kJ
7. 3.340×10^3 kJ
8. $+113$ kJ/mol rxn
9. 9.63 kJ
10. (a) -594 kJ/mol rxn   (b) $+9.2$ kJ/mol rxn   (c) -335 kJ/mol rxn
    (d) -236 kJ/mol rxn   (e) $+192$ kJ/mol rxn   (f) $-10,111.8$ kJ/mol rxn
11. 330.4 kJ/mol
12. -11 kJ/mol rxn
13. -220.1 kJ/mol rxn
14. -3242 kJ/mol rxn
15. -73.6 kJ/mol rxn
16. -4495 kJ/mol rxn
17. from bond energies, -1964 kJ/mol rxn; from $\Delta H°_f$, -2511 kJ/mol rxn
18. from bond energies, -810 kJ/mol rxn (from $\Delta H°_f$, -326 kJ/mol rxn)
19. H-Cl B.E.: calculated 499 kJ/mol; tabulated 431 kJ/mol
20. $+327$ kJ/mol
21. (a) $+100$ J   (b) $+600$ J   (c) -133.9 J   (d) -100 J   (e) -6.50 J
22. (a) 3.90 L·atm or 396 J   (b) 0.157 mol H_2, 0.316 g H_2
23. $w = -229$ J, $q = -457$ J, $\Delta E = -686$ J
24. -3.69 kJ
25. $w = -539$ kJ
26. (a) 378 J/mol rxn·K   (b) $+261.7$ J/mol rxn·K
    (c) $+15.2$ J/mol rxn·K   (d) -12.3 J/mol rxn·K
27. (a) -1617 kJ/mol rxn   (b) $+199.7$ kJ/mol rxn   (c) -29.4 kJ/mol rxn
    (d) -35.9 kJ/mol rxn
    (a), (c) and (d) are spontaneous at 25°C
28. (a) -1617 kJ/mol rxn   (b) $+199.4$ kJ/mol rxn   (c) -29.3 kJ/mol rxn
    (d) -35.7 kJ/mol rxn
29. spontaneous: (a) below 4.58×10^3 K (4.31×10^3 °C)   (b) above 1058 K (785°C)
    (c) all temperatures   (d) below 3.20×10^3 K (2.93×10^3 °C)

30. (a) 371K (98°C), calculated; 114°C, tabulated
    (b) 350K (77°C), calculated; 78.5°C, tabulated
    (c) 347K (74°C), calculated; 76.8°C, tabulated
    (d) 376K (103°C), calculated; 102.8°C, tabulated

31. above 586K (313°C)   32. −123 J/K   33. below 772K (499°C)

34. −185.1 kJ   35. +158 kJ/mol rxn   36. −1.11 × 10³ kJ/mol of CS_2

37. $\Delta H°_{rxn}$ = −355.2 kJ/mol rxn, $\Delta S°_{rxn}$ = +125.7 J/mol rxn·K,
    $\Delta G°_{rxn}$ = −392.7 kJ/mol rxn, spontaneous at all tempteratures

38. q = +14.0 kJ, w = −2.51 kJ, ΔE = 11.5 kJ

39. −56 kJ/mol of NaOH

40. (a) $\Delta H°_{vap}$ = +32 kJ/mol rxn (b) +3.4 kJ/mol rxn (c) 258 kJ/mol rxn (d) 260 kJ/mol rxn
    the results to (c) and (d) are close, indicating that the approximation of PΔV with Δn_{gas}RT is a fairly good one
    (e) $\Delta S°_{vap}$ = +94 J/K
    (f) 340K (67°C), calculated, 61.7°C, tabulated

41. (a) $C_2H_4(g) + 3O_2(g) \rightarrow 2CO_2(g) + 2H_2O(g)$
    (b) from bond energies, −1059 kJ/mol rxn
    (c) from $\Delta H°_f$ data, −1324 kJ/mol rxn

Chapter 16　Chemical Kinetics

16-1 Introduction

Chemical kinetics is the study of the rates of chemical reactions and the mechanisms, or series of steps, by which they occur. The rate of a reaction is independent of the spontaneity of a process. Highly spontaneous reactions with very negative standard free energies can occur too slowly to be observed. Conversely, nonspontaneous processes can occur rapidly for the small extent to which they occur.

The **rate** of a reaction can be defined as the concentration (or amount) of products produced or reactant consumed per unit time. Reaction rates are usually expressed in units of moles per liter per unit time. Hypothetical reactions can be represented as shown below where the capital letters represent formulas and the lower case letters represent coefficients in the balanced equation.

$$aA + bB \rightarrow cC + dD$$

The reaction rate can be represented as the rate of decrease in concentration of a reactant **or** the rate of increase of a product. For example, $-\Delta[A]/\Delta t$ represents the change in concentration of A (the square brackets signify molarity in moles per liter) per unit time interval (Δt). The negative sign signifies that the concentration of A decreases with time. The rates of decrease and increase of reactants and products, respectively, are related to the coefficients in the balanced equation of the overall reaction.

$$\text{Rate} = -\frac{\Delta[A]}{a\Delta t} = -\frac{\Delta[B]}{b\Delta t} = \frac{\Delta[C]}{c\Delta t} = \frac{\Delta[D]}{d\Delta t}$$

The rates of reactions depend upon four factors: (1) the nature of the reactants; (2) the concentrations of reactants; (3) the temperature; and, (4) catalysts.

Chapter 16 Chemical Kinetics

16-2 Nature of Reactants

The rate of a reaction depends upon the chemical identity of the substances mixed, their physical states and what the products are to be. For reactions involving solids, rate increases as particle size decreases and surface area increases because a large surface area exposes more atoms, ions or molecules to other reactants. Reactions between ions are generally more rapid than those between neutral species due to coulombic attractions between ions with opposite charges and because no covalent bonds must be broken.

16-3 Concentrations of Reactants: The Rate-Law Expression

As concentrations change, the rate of a reaction changes. Usually the rate of a forward reaction decreases with time as reactants are consumed. The expressions for rate above describe the **average** rate over some arbitrary time interval. It is often useful to use another kind of expression to describe reaction rate, called the **rate-law expression** or simply the **rate law**.

$$\text{Rate} = k[A]^x[B]^y$$

Here k is the **specific rate constant**, which is unique for each reaction. The specific rate constant increases with increasing temperature. The units of k vary with the form of the rate-law equation. The powers **x** and **y** are the **orders** of the reaction with respect to A and B, respectively. These powers **must be experimentally** determined and bear no necessary relationship to the coefficients in the balanced equation. The sum **x + y** is the **overall order** of the reaction. The rate-law expression describes the instantaneous rate at some arbitrary time after reactants are mixed, usually the initial rate, which occurs immediately after mixing. Initial rates can be found by plotting concentration data and taking the slope of the initial portion of the line. Refer to Table 16-1 and Figure 16-2 in the text.

The following reaction illustrates some of these points.

$$2NO(g) + 2H_2(g) \rightarrow N_2(g) + 2H_2O(g)$$

The rate of the reaction is:

$$\text{Rate} = -\frac{\Delta[NO]}{2\Delta t} = -\frac{\Delta[H_2]}{2\Delta t} = \frac{\Delta[N_2]}{\Delta t} = \frac{\Delta[H_2O]}{2\Delta t}$$

The rate law expression was experimentally determined to be:

$$\text{Rate} = k[NO]^2[H_2]^1$$

The reaction is second order in NO, first order in H_2, and third order overall. If numerical values were provided for concentrations and rates, k could be evaluated. The following examples illustrate how orders and specific rate constants can be determined. Typically, we vary the concentration of one reactant constant while holding the concentrations of any other reactants constant. The data can be analyzed by looking at ratios or by an algebraic approach.

Example 16-1 Method of Initial Rates–Ratio Method

Using the following hypothetical reaction and data at constant temperature, determine the rate-law expression for this reaction and evaluate k, the specific rate constant.

$$2A + 3B \rightarrow 2C + D$$

Trial	Initial [A]	Initial [B]	Initial Rate of Formation of D
1	0.10 M	0.10 M	2.0 x 10^{-3} $M \cdot s^{-1}$
2	0.10 M	0.30 M	6.0 x 10^{-3} $M \cdot s^{-1}$
3	0.30 M	0.30 M	5.4 x 10^{-2} $M \cdot s^{-1}$

Plan

The general form of the rate law expression is: Rate = $k[A]^x[B]^y$. We will inspect the data and see how changing the concentration of a reactant affects the rate of the reaction. In this example, we will use the ratio method of evaluating the changes.

Solution

Experiments 1 and 2 indicate that the change in rate is only due to the change in [B], because the concentration of [A] is held constant. [B] has been multiplied by a factor of:

$$\frac{0.30}{0.10} = 3.0 = [B]\text{ratio}$$

At the same time, the rate has changed by a factor of:

$$\frac{6.0 \times 10^{-3}}{2.0 \times 10^{-3}} = 3.0 = \text{rate ratio}$$

The exponent y can be found from:

rate ratio = $([B]\text{ratio})^y$ so $3.0 = (3.0)^y$ or y = 1 (first order in B)

In experiments 2 and 3, [B] is constant, so the observed change in the rate depends only on [A]. The [A] ratio is found:

$$\frac{0.30}{0.10} = 3.0 = [A]\text{ratio}$$

The rate ratio is:

$$\frac{5.4 \times 10^{-2}}{6.0 \times 10^{-3}} = 9.0 = \text{rate ratio}$$

rate ratio = ([A]ratio)^x so $9.0 = (3.0)^x$ or $x = 2$ (second order in A)

The reaction is first order in [B], second order in [A] and third order overall. The specific rate constant, k, can be evaluated using the general form of the rate-law expression and substituting data from any single experiment. If the first experiment is used:

$$\text{Rate} = k[A]^2[B]^1 \quad \text{so} \quad k = \frac{\text{Rate}}{[A]^2[B]} = \frac{2.0 \times 10^{-3} \, M \cdot s^{-1}}{[0.10 \, M]^2[0.10 \, M]} = 2.0 \, M^{-2} \cdot s^{-1}$$

At this temperature, the rate-law is: $\boxed{\text{Rate} = 2.0 \, M^{-2} \cdot s^{-1}[A]^2[B]}$

Example 16-2 Using the Rate-Law Expression

What will be the rate of the reaction in Example 16-1, if [A] = 0.50 M and [B] = 0.40 M at the same temperature as in that example?

Plan
Since the rate-law is known, the initial rate for any initial concentrations at the same temperature can be determined.

Solution
$$\text{Rate} = 2.0 \, M^{-2} \cdot s^{-1}[A]^2[B])[A]^2[B]) = 2.0 \, M^{-2} \cdot s^{-1}[0.50 \, M]^2[0.40 \, M]) = \boxed{0.20 \, M \cdot s^{-1}}$$

Example 16-3 Method of Initial Rates–Algebraic Method

Derive the rate-law expression for the hypothetical reaction below at the temperature at which the following data were obtained.

$$A + B + 3C \rightarrow \text{Products}$$

Trial	Initial [A]	Initial [B]	Initial [C]	Initial Rate
1	0.20 M	0.10 M	0.30 M	4.0×10^{-4} M·s^{-1}
2	0.20 M	0.20 M	0.30 M	4.0×10^{-4} M·s^{-1}
3	0.20 M	0.30 M	0.60 M	8.0×10^{-4} M·s^{-1}
4	0.40 M	0.40 M	1.20 M	6.4×10^{-3} M·s^{-1}

Plan

The general form of the rate law expression is: Rate = $k[A]^x[B]^y[C]^z$. We will inspect the data and see how changing the concentration of a reactant affects the rate of the reaction. In this example, we will use an algebraic method.

Solution

In experiments 1 and 2, the concentrations of both A and C are the same. Let's look at the ratio of Rate 2/Rate 1.

$$\frac{\text{Rate 2}}{\text{Rate 1}} = \frac{4.0 \times 10^{-4} \text{ M·s}^{-1}}{4.0 \times 10^{-4} \text{ M·s}^{-1}} = \frac{k[0.20]^x[0.20]^y[0.30]^z}{k[0.20]^x[0.10]^y[0.30]^z} = \frac{[0.20]^y}{[0.10]^y}$$

$1.0 = 2.0^y$ so $y = 0$ (zero order in B)

Because $[B]^0 = 1$, B will be omitted from the rate-law expression.

In experiments 2 and 3, [A] is constant, so the change in the rate law is only dependent on [C]. Remember that changes in [B] have no effect on the rate. We can look at Rate 3/Rate 2.

$$\frac{\text{Rate 3}}{\text{Rate 2}} = \frac{8.0 \times 10^{-4} \text{ M·s}^{-1}}{4.0 \times 10^{-4} \text{ M·s}^{-1}} = \frac{k[0.20]^x[0.60]^z}{k[0.20]^x[0.30]^z}$$

$2.0 = 2.0^z$ so $z = 1$ (first order in C)

We now know that Rate = $k[A]^x[C]$. There is no set of data in which [A] changes and [C] is held constant; however, the data from experiments 3 and 4 can be used to evaluate the order with respect to A. We can look at Rate 4/Rate 3.

$$\frac{\text{Rate 4}}{\text{Rate 3}} = \frac{6.4 \times 10^{-3} \, M \cdot s^{-1}}{8.0 \times 10^{-4} \, M \cdot s^{-1}} = \frac{k[0.40]^x[1.20]}{k[0.20]^x[0.60]}$$

$8.0 = [2.0]^x(2.0)$, then $4.0 = [2.0]^x$ so $x = 2$ (second order in A)

The specific rate constant can now be found, using the data of any trial. We will use experiment 4. Rate = $k[A]^2[C]$

$$k = \frac{\text{Rate}}{[A]^2[C]} = \frac{6.4 \times 10^{-3} \, M \cdot s^{-1}}{[0.40]^2[1.20]} = 3.3 \times 10^{-2} M^{-2} \cdot s^{-1}$$

The complete rate law expression at this temperature is:

$$\text{Rate} = \boxed{3.3 \times 10^{-2} M^{-2} \cdot s^{-1} \, [A]^2[C]}$$

16-4 Concentration versus Time: The Integrated Rate Equation

The **half-life**, $t_{1/2}$, of a reactant is the time required for half of the reactant to be consumed. The integrated rate-law relates concentrations at any time to the specific rate constant, k, and to a, the coefficient of A. In each case below:

[A] represents the concentration of reactant A after time t.
$[A]_o$ represents the initial concentration of A at t = 0.

A reaction with Rate = k, is **zero order** in A and **zero order overall**. For this type of reaction, the integrated rate-law and the half-life, $t_{1/2}$, are:

$$[A] = [A]_o - akt \quad \text{and} \quad t_{1/2} = \frac{[A]_o}{2ak}$$

A reaction with Rate = k[A] is **first order** in A and **first order overall**. For this type of reaction, the integrated rate-law and the half-life, $t_{1/2}$, are:

$$\ln\left[\frac{[A]_o}{[A]}\right] = akt \quad \text{or} \quad \log\left[\frac{[A]_o}{[A]}\right] = \frac{akt}{2.303} \quad \text{and} \quad t_{1/2} = \frac{0.693}{ak}$$

A reaction with Rate = $k[A]^2$ is **second order** in A and **second order overall**. The integrated rate law and the half life, $t_{1/2}$, are:

$$\frac{1}{[A]} - \frac{1}{[A]_o} = akt \quad \text{and} \quad t_{1/2} = \frac{1}{ak[A_o]}$$

Example 16-4 Half-Life–Zero Order Reaction

The decomposition of HI on the surface of gold wire is a zero order reaction with a half life of 5.00×10^2 seconds at a certain temperature. If the original concentration of HI is 5.46 M, what is the specific rate constant of this reaction?

$$HI(g) \rightarrow 1/2 H_2(g) + 1/2 I_2(g)$$

Plan

We will use the relationship between k and $t_{1/2}$ for a zero order process.

Solution

$$t_{1/2} = \frac{[A]_0}{2ak} \text{ so } k = \frac{2 a t_{1/2}}{[A]_0} = \frac{2(1)(5.00 \times 10^2 \text{ s})}{5.46 \, M} = \boxed{183 \text{ s} \cdot M^{-1}}$$

Example 16-5 Half-Life–First Order Reaction

Calculate the half-life for the decomposition of CO_2 which is first order with respect to CO_2 at 1000°C with $k = 1.8 \times 10^{-13}$ s^{-1}.

$$CO_2(g) \rightarrow CO(g) + 1/2 O_2(g)$$

Plan

We will use the relationship between $t_{1/2}$ and k for a first order reaction.

Solution

$$t_{1/2} = \frac{0.693}{k} = \frac{0.693}{1.8 \times 10^{-13} \text{ s}^{-1}} = \boxed{3.8 \times 10^{12} \text{ s}}$$

The reaction is very slow. Note that 3.8×10^{12} s is 1.2×10^5 years.

Example 16-6 Concentration versus Time–First Order Reaction

A conversion of one compound to another (A → B) is first order with the specific rate constant $k = 3.3 \times 10^3$ s^{-1} at 1000°C. How long would it take for a 15.0 M sample of the compound to react to the point where only 0.25 M remains?

Plan
We will use the integrated rate law for a first order reaction with:

$[A]_0 = 15.0\ M$, $[A] = 0.25\ M$, and $k = 3.3 \times 10^3\ s^{-1}$

Solution
$$\ln\left[\frac{[A]_0}{[A]}\right] = akt \text{ so } \ln\left[\frac{15.0\ M}{0.25\ M}\right] = \ln(60.) = 4.09 = 1(3.3 \times 10^3\ s^{-1})t$$

$$t = \frac{4.09}{3.3 \times 10^3\ s^{-1}} = \boxed{1.2 \times 10^{-3}\ s}$$

Example 16-7 Half-Life–Second Order Reaction

The rate-law expression for the following reaction is: Rate = $k[NOCl]^2$. At 673°C, $k = 0.70\ M^{-1}s^{-1}$. What is the half-life when the reaction initiated with $0.25\ M$ NOCl?

$$NOCl(g) \rightarrow NO(g) + 1/2\ Cl_2(g)$$

Plan
This is a second order reaction with respect to NOCl and overall so we can use the appropriate equation above.

Solution
$$t_{1/2} = \frac{1}{ak[A]_0} = \frac{1}{(0.70\ M^{-1}s^{-1})(0.25\ M)} = \frac{1}{0.18\ s^{-1}} = \boxed{5.6\ s}$$

Example 16-8 Half-Life–Second Order Reaction

What is the half-life for the reaction of Example 16-7 if the initial concentration of NOCl is $3.75\ M$?

Plan
This is similar to Example 16-7 except that we need to use a different starting concentration.

Solution
$$t_{1/2} = \frac{1}{ak[A]_0} = \frac{1}{(0.70\ M^{-1}s^{-1})(3.75\ M)} = \frac{1}{2.6\ s^{-1}} = \boxed{0.38\ s}$$

Example 16-9 Concentration versus Time–Second Order Reaction

If the reaction in Example 16-7 is carried out at 673°C beginning with 0.250 mole NOCl in a 1.00 liter vessel, how many moles of NOCl will remain after 80.0 seconds? How many grams of NOCl will remain?

Plan
The reaction is second order with respect to NOCl and second order overall. We can use the appropriate integrated rate law with:

$$[NOCl]_0 = \frac{0.250 \text{ mol}}{1 \text{ L}} = 0.250 \, M, \; t = 80.0 \text{ s and } k = 0.70 \, M^{-1}s^{-1}$$

Solution

$$\frac{1}{[NOCl]} - \frac{1}{[NOCl]_0} = akt \text{ so } \frac{1}{[NOCl]} - \frac{1}{[0.250 \, M]_0} = 1(0.70 \, M^{-1}s^{-1})(80.0 \text{ s})$$

Finding a common denominator for the left side yields:

$$\frac{0.250 \, M - [NOCl]}{(0.250 \, M)([NOCl])} = 56 \, M^{-1} \text{ so } 0.250 \, M - [NOCl] = 14[NOCl]$$

$$0.250 \, M = 15[NOCl] \text{ and } [NOCl] = \frac{0.250 \, M}{15} = 0.017 \, M \text{ NOCl}$$

$$? \text{ mol NOCl remaining} = 0.017 \frac{\text{mol NOCl}}{L} \times 1.00 \, L = \boxed{0.017 \text{ mol NOCl}}$$

$$? \text{ g NOCl remaining} = 0.017 \text{ mol NOCl} \times \frac{65.5 \text{ g NOCl}}{1 \text{ mol NOCl}} = \boxed{1.1 \text{ g NOCl}}$$

16-5 Reaction Mechanisms and the Rate-Law Expression

A **reaction mechanism** consists of the actual steps by which a process is hypothesized to occur. A reaction mechanism consists of a series of **elementary steps**. Elementary steps are different from a normal balanced equation in that the order of each reactant or product in an elementary step is given by the coefficient of that substance in that step. Substances that are produced in one step and consumed in another are called **intermediates**. A reaction mechanism must do the following:

1. Add up to the balanced chemical equation.

2. Agree with the observed rate law found from experimental data.
3. Account for any known intermediates.

the series of elementary steps, there always must be one **slow** or **rate determining** step that nctions as a bottleneck in the production of products. If a **fast** step comes before a slow step it ust be a fast equilibrium, since "products" will build up and reach a high enough concentration ich that the reverse process will begin to occur to an appreciable extent. Remember that an quilibrium is reached when the rate of the forward process is equal to the rate of the reverse rocess. We never look beyond the reactants in the slow step of the mechanism for the rate law, ecause it is assumed that, once the reactants get over the highest activation energy barrier, they an then "coast down" over smaller subsequent barriers to become products. The rate law may nclude reactants, products, catalysts, inhibitors, etc., that have concentrations that can be monitored. Any intermediates must be expressed in terms of these other quantities by considering how they are made.

For the reaction: $2NO(g) + 2H_2(g) \rightarrow 2H_2O(g) + N_2(g)$, the rate-law expression is: Rate = $k[NO]^2[H_2]$. Many mechanisms are consistent with this rate-law, including the following:

Step 1. $2NO + H_2 \rightarrow H_2N_2O_2$ (slow)
Step 2. $H_2N_2O_2 + H_2 \rightarrow 2H_2O + N_2$ (fast)
          ─────────────────────────────────
          $2NO + 2H_2 \rightarrow 2H_2O + N_2$ (overall)

This would agree with the observed rate-law, since two molecules of NO and 1 molecule of H_2 react in the slow step, however, the simultaneous collision of three particles in the proper orientation is very unlikely and would lead to a very slow reaction. Because of probability considerations, most mechanisms are comprised of either **bimolecular** collisions or **unimolecular** decompositions. The mechanism believed correct for the reaction is:

Step 1. $2NO \rightleftharpoons N_2O_2$ (fast, equilibrium)
Step 2. $N_2O_2 + H_2 \rightarrow N_2O + H_2O$ (slow)
Step 3. $N_2O + H_2 \rightarrow N_2 + H_2O$ (fast)
          ─────────────────────────────────
          $2NO + 2H_2 \rightarrow 2H_2O + N_2$ (overall)

The slow step determines the overall rate of the reaction, which would be:

Rate = $k_2[N_2O_2][H_2]$

where k_2 is the specific rate constant for that step. However, the rate-law expression must be in terms of original reactants. The N_2O_2 is produced by the rapid equilibrium in the first step. In an equilibrium, the rate of the forward and reverse process are the same so:

$$k_{1f}[NO]^2 = k_{1r}[N_2O_2] \text{ and } [N_2O_2] = \frac{k_{1f}}{k_{1r}}[NO]^2$$

Here k_{1f} and k_{1r} are the specific rate constants for the forward and reverse reactions, respectively, of the first step. Substituting this into the previous equation gives:

$$\text{Rate} = k_2[N_2O_2][H_2] = k_2 \times \frac{k_{1f}}{k_{1r}}[NO]^2[H_2] = k[NO]^2[H_2]$$

where **k** is the observed overall specific rate constant. In order for a mechanism to be valid, its steps must add to the overall reaction and it must account for the observed rate-law. Studies about the intermediates formed are necessary to choose between postulated mechanisms.

Example 16-10 Mechanism of a Reaction

The reaction below is found to obey the indicated rate-law expression. Which of the mechanisms that follow are consistent with this expression?

$$2N_2O_5 \rightarrow 4NO_2 + O_2 \qquad \text{Rate} = k[N_2O_5]$$

Possible mechanism 1:

Step 1. $N_2O_5 \rightarrow N_2O + 2O_2$ (slow)
Step 2. $N_2O_5 + N_2O \rightarrow 2NO_2 + O_2 + N_2$ (fast)
Step 3. $N_2 + 2O_2 \rightarrow 2NO_2$ (fast)

Possible mechanism 2:

Step 1. $N_2O_5 \rightarrow NO_2 + NO_3$ (slow)
Step 2. $N_2O_5 + NO_2 \rightarrow N_2O_4 + NO_3$ (fast)
Step 3. $NO_3 + N_2O_4 \rightarrow 2NO_2 + NO_3$ (fast)
Step 4. $NO_3 + NO_3 \rightarrow 2NO_2 + O_2$ (fast)

Possible mechanism 3:

Step 1. $2N_2O_5 \rightleftharpoons 2NO_2 + 2NO_3$ (fast, equilibrium)
Step 2. $NO_3 \rightarrow NO + O_2$ (slow)
Step 3. $NO + NO_3 \rightarrow 2NO_2$ (fast)

Plan

We will inspect these possible mechanisms as in the example above and compare the rate law indicated by the mechanisms to the actual rate law.

Solution
All three mechanisms sum to give the overall reaction expected and, therefore, meet that requirement. Possible mechanisms 1 and 2 have a slow first step that would give the indicated rate-law. Mechanism 1 postulates a termolecular collision in Step 3 and for that reason is highly unlikely. Mechanism 2, although it has many steps, has only bimolecular collisions after the first decomposition and would be possible. Mechanism 3 is not possible because the equilibrium in the first step, analyzed in the manner done earlier, would cause the rate-law to show a dependence on the NO_2 intermediate:

Rate = $k_2[NO_3]$ would be the rate-law from slow Step 2:

$$k_{1f}[N_2O_5]^2 = k_{1r}[NO_2]^2[NO_3]^2 \text{ so } [NO_3] = \frac{(k_{1f})^{1/2}[N_2O_5]}{(k_{1r})^{1/2}[NO]}$$

Substituting for the NO_3 yields: Rate = $k\frac{[N_2O_5]}{[NO_2]}$ (not the known rate law)

16-6 Temperature: The Arrhenius Equation

In order for a reaction to occur, colliding particles must have the proper orientation relative to each other and they must have sufficient energy to allow their own bonds, if any, to be broken. The amount of energy necessary to initiate a reaction is called the energy of activation or the **activation energy**, E_a. Slow reactions have relatively high activation energies, while fast reactions have low activation energies. Because particles have higher average kinetic energy as the temperature increases, increasing the temperature increases the rate of a reaction. This occurs because a higher percentage of the particles possess an energy that is greater than the activation energy. Arrhenius found the relationship below, which relates the specific rate constant, k, at a temperature, T, and the activation energy, E_a:

$$k = A\, e^{-E_a/RT}$$

The **A** is a collision frequency factor, related to the number of colliding particles with proper orientations. **R** is the ideal gas constant. The units of R must agree with those of E_a. **T** is the temperature in kelvins and **e** is Euler's number, the basis of the natural logarithm system. For the same reaction at different temperatures with a specific rate constant k_1 at T_1 and a specific rate constant k_2 at T_2, the following relationship can be derived from the Arrhenius equation:

$$\ln\left[\frac{k_2}{k_1}\right] = \frac{E_a}{R}\left[\frac{T_2 - T_1}{T_2 T_1}\right] \text{ or } \log\left[\frac{k_2}{k_1}\right] = \frac{E_a}{2.303R}\left[\frac{T_2 - T_1}{T_2 T_1}\right]$$

Example 16-11 Arrhenius Equation

For the reaction, $N_2O_5 \rightarrow NO_2 + NO_3$, Ea = 88 kJ/mol. The specific rate constant at 0°C is 9.16×10^{-3} s^{-1} What is the specific rate constant at 100°C?

Plan

We can use the formula derived from the Arrhenius equation with:
$k_1 = 9.16 \times 10^{-3}$ s^{-1}, $T_1 = 0°C + 273 = 273K$, $k_2 = ?$,
$T_2 = 100°C + 273 = 373K$, $E_a = 88$ kJ/mol, $R = 8.314 \times 10^{-3}$ kJ/mol·K

Solution

$$\ln\left[\frac{k_2}{9.16 \times 10^{-3} \text{ s}^{-1}}\right] = \frac{88 \text{ kJ/mol}}{8.314 \times 10^{-3} \text{ kJ/mol·K}} \left[\frac{373K - 273K}{373K \times 273K}\right]$$

$$\frac{k_2}{9.16 \times 10^{-3} \text{ s}^{-1}} = e^{10.394} = 3.27 \times 10^4 \text{ (use } e^x \text{ or inv lnx on calculator)}$$

$k_2 = 9.16 \times 10^{-3}$ s^{-1}(3.27 × 10^4) = $\boxed{300 \text{ s}^{-1}}$

Example 16-12 Activation Energy

A reaction is found to have a specific rate constant of 14.4 min^{-1} at 298K and a specific rate constant of 94.2 min^{-1} at 308K. What is the activation energy of this reaction?

Plan

We will use the same formula as in Example 16-11 with:
$k_1 = 14.4 \, M$ min^{-1}, $T_1 = 298K$, $k_2 = 94.2 \, M$ min^{-1}, $T_2 = 308K$, $E_a = ?$
$R = 8.314 \times 10^{-3}$ kJ/mol·K

Solution

$$\ln\left[\frac{94.2 \, M \text{ min}^{-1}}{14.4 \, M \text{ min}^{-1}}\right] = \frac{E_a}{8.314 \times 10^{-3} \text{ kJ/mol·K}} \left[\frac{308K - 298K}{308K \times 298K}\right]$$

$\ln(6.54) = 1.88 = E_a(1.31 \times 10^{-2}$ mol/kJ)

$E_a = \dfrac{1.88}{1.31 \times 10^{-2} \text{ mol/kJ}} = \boxed{144 \text{ kJ/mol}}$

EXERCISES

Nature of Reactants

1. For each of the following pairs of reactions, which reaction is likely to occur more rapidly at room temperature? Why?

   (a) $K_2SO_4(s) + Ba(CH_3COO)_2(s) \rightarrow BaSO_4(s) + 2KCH_3COO(s)$

   $K_2SO_4(aq) + Ba(CH_3COO)_2(aq) \rightarrow BaSO_4(s) + 2KCH_3COO(aq)$

   (b) $PCl_3(g) + Cl_2(g) \rightarrow PCl_5(g)$

   $AgNO_3(aq) + NaCl(aq) \rightarrow AgCl(s) + NaNO_3(aq)$

Concentration of Reactants and the Rate-Law Expression

2. Use the data below to write the rate-law expression, including the specific rate constant, for the reaction below at the temperature at which the reaction was performed.

$$NO_2 + CO \rightarrow NO + CO_2$$

Trial	Initial $[NO_2]$ (M)	Initial $[CO]$ (M)	Initial Rate ($M \cdot s^{-1}$)
1	0.300	0.200	1.35×10^{-3}
2	0.300	0.600	1.35×10^{-3}
3	0.600	0.600	5.40×10^{-3}

3. (a) What would be the rate of the reaction in Exercise 2 at the same temperature if the initial concentrations of NO_2 and CO were 0.750 M and 0.265 M, respectively?

   (b) What would be the rate of the reaction at the same temperature if the initial concentrations of NO_2 and CO were 0.173 M and 0.228 M, respectively?

4. Determine the rate-law expression for the reaction below at the temperature at which the tabulated initial rate data for this hypothetical reaction were obtained.

$$3A + 2B + 3C \rightarrow Products$$

Trial	Initial [A] (M)	Initial [B] (M)	Initial [C] (M)	Initial Rate ($M \cdot s^{-1}$)
1	0.30	0.20	0.10	4.0×10^{-2}
2	0.50	0.20	0.10	4.0×10^{-2}
3	0.50	0.20	0.25	1.0×10^{-1}
4	0.30	0.40	0.10	1.6×10^{-1}

5. A reaction is third order in A and first order in B. If the concentration of A is reduced to one-fourth of its original value and the concentration of B is doubled, the rate of the reaction will (increase or decrease) by a factor of _____.

6. Determine the rate-law expression for the hypothetical reaction below at the temperature which the initial data were taken.

$$3A(g) + 4B(g) \rightarrow 2C(g) + D(g)$$

Trial	Initial [A] (M)	Initial [B] (M)	Initial Rate (M·min^{-1})
1	0.200	0.300	7.42
2	0.250	0.300	7.42
3	0.200	0.250	4.29

7. Consider the initial rate data obtained at a certain temperature for the reaction below in aqueous solution. Write the rate-law expression for this reaction at that temperature.

$$H_2O_2 + 3I^- + 2H^+ \rightarrow 2H_2O + I_3^-$$

Trial	Initial [H$_2$O$_2$] (M)	Initial [H$^+$] (M)	Initial [I$^-$] (M)	Initial Rate (M·s^{-1})
1	0.200	0.300	0.100	6.40 x 10^{-3}
2	0.300	0.300	0.250	2.40 x 10^{-2}
3	0.300	0.600	0.250	2.40 x 10^{-2}
4	0.200	0.900	0.400	2.56 x 10^{-2}

8. What would be the rate in Exercise 7 if initially 0.150 mol of H$_2$O$_2$, 0.180 mol of H$^+$ and 0.260 mol of I$^-$ were placed in a 5.00 liter flask?

Integrated Rate-Laws and Half-Lives of Reactants

9. The decomposition (shown below) of the flammable gas dimethyl ether, (CH$_3$)$_2$O, follows first order kinetics At 500 °C, the rate law is: Rate = 0.0246 min^{-1} [(CH$_3$)$_2$O]. Suppose we place 352 grams of (CH$_3$)$_2$O in a 3.00 liter container at 500°C.

$$(CH_3)_2O(g) \rightarrow CH_4(g) + H_2(g) + CO(g)$$

(a) What is the half life of this process?

(b) How many grams of (CH$_3$)$_2$O remain unreacted after 75.0 minutes?

(c) Refer to (b). How many grams of CH$_4$ are produced?

(d) How long does it take to reduce the mass of unreacted (CH$_3$)$_2$O to 14.0 grams?

10. The half-life of the first order gas phase reaction below is 9.6×10^{28} seconds at 300 K.

$$C_2H_5Cl(g) \rightarrow C_2H_4(g) + HCl(g)$$

   (a) Evaluate the specific rate constant at 300 K.

   (b) How long would it take for 1.00% of a given sample of C_2H_5Cl at 300K to decompose into C_2H_4 and HCl?

   (c) What percentage of a 2.00 g sample remains after one billion (10^9) years?

11. At 300K, the decomposition reaction $2NOCl \rightarrow 2NO + Cl_2$ has a second order specific rate constant of $2.8 \times 10^{-5}\ M^{-1}s^{-1}$. Suppose 1.0 mole of NOCl is introduced into a 2.0 liter flask at 300K.

   (a) Evaluate the half-life of the reaction.

   (b) How much NOCl will remain after 30 minutes?

   (c) After what time period will 0.9 mole of NOCl remain?

   (d) Refer to (c). How many moles of Cl_2 will be present at that time?

12. The decomposition of HI on gold wire was cited in Example 16-4 as being a zero order reaction. The specific rate constant has a value of 15 Ms^{-1} at some temperature and starting concentration of 3.0 M.

   (a) How long does it take the 3.0 M sample of HI to decompose so that only 0.20 M remains?

   (b) What is the half-life of the reaction under these conditions?

13. For the reaction below, $k = 5.00 \times 10^{-4}\ s^{-1}$ at 45°C.

$$N_2O_5(g) \rightarrow NO_2(g) + 1/2 O_2(g)$$

   (a) What is the half-life of this reaction at this temperature?

   (b) If the $[N_2O_5]$ is originally 4.50 M, what molarity of N_2O_5 remains unreacted in 1.50 hours?

Mechanisms of Reactions

14. Propose two possible mechanisms that agree with the rate-law expression for the reaction below.

$$SO_2Cl_2(g) \rightarrow SO_2(g) + Cl_2(g). \quad \text{Rate} = k[SO_2Cl_2]$$

15. Propose two possible mechanisms that agree with the rate-law expression for the reaction below.

$$2N_2O \rightarrow 2N_2 + O_2 \quad \text{Rate} = k[N_2O]^2$$

16. Decide which of the suggested possible mechanisms are consistent with the observed rate-law expression.

$$IBr(g) \rightarrow I_2(g) + Br_2(g) \quad \text{Rate} = k[IBr]^2$$

(a) $IBr \rightarrow I + Br$ (slow)
$I + IBr \rightarrow I_2 + Br$ (fast)
$Br + Br \rightarrow Br_2$ (fast)

(b) $IBr \rightleftharpoons I + Br$ (fast, equilibrium)
$I + IBr \rightarrow I_2 + Br$ (slow)
$Br + Br \rightarrow Br_2$ (fast)

(c) $IBr + IBr \rightarrow I_2Br^+ + Br^-$ (slow)
$I_2Br^+ \rightarrow I_2 + Br^+$ (fast)
$Br^- + Br^+ \rightarrow Br_2$ (fast)

(d) $IBr \rightleftharpoons I^+ + Br^-$ (fast, equilibrium)
$I^+ + IBr \rightarrow I_2Br^+$ (slow)
$I_2Br^+ + Br^- \rightarrow I_2 + Br_2$ (fast)

(e) $IBr + IBr \rightarrow I_2 + Br_2$ (one step)

17. Write the rate-law expression that is consistent with the following mechanism.

$H_3AsO_4(aq) + H^+(aq) \rightleftharpoons H_4AsO_4(aq)$ fast, equilibrium

$H_4AsO_4(aq) + I^-(aq) \rightarrow H_3AsO_3(aq) + HIO(aq)$ slow

$HIO(aq) + I^-(aq) + H^+(aq) \rightarrow I_2(s) + H_2O(\ell)$ fast

Effect of Temperature on Reaction Rate

18. The reaction below has specific rate constants as follows: $k = 2.0 \times 10^{-2}$ M$^{-1} \cdot$s^{-1} at 35°C, $k = 7.2 \times 10^{-2}$ M$^{-1} \cdot$s^{-1} at 55°C. What is the activation energy of the reaction?

$$S_2O_8^{2-}(aq) + 2I^-(aq) \rightarrow I_2(aq) + 2SO_4^{2-}(aq)$$

19. Refer to Exercise 18. What would be the specific rate constant of the reaction at 0°C?

20. Refer to Exercise 18. How much would the activation energy need to be lowered to triple the rate of the reaction at 35°C?

21. Refer to Exercise 18. To what temperature would the reaction mixture need to be heated to increase the rate of the reaction observed at 35°C by a factor of 100?

Miscellaneous Exercises

22. Nuclear decay processes (Chapter 30 of the text) follow first order kinetics. ^{33}P is used to determine blood volume and, when incorporated into phosphate ion, to study brain and breast tumors. The half-life of ^{33}P is 14.282 days.

    (a) What is k, the specific rate constant, in day^{-1}?

    (b) How many days are required for 90.0% of the ^{33}P to be converted to products?

23. The hypothetical reaction has the following data. Determine the rate-law expression, including the value of the specific rate constant, for this reaction.

    $$2A(g) + 3B(g) \rightarrow \text{Products}$$

Trial	Initial [A] (M)	Initial [B] (M)	Initial Rate (M·s^{-1})
1	2.75	1.61	0.338
2	3.66	1.61	0.450
3	4.68	1.22	0.251

24. What is the rate of the reaction in Exercise 23 if initially [A] = 2.00 M and [B] = 3.00 M?

25. For a second order reaction, the specific rate constant has a value of 8.66 x 10^{-5} M^{-1}·s^{-1} at 16°C. The activation energy is 85.4 kJ/mol. What is the specific rate constant at 146°C?

26. The concentration of a reactant that operates according to first order kinetics and has a coefficient of 1 drops from 2.04 M to 0.300 M in 425 seconds. What are (a) the specific rate constant in s^{-1} and (b) the half-life, in seconds, of this process?

27. For CH$_3$CHO(g) + I$_2$(g) → CH$_3$I(g) + HI(g) + CO(g), E$_a$ = 108 kJ/mol. How much faster is this reaction at 55°C than at 15°C?

28. What rate-law expression is consistent with the mechanism below?

    Cl$_2$(g) ⇌ 2Cl(g)   fast, equilibrium

    CHCl$_3$(g) + Cl(g) → CCl$_3$(g) + HCl(g)   slow

    CCl$_3$(g) + Cl(g) → CCl$_4$(g)   fast

ANSWERS TO EXERCISES

1. In each pair, the reaction in aqueous solution will be more rapid because there are no solids to break up, nor any covalent bonds to break.

2. Rate = $1.50 \times 10^{-2} M^{-1} \cdot s^{-1} [NO_2]^2$

3. (a) Rate = $8.44 \times 10^{-3} M \cdot s^{-1}$ (b) Rate = $4.49 \times 10^{-4} M \cdot s^{-1}$

4. Rate = $10 M^{-2} \cdot s^{-1} [B]^2[C]$ 5. decrease by a factor of 8

6. Rate = $275 M^{-2} \cdot min^{-1} [B]^3$

7. Rate = $0.320 M^{-1} \cdot min^{-1} [H_2O_2][I^-]$ 8. Rate = $4.99 \times 10^{-4} M \cdot s^{-1}$

9. (a) $t_{1/2}$ = 28.2 min (b) 55.2 g $(CH_3)_2O$ (c) 103 g CH_4 (d) 131 min

10. (a) $7.2 \times 10^{-30} s^{-1}$ (b) 1.4×10^{27} s (4.4×10^{19} years) (c) nearly 100%

11. (a) 7.1×10^4 s (≈20 hours) (b) 0.98 mol (0.49 M)
    (c) 7.9×10^3 s (2.2 hours) (d) 0.050 mol Cl_2

12. (a) 0.19 s (b) 0.10 s   13. (a) 1.39×10^3 s (b) 0.306 M

14. $SO_2Cl_2 \rightarrow SO_2 + Cl_2$ (one step) | $SO_2Cl_2 \rightarrow SO_2Cl^+ + Cl^-$ (slow)
                                                  | $SO_2Cl^+ + Cl^- \rightarrow SO_2 + Cl_2$ (fast)

15. $N_2O + N_2O \rightarrow 2N_2 + O_2$ (one step) | $N_2O + N_2O \rightarrow N_2O_2 + N_2$ (slow)
                                                   | $N_2O_2 \rightarrow N_2 + O_2$ (fast)

16. c, e   17. Rate = $k[H_3AsO_4][H^+][I^-]$ 18. 54 kJ/mol   19. $1.3 \times 10^{-3} M^{-1} \cdot s^{-1}$

20. lower by 2.8 kJ   21. 394 K (121°C)

22. (a) k = 0.0485 day^{-1} (b) $t_{1/2}$ = 47.5 days   23. $2.95 \times 10^{-2} M^{-3} \cdot s^{-1} [A][B]^3$

24. $1.59 M \cdot s^{-1}$ 25. $6 s^{-1}$ 26. (a) k = $0.00453 s^{-1}$ (b) $t_{1/2}$ = 153 s

27. about 200 times ($e^{5.5}$) faster   28. Rate = $k[CHCl_3][Cl_2]^{1/2}$

Chapter 17: Chemical Equilibrium

17-1 The Equilibrium Constant

Most chemical processes are **reversible**, that is, they occur in both forward and reverse directions and do not go to completion. Double arrows indicate reversibility. For such processes, a **dynamic equilibrium** is eventually established in which the forward and reverse reactions occur simultaneously at the same rate and some of each species is present. Unless the equilibrium is then disturbed, no further changes in concentration occur. The equilibrium may be approached by starting with either "reactants" or with "products" or with a mixture of each with "products" and "reactants" designated as indicated in the generalized reaction below.

$$aA + bB \rightleftharpoons cC + dD \qquad \text{"Reactants"} \rightleftharpoons \text{"Products"}$$

The position of the equilibrium is related to the magnitude of the **equilibrium constant** (K_c). For the generalized reaction above, the **mass action expression** for the equilibrium constant is:

$$K_c = \frac{[C]^c[D]^d}{[A]^a[B]^b} \qquad \left\{\frac{\text{"Products"}}{\text{"Reactant"}}\right\}$$

The square brackets, [ ], refer to concentration in molarity, M. (More strictly, values used are shown without units because the equilibrium constant is thermodynamically defined in terms of activities, which are **ratios** of values to unit molarity or unit pressure.) The powers to which the concentrations are raised are the coefficients in the balanced equation. By convention, the species on the right are called the "products" and appear in the numerator, while the species on the left, the "reactants" are always in the denominator of this mass action expression. Values of equilibrium constants must be experimentally determined and only on the system and the temperature. The larger the numerical value of the equilibrium constant, the more the forward reaction is favored; therefore, a higher percentage of the reactants will be converted to products. Conversely, the smaller the equilibrium constant, the more the reverse reaction is favored; therefore, a smaller percentage of the reactants will be converted to products.

Example 17-1 Calculation of K_c

The following amounts were found at equilibrium in a 3.00 liter container at 400°C: 0.0420 mole N_2, 0.516 mole H_2, and 0.0357 mole NH_3. Evaluate K_c.

$$N_2(g) + 3H_2(g) \rightleftharpoons 2NH_3(g)$$

Plan

K_c is defined in terms of activities, which are related to the equilibrium concentrations in moles per liter. We find these by dividing the number of moles by the volume in liters. We can then use the equilibrium molarities, without units, in the mass action expression.

Solution

$$? [N_2] = \frac{0.0420 \text{ mol } N_2}{3.00 \text{ L}} = 0.0140 \, M \, N_2$$

$$? [H_2] = \frac{0.516 \text{ mol } H_2}{3.00 \text{ L}} = 0.172 \, M \, H_2$$

$$? [NH_3] = \frac{0.0357 \text{ mol } NH_3}{3.00 \text{ L}} = 0.0119 \, M \, NH_3$$

$$? K_c = \frac{[NH_3]^2}{[N_2][H_2]^3} = \frac{(0.0119)^2}{(0.0140)(0.172)^3} = \boxed{1.99}$$

17-2 Variation of K_c with the Form of the Balanced Equation

The form of the mass action expression depends upon the coefficients of the balanced equation. If the equation in Example 17-1 had been written with the coefficients multiplied by 1/2 as follows: $1/2 N_2(g) + 3/2 H_2(g) \rightleftharpoons NH_3(g)$, then the equilibrium constant, K'_c, would be:

$$K'_c = \frac{[NH_3]}{[N_2]^{1/2}[H_2]^{3/2}} = \frac{(0.0119)}{(0.0140)^{1/2}(0.172)^{3/2}} = 1.41 = K_c^{1/2}$$

If the equation is reversed, $2NH_3(g) \rightleftharpoons N_2(g) + 3H_2(g)$, which is similar to multiplying the coefficients multiplied by -1, then the equilibrium constant, K''_c, would be:

$$K''_c = \frac{[N_2][H_2]^3}{[NH_3]^2} = \frac{(0.0140)(0.172)^3}{(0.0119)^2} = 0.530 = K_c^{-1}$$

In summary, if the coefficients are multiplied by a number, **n**, the value of the original equilibrium constant is raised to the **n**th power.

17-3 The Reaction Quotient

The mass action expression for the **reaction quotient, Q**, has the same form as the equilibrium constant; however, the concentrations are not necessarily those at equilibrium.

If	Then
$Q_c > K_c$	Reverse reaction favored until equilibrium is established
$Q_c = K_c$	Forward and reverse reaction are in equilibrium
$Q_c < K_c$	Forward reaction favored until equilibrium is established

Evaluation of Q_c for a reacting system at any instant allows determination of which concentrations must increase and which must decrease in order for equilibrium to be established.

Example 17-2 The Reaction Quotient

Initial concentrations as listed below are placed into a container at 400°C for the reaction in Example 17-1. What happens under these initial conditions?

$$[N_2] = 0.142 \text{ M}, [H_2] = 0.0265 \text{ M}, \text{ and } [NH_3] = 0.0384 \text{ M}$$

Plan
We can find Q_c for these starting concentrations. Q_c can then be compared to K_c to see if the reaction is at equilibrium or if reactants or products will be favored.

Solution

$$? \; Q_c = \frac{[NH_3]^2}{[N_2][H_2]^3} = \frac{(0.0384)2}{(0.142)(0.0.0265)3} = 558$$

$Q_c \gg K_c$, therefore, NH_3 (product) will decompose to produce more N_2 and H_2 until equilibrium is established. The reactants are favored.

17-4 Uses of the Equilibrium Constant, K_c

Once the equilibrium constant for a given reaction has been determined at a certain temperature, it may be used to determine equilibrium concentrations at that temperature for any combination of initial concentrations.

Example 17-3 Finding Equilibrium Concentrations

Phosgene, $COCl_2$, is a poisonous gas that decomposes into carbon monoxide and chlorine according to the equation below. For this decomposition, $K_c = 0.083$ at 900°C. If the reaction is initiated with 0.450 mole of $COCl_2$ at 900°C in a 5.00 liter container, what will be the concentrations of all species present at equilibrium?

$$COCl_2(g) \rightleftharpoons CO(g) + Cl_2(g)$$

Plan

Intuitively, we expect that the reaction will proceed to the right to form products. We can check this by evaluating Q_c, remembering to convert the number of moles to molarities by dividing by the volume. We then construct a table showing: (1) initial concentrations; (2) changes in concentrations due to the net reaction; and (3) equilibrium concentrations, in which unknown concentrations are algebraically represented. In this case, $x = M\ COCl_2$ that decomposes. We substitute equilibrium concentrations in terms of unknowns into the mass action expression and set the expression equal to K_c. The resulting equation can then be algebraically solved for x and the equilibrium concentrations can be found.

Solution

$$?\ \text{initial}\ [COCl_2] = \frac{0.450\ \text{mol}\ COCl_2}{5.00\ L} = 0.0900\ M$$

$$?\ Q_c = \frac{[CO][Cl_2]}{[COCl_2]} = \frac{(0\ M)(0\ M)}{(0.0900\ M)} = 0$$

Because $Q_c < K_c$, products will be made

	$COCl_2$	$\rightleftharpoons$	CO	+	Cl_2
initial:	0.0900 M		0 M		0 M
change:	-x M		+x M		+x M
equilibrium	(0.0900-x) M		x M		x M

We substitute these equilibrium values are substituted into the K_c expression:

$$? \ K_c = \frac{[CO][Cl_2]}{[COCl_2]} = \frac{(x)(x)}{(0.0900 - x)} = 0.083$$

$$x^2 = 0.0075 - 0.083x \text{ or } x^2 + 0.083x - 0.0075 = 0$$

This is a quadratic equation in the standard form of $ax^2 + bx + c = 0$. In this case, $a = 1$, $b = 0.083$, and $c = -0.0075$. It may be solved by the quadratic formula, which gives two solutions, only one of which will usually have physical significance.

$$x = \frac{-b \pm \sqrt{b^2 - 4ac}}{2a} = \frac{-0.083 \pm \sqrt{(0.083)^2 - 4(1)(-.0075)}}{2(1)}$$

$$x = \frac{-0.083 \pm \sqrt{0.037}}{2} = \frac{-0.083 \pm 0.19}{2} \text{ therefore, } x = +0.054 \ M \text{ and } -0.14 \ M$$

Since x represents the amount of $COCl_2$ that decomposes, it has a maximum value of 0.0900 M (the initial concentration). It also has a minimum value of 0 M because the concentrations of the products cannot be negative, ie. $0 < x < 0.0900$. Therefore, $x = -0.14$ is the extraneous root and $x = 0.054$ M is the root that has physical meaning. The equilibrium concentrations are:

$$[CO] = [Cl_2] = x = \boxed{0.054 \ M}$$

$$[COCl_2] = (0.0900 \ M - x) = (0.0900 \ M - 0.054 \ M) = \boxed{0.036 \ M}$$

These values can be checked by substitution into the original mass action expression. This verification can be done for all answers.

$$K_c = \frac{[CO][Cl_2]}{[COCl_2]} = \frac{(0.054)(0.054)}{(0.036)} = 0.081$$

This value is within roundoff error of 0.083.

Example 17-4 Finding Equilibrium Concentrations

At 750°C, the equilibrium constant for the following reaction is 0.771. If 0.200 mol of H_2O and 0.200 mol of CO are placed in a 1.00 liter flask, what will be the equilibrium concentrations of all species at 750°C?

Plan

$$H_2(g) + CO_2(g) \rightleftharpoons H_2O(g) + CO(g)$$

We need to find the starting concentrations of H_2O and CO. Without calculating the reaction quotient (which would have a value of ∞), it is apparent that the normal "products" must react to produce "reactants". We can let x be the concentration of H_2O (or CO, since they react 1:1), which reacts. We set up a grid of changes and solve as in the previous example.

Solution

	$H_2(g)$	+	$CO_2(g)$	$\rightleftharpoons$	$H_2O(g)$	+	$CO(g)$
initial:	0 M		0 M		0.200 M		0.200 M
change:	+x M		+x M		-x M		-x M
equilibrium:	x M		x M		(0.200-x) M		(0.0200-x) M

$$K_c = \frac{[H_2O][CO]}{[H_2][CO_2]} = \frac{(0.200-x)(0.200-x)}{(x)(x)} = \frac{(0.200-x)^2}{x^2} = 0.771$$

The unknown side is a perfect square. Taking the square roots:

$$\frac{0.200-x}{x} = 0.878 \text{ therefore } 0.200-x = 0.878x \text{ or } 1.878x = 0.200$$

$$x = \frac{0.200}{1.878} = 0.106 \, M$$

The equilibrium concentrations are:

$[H_2] = [CO_2] = x = \boxed{0.106 \, M}$

$[H_2O] = [CO] = (0.200-x) \, M = (0.200 - 0.106) = \boxed{0.094 \, M}$

Check: $\dfrac{[H_2O][CO]}{[H_2][CO_2]} = \dfrac{(0.094)(0.094)}{(0.106)(0.106)} = 0.79$

This is within roundoff error of 0.771.

17-5 Factors that Affect Equilibria

Le Chatelier's Principle states that when a stress is applied to a system at equilibrium, the system responds to relieve the stress and reestablishes equilibrium. The stresses to be considered are: (1) changes in concentration; (2) changes in volume (and pressure); (3) changes in temperature; and, (4) addition of a catalyst (**not** a stress of the equilibrium position).

17-6 Changes in Concentration

If the concentration of one of the species in a reaction is changed by removing that species, the system responds by temporarily producing that species. Addition of a species causes some of its excess to be used and favors the other side of the reaction. In other words, the system tries to undo the stress.

Example 17-5 Applying a Stress to a System at Equilibrium

If 1.500 mole PCl_3 and 0.500 mole Cl_2 are placed in a 5.00 liter container, 0.390 mole PCl_5 is present after equilibrium has been established. If an additional 0.250 mole of Cl_2 is then added, what will be the concentrations of all species when equilibrium is reestablished?

$$PCl_5(g) \rightleftharpoons PCl_3(g) + Cl_2(g)$$

Plan

We must first find the initial concentrations of PCl_3 and Cl_2 and the equilibrium concentration of PCl_5. Then we must analyze the data stoichiometrically to determine the equilibrium concentrations of PCl_3 and Cl_2. These equilibrium concentrations are used with the mass action expression to find the value of K_c. We must then find the molarity of Cl_2 added and evaluate Q_c to determine which direction the reaction will shift. We set up a grid with **x** as the amount by which the concentrations change and proceed as in previous examples.

Solution

$$? \text{ initial } [PCl_3] = \frac{1.500 \text{ mol } PCl_3}{5.00 \text{ L}} = 0.300 \, M \, PCl_3$$

$$? \text{ initial } [Cl_2] = \frac{0.500 \text{ mol } Cl_2}{5.00 \text{ L}} = 0.100 \, M \, Cl_2$$

$$? \text{ equilibrium } [PCl_5] = \frac{0.390 \text{ mol } PCl_5}{5.00 \text{ L}} = 0.0780 \, M \, PCl_5$$

Because all the coefficients are 1, the change in concentration of PCl_3 and Cl_2 will be the same as the increase in PCl_5 concentration.

	$PCl_5(g)$	⇌	$PCl_3(g)$	+	$Cl_2(g)$
initial:	0 M		0.300 M		0.100 M
change	+0.0780 M		-0.0780 M		-0.0780 M
equilibrium:	0.0780 M		0.222 M		0.022 M

$$K_c = \frac{[PCl_3][Cl_2]}{[PCl_5]} = \frac{(0.222)(0.022)}{(0.0780)} = 0.063$$

The stress applied is the addition of Cl_2:

$$\text{? added } [Cl_2] = \frac{0.250 \text{ mol } Cl_2}{5.00 \text{ L}} = 0.0500 \text{ } M \text{ } Cl_2$$

LeChatelier's Principle indicates that the reaction should shift to the left. The reaction quotient confirms this:

$$Q_c = \frac{[PCl_3][Cl_2]}{[PCl_5]} = \frac{(0.222)(0.022 + 0.0500)}{(0.0780)} = 0.20$$

$Q_c > K_c$ means we must make more reactants (shift left).

We can now proceed with the complete analysis:

	$PCl_5(g)$	⇌	$PCl_3(g)$	+	$Cl_2(g)$
equilibrium:	0.0780 M		0.222 M		0.022 M
stress:	+0.0500 M				+0.0500 M
new initial:	0.0780 M		0.222 M		0.072 M
shift:	+x M		-x M		-x M
new equil:	(0.0780+x) M		(0.222-x) M		(0.072-x) M

$$K_c = \frac{[PCl_3][Cl_2]}{[PCl_5]} = \frac{(0.222-x)(0.072-x)}{(0.0780+x)} = 0.063$$

$0.00491 + 0.063x = 0.016 - 0.294x + x^2$

$x^2 - 0.357x + 0.011 = 0$

Solving with the quadratic formula gives: x = 0.323 and 0.034. The value of x must be between 0 M and 0.072 M to be physically significant, so x = 0.034.

$[PCl_5] = 0.0780 + x = 0.0780 + 0.034 = \boxed{0.112\ M}$

$[PCl_3] = 0.222 - x = 0.222 - 0.034 = \boxed{0.188\ M}$

$[Cl_2] = 0.072 - x = 0.072 - 0.034 = \boxed{0.038\ M}$

Checking with the K_c expression gives:

$$K_c = \frac{[PCl_3][Cl_2]}{[PCl_5]} = \frac{(0.18)(0.038)}{(0.112)} = 0.064$$

This is within roundoff error of 0.063.

Example 17-6 Applying a Stress to a System at Equilibrium

Refer back to Example 17-5. If 0.050 mol of Cl_2 (0.010 M) had been removed, what equilibrium concentrations would have resulted?

Plan
We already know that $K_c = 0.063$ from Example 17-5. We follow a similar series of steps to determine the equilibrium concentrations.

Solution
We expect the equilibrium to shift right since a product has been removed. The evaluation of Q_c confirms this.

$$Q_c = \frac{[PCl_3][Cl_2]}{[PCl_5]} = \frac{(0.222)(0.022 - 0.0100)}{(0.0780)} = 0.034$$

$Q_c < K_c$ means we must make more products or shift right.

We now proceed with the analysis

	$PCl_5(g)$		$\rightleftharpoons$	$PCl_3(g)$	+	$Cl_2(g)$	
equilibrium:	0.0780	M		0.222	M	0.022	M
stress:						-0.010	M
new initial:	0.0780	M		0.222	M	0.012	M
shift:	$-x$	M		$+x$	M	$+x$	M
new equil:	$(0.0780-x)$	M		$(0.222+x)$	M	$(0.012+x)$	M

$$K_c = \frac{[PCl_3][Cl_2]}{[PCl_5]} = \frac{(0.222+x)(0.012+x)}{(0.0780-x)} = 0.063$$

$$0.0049 - 0.063x = 0.0027 + 0.234x + x^2 \quad \text{or} \quad x^2 + 0.297x - 0.0022 = 0$$

Solving with the quadratic equation yields: $x = +0.007$ and -0.304. The latter is not physically significant because the use of $x = -0.304$ would lead to some negative molarities; therefore, $x = 0.007$ M. The equilibrium concentrations are:

$[PCl_5] = 0.0780 - x = 0.0780 - 0.007 = \boxed{0.071 \, M}$

$[PCl_3] = 0.222 + x = 0.222 + 0.007 = \boxed{0.229 \, M}$

$[Cl_2] = 0.012 + x = 0.012 + 0.007 = \boxed{0.019 \, M}$

To check: $K_c = \dfrac{[PCl_3][Cl_2]}{[PCl_5]} = \dfrac{(0.229)(0.019)}{(0.071)} = 0.061$

This result is within reasonable roundoff error.

17-7 Changes in Volume (Pressure)

At constant temperature, a change in the volume of a gas causes a change in its pressure and vice versa. The pressure of a gas is directly related to its concentration. From the ideal gas law:

$$P = \left(\frac{n}{V}\right)RT = MRT$$

As a result, pressure or volume changes can shift equilibria. For a system that has different numbers of moles of gases for the reactants and products, an increase in volume causes a shift in the direction that produces a larger number of moles of gas. Since volume and pressure are inversely proportional to each other, an increase in pressure that causes a volume change, causes a shift toward the side with the smaller number of moles of gas. (Pressure increases caused by addition of a non-reactive gas to the system, with no subsequent volume change, do not disturb the equilibrium. This is because the partial pressures and concentrations of the reactant gases do not change when there is no volume change.) Equilibria involving only solids and liquids and those in which there are equal numbers of moles of gaseous reactants and products are not affected by pressure/volume changes.

Example 17-7 Applying a Stress to a System at Equilibrium

Refer to Example 17-5. What would be the new equilibrium concentrations if the volume of the original system were suddenly reduced to 2.00 liters with no change in temperature?

Plan

We know that a decrease in volume should favor the side with PCl_5 because it has a smaller number of total moles in the balanced equation. We must calculate the new concentrations. Concentration is inversely related to volume, so that: $M_1V_1 = M_2V_2$. We can find the new concentrations, after dilution, but before the system has equilibrated. We will check Q_c to confirm that more PCl_5 will be made. We will then analyze the data and solve for equilibrium concentrations as in previous examples.

Solution

The initial volume is 5.00 L and the final volume is 2.00 L, so in each case, the final concentration, immediately after the volume reduction will be:

$$M_2 = \frac{M_1V_1}{V_2} = \frac{M_1(5.00 \text{ L})}{(2.00 \text{ L})} = 2.50 M_1$$

$? [PCl_3] = 2.50(0.222\ M) = 0.555\ M$

$? [Cl_2] = 2.50(0.022\ M) = 0.055\ M$

$? [PCl_5] = 2.50(0.0780\ M) = 0.195\ M$

$? Q_c = \dfrac{[PCl_3][Cl_2]}{[PCl_5]} = \dfrac{(0.555)(0.055)}{(0.195)} = 0.16$

Because $Q_c > K_c$, the reaction does shift left, making the reactant PCl_5. Now we can analyze the effect of the stress, using the new starting concentrations and noting that PCl_5 is produced.

	$PCl_5(g)$	$\rightleftharpoons$	$PCl_3(g)$	$+$	$Cl_2(g)$
initial:	0.195 M		0.555 M		0.055 M
shift:	+x M		-x M		-x M
new equil:	(0.195+x) M		(0.555-x) M		(0.055-x) M

$$K_c = \frac{[PCl_3][Cl_2]}{[PCl_5]} = \frac{(0.555-x)(0.055-x)}{(0.195+x)} = 0.063$$

$0.012 + 0.063x = 0.031 - 0.610x + x^2$ or $x^2 - 0.673x + 0.019 = 0$

The two possible roots are: x = 0.643 and 0.030, but since 0 < x < 0.055 to be physically significant, x = 0.030 M

When equilibrium is reestablished, the concentrations are:

$[PCl_5] = 0.195 + x = 0.195 + 0.030 = \boxed{0.225 \text{ M}}$

$[PCl_3] = 0.555 - x = 0.555 - 0.030 = \boxed{0.525 \text{ M}}$

$[Cl_2] = 0.055 - x = 0.055 - 0.030 = \boxed{0.025 \text{ M}}$

To check: $K_c = \dfrac{[PCl_3][Cl_2]}{[PCl_5]} = \dfrac{(0.525)(0.025)}{(0.225)} = 0.0058$

If the volume had been increased instead (or the pressure decreased), the forward reaction would have been favored (Q < K). The equilibrium position would have shifted to the right.

17-8 Changes in Temperature

The magnitude of an equilibrium constant is temperature dependent and will be mathematically treated in Section 17-13 of this book. At this point, LeChatelier's Principle may be used to predict the direction of shift due to temperature changes. An exothermic reaction releases heat. For example:

$$2H_2(g) + O_2(g) \rightleftharpoons 2H_2O(g) + 483.6 \text{ kJ}$$

An increase of temperature, at constant volume and pressure, increases the amount of heat available to the system. Since the reaction to the left is endothermic (requires heat), the equilibrium shifts to that side to consume the heat. The numerical value of the equilibrium constant actually decreases. A decrease in temperature of an exothermic process will cause an increase in equilibrium constant and a shift to the right. Temperature changes affect endothermic reactions in exactly the opposite way.

17-9 Adding a Catalyst

Catalysts equally affect the rates at which both the forward and reverse reactions occur and, generally, diminish the time needed to establish an equilibrium. A catalyst does not cause a shift in the position of an equilibrium.

Example 17-8 Applying a Stress to a System at Equilibrium

The following systems are at equilibrium at a given temperature. Predict the effects of the following stresses on the equilibrium position and on the relative amounts of each substance.

(1) $2SO_2(g) + O_2(g) \rightleftharpoons 2SO_3(g) + 197.6$ kJ
(2) $N_2(g) + 2O_2(g) + 66.4$ kJ $\rightleftharpoons 2NO_2(g)$
(3) $2N_2O_5(g) \rightleftharpoons 4NO_2(g) + O_2(g) + 110.8$ kJ

(a) increasing the amount of O_2, (b) decreasing the amount of O_2, (c) raising T at constant P and V, (d) lowering T at constant P and V, (e) increasing V at constant T, (f) decreasing P at constant T, (g) decreasing V at constant T, (h) adding a catalyst

Plan

We will follow the guidelines discussed in the preceding sections. Arrows ($\rightarrow$ or $\leftarrow$) indicate shifts in position of equilibrium to the right or left, + indicates an increase, – a decrease and 0 no change in equilibrium amount.

Solution

(1)

	Shift	SO_2	O_2	SO_3
(a)	$\rightarrow$	–	+	+
(b)	$\leftarrow$	+	–	–
(c)	$\leftarrow$	+	+	–
(d)	$\rightarrow$	–	–	+
(e)	$\leftarrow$	+	+	–
(f)	$\leftarrow$	+	+	–
(g)	$\rightarrow$	–	–	+
(h)	0	0	0	0

(2)

	Shift	N_2	O_2	NO_2
(a)	$\rightarrow$	–	–	+
(b)	$\leftarrow$	+	+	–
(c)	$\rightarrow$	–	–	+
(d)	$\leftarrow$	+	+	–
(e)	$\leftarrow$	+	+	–
(f)	$\leftarrow$	+	+	–
(g)	$\rightarrow$	–	–	+
(h)	0	0	0	0

(3)

	Shift	N_2O_5	NO_2	O_2
(a)	←	+	−	+
(b)	→	−	+	−
(c)	←	+	−	−
(d)	→	−	+	+
(e)	→	−	+	+
(f)	→	−	+	+
(g)	←	+	−	−
(h)	0	0	0	0

17-10 Relationship Between Partial Pressure and the Equilibrium Constant

It is often convenient to measure the partial pressures of gases rather than their concentrations. Since the partial pressure is directly proportional to its concentration, the equilibrium constant for a reaction involving gases can also be defined in terms of partial pressures. For the system:

$$aA(g) + bB(g) \rightleftharpoons cC(g) + dD(g)$$

$$K_c = \frac{[C]^c[D]^d}{[A]^a[B]^b} \text{ and } K_p = \frac{(P_C{}^c)(P_D{}^d)}{(P_A{}^a)(P_B{}^b)}$$

For an ideal gas, $P = [M]RT$ (with the value of $R = 0.0821 \frac{L \cdot atm}{mol \cdot K}$). At any constant temperature, R and T are constant, so the relationship between K_c and K_p is:

$$K_p = K_c(RT)^{\Delta n} \quad (\Delta n = (n_{gas\ products}) - (n_{gas\ reactants}))$$

When the numbers of moles of product and reactant gases are equal, $K_c = K_p$, for the reaction.

Example 17-9 Relationship Between K_p and K_c

We found in Example 17-1 that at 400°C, K_c for the reaction below with equilibrium concentrations of $[N_2] = 0.0140\ M$, $[H_2] = 0.172\ M$ and $[NH_3] = 0.0119\ M$ was 1.99. Calculate (a) K_p at 400°C, (b) the partial pressure of each gas, and (c) the total pressure in this system.

Plan

(a) We will use the formula above to evaluate K_p. (b) From the ideal gas law, we know the relationship between pressure and concentration and can solve for the pressures. (c) From Dalton's law, we know that the total pressure in the system is the sum of the partial pressures of the gases. Note that it is helpful to consider the apparent units of K_c and K_p to recognize which value of R must be used for these conversions.

Solution

(a) The change in moles of gas is: $\Delta n = 2 - (1 + 3) = -2$

$$? \; K_p = K_c(RT)^{\Delta n} = 1.99 \; M^{-2} \left[\left(0.0821 \frac{L \cdot atm}{mol \cdot K} \right)(673K) \right]^{-2}$$

$$= 1.99 \; M^{-2} \left[55.3 \frac{L \cdot atm}{mol} \right]^{-2} = 1.99(3.27 \times 10^{-4}) \, atm^{-2}$$

$$= \boxed{6.51 \times 10^{-4}}$$

(b) In each case, the partial pressure is $P = $ (concentration)RT. From part (a), we see that $RT = 55.3 \frac{L \cdot atm}{mol}$. Therefore,

$$P_{N_2} = 0.0140 \frac{mol}{L} \times 55.3 \frac{L \cdot atm}{mol} = \boxed{0.774 \; atm}$$

$$P_{H_2} = 0.172 \frac{mol}{L} \times 55.3 \frac{L \cdot atm}{mol} = \boxed{9.51 \; atm}$$

$$P_{NH_3} = 0.0119 \frac{mol}{L} \times 55.3 \frac{L \cdot atm}{mol} = \boxed{0.658 \; atm}$$

(c) $P_{total} = 0.774 \, atm + 9.51 \, atm + 0.658 \, atm = \boxed{10.9 \; atm}$.

The K_p also could have been evaluated by finding the partial pressures first and substituting into the K_p expression:

$$K_p = \frac{P_{NH_3}^2}{P_{N_2} P_{H_2}^3} = \frac{(0.658)^2}{(0.774)(9.51)^3} = \boxed{6.50 \times 10^{-4}}$$

This agrees with the previous result within roundoff error.

17-11 Heterogeneous Equilibria

Heterogeneous equilibria involve species in more than one phase. In the fundamental thermodynamic definition of equilibrium constants, activities are used. Pure solids and liquids have activity equal to 1 and, therefore, do not appear in mass action expressions. Solvents in dilute solution also have an activity of 1. Only gases appear in the expression for K_p.

Example 17-10 K_c and K_p for Heterogeneous Equilibria

Write mass action expressions for the equilibrium constants, K_p and K_c, for the following reactions. (All solutions are dilute.)

(1) $2NaNO_3(s) \rightleftharpoons 2NaNO_2(s) + O_2(g)$
(2) $N_2H_4(\ell) + 2H_2O_2(\ell) \rightleftharpoons N_2(g) + 4H_2O(g)$
(3) $2NH_3(g) + 3CuO(s) \rightleftharpoons 3H_2O(\ell) + N_2(g) + 3Cu(s)$
(4) $HF(aq) + H_2O(\ell) \rightleftharpoons H_3O^+(aq) + F^-(aq)$
(5) $3H_3PO_4(\ell) + PH_3(g) \rightleftharpoons 4H_3PO_3(\ell)$

Plan

We will follow the form indicated above, remembering that pure solids, pure liquids, and the solvents in dilute solutions all have activity of 1.

Solution

(1) $K_c = [O_2]$ $\qquad K_p = P_{O_2}$

(2) $K_c = [N_2][H_2O]^4$ $\qquad K_p = P_{N_2} \times P_{H_2O}^4$

(3) $K_c = \dfrac{[N_2]}{[NH_3]^2}$ $\qquad K_p = \dfrac{P_{N_2}}{P_{NH_3}^2}$

(4) $K_c = \dfrac{[H_3O^+][F^-]}{[HF]}$ $\qquad K_p$ is undefined   (Water is the solvent.)

(5) $K_c = \dfrac{1}{[PH_3]}$ $\qquad K_p = \dfrac{1}{P_{PH_3}}$

7-12 Relationship Between $\Delta G°$ and the Equilibrium Constant

The value of $\Delta G°$ indicates the free energy transferred in establishing equilibrium at standard temperature (usually 25°C) and one atmosphere pressure when **all** substances, reactants and products, are mixed in unit molar or pressure quantities. The value of ΔG indicates the free energy transfer under any specific set of conditions except standard thermodynamic conditions. The two quantities are related to each other by the relationship:

$$\Delta G = \Delta G° + RT \ln Q \quad \text{or} \quad \Delta G = \Delta G° + 2.303\, RT \log Q$$

where Q is the reaction quotient, T is the absolute temperature, and R is the ideal gas constant (8.314×10^{-3} kJ/mol·K when $\Delta G°$ is in kJ/mol).

At equilibrium, $\Delta G = 0$, and $Q = K$, so

$$\Delta G° = -RT \ln K \quad \text{or} \quad \Delta G° = -2.303\, RT \log K$$

where $K = K_p$ for gaseous systems, $K = K_c$ for reactions in aqueous solution. In heterogeneous equilibria, gaseous substances are shown in terms of partial pressures, while dissolved species are reported in terms of molarities; therefore, K is a combination of K_p and K_c.

Example 17-11 K versus $\Delta G°$

Evaluate K_p and K_c at 25°C for the following reaction from thermodynamic data in Appendix K of the text.

$$I_2(g) + Cl_2(g) \rightleftharpoons 2ICl(g)$$

Plan

We can calculate $\Delta G°$ directly from $\Delta G°_f$ values (review Chapter 15 if necessary). We can then use the relationship $\Delta G°_{298} = -2.303\, RT \log K_p$ to evaluate K_p. Finally, we can evaluate K_c using the relationship that $K_p = K_c(RT)^{-\Delta n}$.

Solution

$$\Delta G°_{298} = 2\Delta G°_f\, ICl(g) - \Delta G°_f\, I_2(g) - \Delta G°_f\, Cl_2(g)$$

$$\Delta G°_{298} = 2(-5.52 \text{ kJ}) - (19.36 \text{ kJ}) - (0 \text{ kJ}) = -30.4 \text{ kJ/mol}$$

At equilibrium $\Delta G°_{298} = -RT \ln K_p$ (since all are gases). Rearranging yields:

$$\ln K_p = \frac{\Delta G°_{298}}{-RT} = \frac{-30.4 \text{ kJ/mol}}{-(8.314 \times 10^{-3} \text{ kJ/mol·K})(298\text{K})} = +12.3$$

$$K_p = e^{+12.3} = \boxed{2.2 \times 10^5} \text{ at 298K}$$

Since equal numbers of moles of gases are produced and consumed ($\Delta n = 0$),

$$K_c = K_p(RT)^{-\Delta n} = K_p(RT)^0 = K_p = \boxed{2.2 \times 10^5} \text{ at 298 K}$$

Example 17-12 K versus $\Delta G°$

The equilibrium constant, K_c is 4.63×10^{-3} for the reaction below at 25°C and 1 atm pressure. Evaluate $\Delta G°_{298}$.

Plan

$$N_2O_4(g) \rightleftharpoons 2NO_2(g)$$

This is essentially the opposite of the previous example. We will use the steps in reverse order.

Solution

K_p can be calculated from K_c ($\Delta n = 2 - 1 = 1$).

$$?\ K_p = K_c(RT)^{\Delta n} = 4.63 \times 10^{-3} \left\{ \left[0.0821 \frac{\text{L·atm}}{\text{mol·K}}\right](298\text{K}) \right\}^1$$

$$K_p = 4.63 \times 10^{-3}[24.5] = 0.113 \text{ at 298 K}$$

$$?\ \text{kJ/mol} = \Delta G°_{298} = -RT \ln K_p = -\left(8.314 \times 10^{-3} \frac{\text{kJ}}{\text{mol·K}}\right)(298\text{K})\ln(0.113)$$

$$\Delta G°_{298} = (-8.314 \times 10^{-3})(298)(-2.180) = \boxed{5.40 \text{ kJ/mol}}$$

17-13 Evaluation of Equilibrium Constants at Different Temperatures

If the assumption (an inexact, but reasonable approximation) is made that $\Delta H°$ is temperature independent, the following relationship may be derived:

$$\ln\left[\frac{K_{T_2}}{K_{T_1}}\right] = \frac{\Delta H°}{R}\left[\frac{T_2 - T_1}{T_2 T_1}\right] \text{ or } \log\left[\frac{K_{T_2}}{K_{T_1}}\right] = \frac{\Delta H°}{2.303R}\left[\frac{T_2 - T_1}{T_2 T_1}\right]$$

If $\Delta H°$ and the equilibrium constant are known at any particular temperature, T_1, the value of the equilibrium constant at another temperature, T_2, can be estimated.

Example 17-13 Evaluating K_p at a Different Temperature

For the following reaction, $K_p = 1.39 \times 10^{-23}$ at 25°C. $\Delta H°$ for the reaction is +178 kJ/mol. Calculate K_p at 800°C.

$$CaCO_3(s) \rightleftharpoons CaO(s) + CO_2(g)$$

Plan
We will use the equation above to solve for the constant at the higher temperature.

Solution
$K_{T_1} = 1.39 \times 10^{-23}, T_1 = 273 + 25 = 298K, K_{T_2} = ?, T_2 = 273 + 800 = 1073K$

$$\ln\left[\frac{K_{T_2}}{1.39 \times 10^{-23}}\right] = \frac{+178 \text{ kJ/mol}}{8.314 \times 10^{-3} \frac{kJ}{mol \cdot K}}\left[\frac{1073K - 298K}{(1073K)(298K)}\right] = 51.9$$

$$\frac{K_{T_2}}{1.39 \times 10^{-23}} = e^{51.9} = 3.47 \times 10^{22}$$

$K_{T_2} = 1.39 \times 10^{-23}(3.47 \times 10^{22}) = \boxed{0.482}$

Notice that when the temperature was raised on this endothermic reaction, the value of the equilibrium constant increased considerably. The position of equilibrium shifted far to the right, just as had been predicted earlier in the LeChatelier's Principle qualitative discussion of the effect of temperature.

EXERCISES

Mass Action Expressions

1. Write the mass action expressions for K_c for the following equations.
   (a) $2CCl_4(g) + Br_2(\ell) \rightleftharpoons 2CCl_3Br(g) + Cl_2(g)$
   (b) $4NH_3(g) + 3O_2(g) \rightleftharpoons 6H_2O(g) + 2N_2(g)$
   (c) $CH_3NH_2(aq) + H_2O(\ell) \rightleftharpoons CH_3NH_3^+(aq) + OH^-(aq)$
   (d) $N_2O_5(g) \rightleftharpoons N_2O(g) + 2O_2(g)$
   (e) $[Co(CO)_4]_2(s) \rightleftharpoons 2Co(s) + 8CO(g)$
   (f) $2CHCl_3(g) + 3H_2(g) \rightleftharpoons 2CH_4(g) + 3Cl_2(g)$
   (g) $2HgO(s) \rightleftharpoons 2Hg(\ell) + O_2(g)$
   (h) $P_4(s) + 6H_2(g) \rightleftharpoons 4PH_3(g)$

2. Write the mass action expression for K_p for the reactions in Exercise 1.

Uses of the Equilibrium Constant

3. At some temperature, the following amounts were found in a closed 2.00 liter container at equilibrium: 0.160 mol H_2, 0.160 mol I_2 and 0.0332 mol HI.

   $$2HI(g) \rightleftharpoons H_2(g) + I_2(g)$$

   (a) Evaluate K_c at this temperature.
   (b) What is K_p for this reaction at the same temperature?
   (c) If 0.300 mol HI is added to this system at this temperature, what are the new equilibrium concentrations of all species?

4. Refer to Exercise 3. Evaluate K_c and K_p at the same temperature for the following reactions.
   (a) $HI(g) \rightleftharpoons 1/2H_2(g) + 1/2I_2(g)$
   (b) $H_2(g) + I_2(g) \rightleftharpoons 2HI(g)$
   (c) $1/2H_2(g) + 1/2I_2(g) \rightleftharpoons HI(g)$

5. For the following reaction $K_c = 3.52$ at 25°C.

$$I_2(s) + Cl_2(g) \rightleftharpoons 2ICl(g)$$

(a) Calculate the equilibrium concentrations of Cl_2 and ICl in a 2.00 liter flask with excess solid I_2 at 25°C, if the reaction is initiated with the following amounts.

   (i) 0.600 mol Cl_2 (ii) 0.500 mol ICl.   (iii) 0.600 mol Cl_2 and 0.500 mol ICl.

(b) What is K_p for this reaction at 25°C?

(c) What are the partial pressures of ICl and Cl_2 at equilibrium if the flask is filled with 2.00 atm pressure of Cl_2 in contact with the solid I_2.

(d) What are the equilibrium partial pressures if the reaction is initiated with 2.00 atm pressure of ICl?

6. At 530K, $K_c = 1.50 \times 10^{-2}$ for the reaction:

$$N_2O(g) + O_2(g) \rightleftharpoons NO_2(g) + NO(g)$$

What are the concentrations of all species at equilibrium in a 2.00 liter container if the reaction is initiated with:

(a) 0.600 mol N_2O and 0.600 mol O_2?

(b) 0.600 mol NO_2 and 0.600 mol NO?

(c) 0.800 mol N_2O and 0.400 mol O_2?

(d) 0.600 mol NO_2, 0.600 mol NO, 0.800 mol N_2O, and 0.800 mol O_2?

(e) 0.800 mol NO_2, 0.800 mol N_2O, 0.400 mol NO, and 0.400 mol O_2?

(f) What is K_p for this reaction at 530 K?

7. At 25°C, $K_c = 4.62 \times 10^{-3}$ for the decomposition of N_2O_4 to NO_2. A 1.50 gram sample of N_2O_4 is placed in a 3.00 liter container at 25°C and allowed to reach equilibrium.

$$N_2O_4(g) \rightleftharpoons 2NO_2(g)$$

(a) What masses of N_2, N_2O_4, and NO_2 are present at equilibrium?

(b) What is the percent dissociation of N_2O_4?

(c) What is K_p at 25°C?

(d) What are the partial pressures of N_2O_4 and NO_2 and what is the total pressure?

8. The reaction below is initiated by adding 3.00 atm $PH_3(g)$ and 10.50 atm Cl_2 to a closed container. At equilibrium, it is found that 1.40 atm PCl_3 have been formed. Compute the partial pressure of each gas at equilibrium and determine K_p for this process at this temperature. (8)

$$PH_3(g) + 3Cl_2(g) \rightleftharpoons PCl_3(g) + 3HCl(g)$$

Application of a Stress to a System at Equilibrium

9. The following system is at equilibrium. In which direction with the equilibrium position shift and how will the equilibrium amount of HCl be affected by the following changes?

$$SiH_4(g) + 4Cl_2(g) \rightleftharpoons SiCl_4(\ell) + 4HCl(g) + 1089 \text{ kJ/mol}$$

(a) lowering the temperature
(b) increasing the volume
(c) adding more SiH_4
(d) adding more $SiCl_4(\ell)$
(e) removing some Cl_2
(f) adding a catalyst
(g) adding an inert gas to increase the pressure with no volume change

10. The following system is at equilibrium. In which direction will the equilibrium shift, and how will the equilibrium amount of H_2 be affected by the change?

$$131 \text{kJ} + C(s,\text{graphite}) + H_2O(g) \rightleftharpoons CO(g) + H_2(g)$$

(a) adding more $C(s,\text{graphite})$
(b) adding more CO
(c) removing some $H_2O(g)$
(d) adding a catalyst
(e) lowering the temperature
(f) decreasing the container volume
(g) increasing the pressure by adding neon, no volume change

11. The following system is at equilibrium at 25°C. In which direction will the equilibrium shift and how will the equilibrium amount of Cl_2 be changed by the following stresses.

$$2PCl_5(g) + 6HCl(g) + 1362.5 \text{ kJ} \rightleftharpoons 2PH_3(g) + 8Cl_2(g)$$

(a) removing some PH_3
(b) adding some PCl_5
(c) removing some HCl
(d) increasing the volume
(e) adding He to change pressure, with no volume change
(f) raising the temperature
(g) adding catalyst

12. At a certain temperature at equilibrium, 0.380 mol H_2, 0.380 mol F_2 and 4.09 mol HF are found in a 10.00 L flask.

$$H_2(g) + F_2(g) \rightleftharpoons 2HF(g)$$

   (a) What is K_c for this reaction?
   (b) What are the new equilibrium concentrations of all species if 3.75 mol BrCl is added to the container?

13. Refer to the reaction and constants in Exercise 7.
   (a) What would be the equilibrium concentrations of N_2O_4 and NO_2 if we began with 1.50 g of N_2O_4 in a 5.00 liter, instead of a 3.00 liter container, at 25°C?
   (b) Determine the percent dissociation of N_2O_4.
   (c) What masses of N_2O_4 and NO_2 will be present?
   (d) What will be the total pressure of the system at equilibrium in the 5.00 liter container?

Relationship between Equilibrium Constants and $\Delta G°$

14. For the system below $K_p = 6.9 \times 10^{-3}$ at 40°C.

$$NH_4CO_2NH_2(s) \rightleftharpoons 2NH_3(g) + CO_2(g)$$

   (a) Evaluate K_c at 40°C.
   (b) What are the partial pressures of each gas and the total pressure at equilibrium?
   (c) What are the concentrations of each gas at equilibrium?
   (d) If the equilibrium partial pressure of CO_2 is increased to 2.40 atm, what is the equilibrium partial pressure of NH_3?
   (e) If the equilibrium partial pressure of NH_3 is increased to 6.45 atm, what is the equilibrium partial pressure of CO_2?
   (f) What is the value of $\Delta G°$ at 40°C?

15. Using data in Appendix K, calculate $\Delta G°$ for the following reactions at 25°C and then compute K_p for each at 25 °C.
   (a) $2S(s, \text{rhombic}) + Cl_2(g) \rightleftharpoons S_2Cl_2(g)$
   (b) $C_2H_2(g) + 2H_2(g) \rightleftharpoons C_2H_6(g)$
   (c) $CuO(s) + H_2(g) \rightleftharpoons Cu(s) + H_2O(g)$

(d) $2CHCl_3(g) + Cl_2(g) \rightleftharpoons 2CCl_4(g) + H_2(g)$

(e) $2BrF_3(g) \rightleftharpoons Br_2(g) + 3F_2(g)$

16. Find K_c at 25°C for the reactions in Exercise 15.

17. Given the following reactions at 1000K and their equilibrium constants, calculate $\Delta G°$ at 1000K for each.

(a) $FeO(s) \rightleftharpoons Fe(s) + CO(g)$ $K_p = 0.403$

(b) $CO_2(g) + CF_4(g) \rightleftharpoons 2COF_2(g)$ $K_p = 0.472$

Temperature Dependence of Equilibrium Constants

18. For the following reaction, $K_p = 6.25 \times 10^3$ at 1000K. Estimate K_p at 440°C.

$$2NOBr(g) + 46.4 \text{ kJ} \rightleftharpoons 2NO(g) + Br_2(g)$$

19. Use the data from Appendix K to calculate $\Delta H°$ at 25°C for all the reactions in Exercise 15. Then, for each, estimate the K_p at 500°C.

Miscellaneous Exercises

20. The reaction below is initiated with 2.00 atm $BrF_3(g)$ and 6.00 atm $H_2(g)$ at some temperature. When the process has come to equilibrium, the partial pressure of the HBr is found to be 1.50 atm. What is K at this temperature?

$$BrF_3(g) + 2H_2(g) \rightleftharpoons HBr(g) + 3HF(g)?$$

21. Use $\Delta G°_f$ values at 25°C to determine the heat of vaporization of $TiCl_4(\ell)$ at this temperature.

(a) What is K_p at this temperature?

(b) What is the vapor pressure, in torr, of $TiCl_4$ at this temperature?

(c) What is K_c at this temperature?

22. Ozone, O_3, can be formed by lightning strikes or other electrical discharges.

$$3/2 O_2(g) \rightleftharpoons O_3(g)$$

(a) Use Appendix K to calculate $\Delta G°$ for this process at 25°C..

(b) Calculate K_p for this reaction at 25°C.

(c) Calculate K_c for this reaction at 25°C.

(d) Use Appendix K to determine $\Delta H°$ for this reaction.

(e) Estimate K_p of this reaction at 750°C.

23. The following system is at equilibrium at 25°C. In which direction will the equilibrium shift and how will the equilibrium amount of CO be affected by the following changes.

$$Fe_2O_3(s) + 3CO(g) \rightleftharpoons 2Fe(s) + 3CO_2(g) + 21.5 \text{ kJ}$$

(a) cooling
(b) adding Fe_2O_3
(c) removing some CO_2
(d) adding more solid Fe
(e) decreasing the volume
(f) decreasing the pressure, with volume change
(g) increasing the pressure by adding argon, no volume change

24. For the reaction below, $K_p = 6.70 \times 10^{-3}$ at 380°C.

$$COCl_2(g) \rightleftharpoons CO(g) + Cl_2(g)$$

(a) A sample of $COCl_2$ is placed in an enclosed 15.0 liter container at 380°C and it exerts a pressure of 4.65 atm before decomposition begins. What will be the partial pressures of all species when equilibrium is established and what will the total pressure be?

(b) If the volume is increased to 45.0 liters with no change in temperature, what will be the new equilibrium partial pressures and total pressure?

(c) Refer to (b). What will be the equilibrium concentrations of the three gases in the 45.0 liter container?

(d) If the volume of the original system is decreased to 3.00 liters, what will be the new equilibrium partial pressures of all species and the total pressure?

(e) Refer to (d). What will be the equilibrium concentrations of the three gases?

(f) Refer to (e). What mass in grams of each of the species will be present?

25. $P_4(s) + 5O_2(g) \rightleftharpoons P_4O_{10}(s)$. For this reaction at 3070K, $K_p = 4.10$. Solid P_4O_{10} is added to a container at this temperature and the system is allowed to come to equilibrium. What is the partial pressure of O_2 at equilibrium?

26. At 25°C, $K_p = 4.46 \times 10^{-30}$ for the process below. Use this information to determine $\Delta G°_f$ for N_2O_3 at this temperature.

$$2N_2(g) + 3 O_2(g) \rightleftharpoons 2N_2O_3(g)$$

ANSWERS TO EXERCISES

1. (a) $K_c = \dfrac{[CCl_3Br]^2[Cl_2]}{[CCl_4]^2}$

   (b) $K_c = \dfrac{[H_2O]^6[N_2]^2}{[NH_3]^4[O_2]^3}$

   (c) $K_c = \dfrac{[CH_3NH_3^+][OH^-]}{[CH_3NH_2]}$

   (d) $K_c = \dfrac{[N_2O][O_2]^2}{[N_2O_5]}$

   (e) $K_c = [CO]^8$

   (f) $K_c = \dfrac{[CH_4]^2[Cl_2]^3}{[CHCl_3]^2[H_2]^3}$

   (g) $K_c = [O_2]$

   (h) $K_c = \dfrac{[PH_3]^4}{[H_2]^6}$

2. (a) $K_p = \dfrac{P_{CCl_3Br}^2 P_{Cl}}{P_{CCl_4}^2}$

   (b) $K_p = \dfrac{P_{H_2O}^6 P_{N_2}^2}{P_{NH_3}^4 P_{O_2}^3}$

   (c) K_p is undefined

   (d) $K_p = \dfrac{P_{N_2O} P_{O_2}^2}{P_{N_2O_5}}$

   (e) $K_p = P_{CO}^8$

   (f) $K_p = \dfrac{P_{CH_4}^2 P_{Cl_2}^3}{P_{CHCl_3}^2 P_{H_2}^3}$

   (g) $K_p = P_{O_2}$

   (h) $K_p = \dfrac{P_{PH_3}^4}{P_{H_2}^6}$

3. (a) $K_c = 23.2$ (b) $K_p = 23.2$ (c) $[H_2] = [I_2] = 0.148\ M$, $[HI] = 0.031\ M$

4. (a) $K_c = 4.82$; $K_p = 4.82$ (b) $K_c = 0.0429$; $K_p = 0.0429$ (c) $K_c = 2.19$; $K_p = 2.19$

5. (a) (i) $[Cl_2] = 0.064\ M$, $[ICl] = 0.472\ M$
       (ii) $[Cl_2] = 0.014\ M$, $[ICl] = 0.222\ M$
       (iii) $[Cl2] = 0.112\ M$, $[ICl] = 0.626\ M$
   (b) $K_p = 86.1$ (c) $P_{Cl_2} = 0.16$ atm, $P_{ICl} = 3.68$ atm
   (d) $P_{Cl_2} = 0.426$ atm, $P_{ICl} = 1.91$ atm

6. (a) $[NO_2] = [NO] = 0.0326\ M$, $[N_2O] = [O_2] = 0.0267M$
   (b) $[NO_2] = [NO] = 0.033\ M$, $[N_2O] = [O_2] = 0.0267M$
   (c) $[N_2O] = 0.369\ M$, $[O_2] = 0.169\ M$, $[NO_2] = [NO] = 0.031\ M$
   (d) $[N_2O] = [O_2] = 0.624\ M$, $[NO_2] = [NO] = 0.076\ M$
   (e) $[N_2O] = 0.585\ M$, $[O_2] = 0.385\ M$, $[NO_2] = 0.215\ M$,

[NO] = 0.015 M
(f) $K_p = 1.50 \times 10^{-2}$

7. (a) $NO_2 = 0.550$ g, $N_2O_4 = 0.950$ g (b) 36.6% (c) $K_p = 0.113$
   (d) $P_{NO_2} = 0.0974$ atm, $P_{N_2O_4} = 0.0842$ atm, $P_{total} = 0.1816$ atm

8. $P_{PH_3} = 1.60$ atm; $P_{Cl_2} = 6.30$ atm; $P_{PCl_3} = 1.40$ atm; $P_{HCl} = 4.20$ atm; $K_p = 0.259$

9. (a) right, increases   (b) left, decreases   (c) right, increases
   (d) no change   (e) left, decreases   (f) none, no change
   (g) none, no change

10. (a) none, no change   (b) left, decreases   (c) left, decreases
    (d) none, no change   (e) left, decreases   (f) left, decreases
    (g) none, no change

11. (a) right, increases   (b) right, increases   (c) left, decreases
    (d) right, increases   (e) none, no change   (f) right, increases
    (g) left, decreases   (h) none, no change

12. (a) $K_c = 0.116$ (b) $[H_2] = [F_2] = 0.672$ M; $[HF] = 7.26$ M

13. (a) 2.90×10^{-3} M NO_2, 1.81×10^{-3} M N_2O_4
    (b) 44.5% (c) 0.667 g NO_2, 0.833 g N_2O_4 (d) 0.115 atm

14. (a) $K_c = 4.1 \times 10^{-7}$
    (b) $P_{CO_2} = 0.12$ atm; $P_{NH_3} = 0.24$ atm; $P_{total} = 0.36$ atm
    (c) $[CO_2] = 4.7 \times 10^{-3}$ M, $[NH_3] = 9.3 \times 10^{-3}$ M
    (d) $P_{NH_3} = 5.4 \times 10^{-2}$ atm   (e) $P_{CO_2} = 1.7 \times 10^{-4}$ atm   (f) $\Delta G°_{rxn} = +12.3$ kJ/mol rxn

15. (a) $\Delta G° = -31.8$ kJ/mol, $K_p = 3.75 \times 10^5$
    (b) $\Delta G° = -242.1$ kJ/mol, $K_p = 2.49 \times 10^{42}$
    (c) $\Delta G° = -99$ kJ/mol, $K_p = 2.17 \times 10^{17}$
    (d) $\Delta G° = +10.20$ kJ/mol, $K_p = 1.63 \times 10^{-2}$
    (e) $\Delta G° = +462.1$ kJ/mol, $K_p = 9.96 \times 10^{-82}$

16. (a) $K_c = 3.75 \times 10^5$ (b) $K_c = 1.49 \times 10^{45}$
    (c) $K_c = 2.17 \times 10^{17}$ (d) $K_c = 9.76$
    (e) $K_c = 1.66 \times 10^{-84}$

17. (a) $\Delta G° = +9.62$ kJ/mol (b) $\Delta G° = +6.24$ kJ/mol   18. $K_p = 661$

19. (a) $\Delta H° = -18$ kJ/mol, $K_p = 4.32 \times 10^3$
    (b) $\Delta H° = -141.8$ kJ/mol, $K_p = 1.32 \times 10^{27}$
    (c) $\Delta H° = -84.8$ kJ/mol, $K_p = 1.60 \times 10^8$
    (d) $\Delta H° = -64.6$ kJ/mol, $K_p = 1.79 \times 10^{-9}$
    (e) $\Delta H° = 542.1$ kJ/mol, $K_p = 2.45 \times 10^{-23}$

20. $K_c = 3.0 \times 10^1$

21. (a) $K_p = 1.51 \times 10^{-2}$ (b) VP $= 1.51 \times 10^{-2}$ atm $= 11.5$ torr (c) $K_c = 6.17 \times 10^{-4}$

22. (a) $\Delta G° = 163$ kJ/mol (b) $K_p = 2.68 \times 10^{-29}$ (c) $K_c = 1.32 \times 10^{-28}$
    (d) $\Delta H° = 143$ kJ/mol (e) $K_r = 1.56 \times 10^{-11}$

23. (a) right, decreases (b) none, no change (c) right, decreases
    (d) none, no change   (e) none, no change (f) none, no change
    (g) none, no change

24. (a) $P_{COCl_2} = 4.48$ atm, $P_{CO} = P_{Cl_2} = 0.173$ atm, $P_{total} = 4.82$ atm
    (b) $P_{COCl_2} = 1.45$ atm, $P_{CO} = P_{Cl_2} = 0.0990$ atm, $P_{total} = 1.65$ atm
    (c) $[COCl_2] = 0.0270\,M$, $[CO] = [Cl_2] = 0.0018\,M$
    (d) $P_{COCl_2} = 22.8$ atm, $P_{CO} = P_{Cl_2} = 0.39$ atm, $P_{total} = 23.6$ atm
    (e) $[COCl_2] = 0.425\,M$, $[CO] = [Cl_2] = 0.0073\,M$
    (f) 126 g $COCl_2$, 0.61 g CO and 1.6 g Cl_2

25. 0.754 atm   26. +83.7 kJ mol

Chapter 18

Ionic Equilibria I: Acids and Bases

18-1 Strong Electrolytes

Strong electrolytes dissociate or ionize essentially completely in aqueous solution. (Refer to Chapters Four and Ten.) Review the list of strong acids and bases that is in Table 4-1 of this book. The concentrations of ions in solutions of strong electrolytes can be found by writing the dissociation or ionization reaction and applying normal stoichiometric considerations.

Example 18-1 Calculation of Concentrations of Ions

What are the concentrations of $Sr(OH)_2$, Sr^{2+}, and OH^- in a 0.074 M solution of $Sr(OH)_2$?

Plan
We first write out the equation for the dissociation of the strong soluble base $Sr(OH)_2$. Then we construct a reaction summary.

Solution

	$Sr(OH)_2$	$\rightarrow$	Sr^{2+}	+	$2OH^-$
initial:	0.074 M		0 M		0 M
change:	$-0.074\ M$		$+0.074\ M$		$+2(0.074)\ M$
final:	0 M		0.074 M		0.15 M

18-2 The Autoionization of Water

Pure water ionizes slightly according to the equation:

$$H_2O(\ell) + H_2O(\ell) \rightleftharpoons H_3O^+(aq) + OH^-(aq)$$

In pure water at 25°C, $[H_3O^+] = [OH^-] = 1.0 \times 10^{-7} M$. In all dilute aqueous solutions, the product of the concentrations of these two ions is a constant, called K_w, the ion product of water. At 25°C, the value obtained and which will be used throughout this book is:

$$K_w = [H_3O^+][OH^-] = 1.0 \times 10^{-14}$$

Example 18-2 Calculation of Concentrations of Ions

What is the $[H_3O^+]$ in the 0.074 M $Sr(OH)_2$ solution of Example 18-1?

Plan

We know that $[OH^-]$ is 0.14 M. We can find $[H_3O^+]$ by applying the relationship for K_w.

Solution

$$[H_3O^+][OH^-] = 1.0 \times 10^{-14} = K_w$$

$$[H_3O^+] = \frac{1.0 \times 10^{-14}}{[OH^-]} = \frac{1.0 \times 10^{-14}}{0.14\,M} = \boxed{7.1 \times 10^{-14}\,M}$$

Note that in this solution of a strong base, $[OH^-] > 10^{-7} M$ and also $[OH^-] > [H_3O^+]$. The opposite is true for solutions of acids, i.e., in acids, $[H_3O^+] > 10^{-7}$ and $[H_3O^+] > [OH^-]$. The $[OH^-]$ contribution due to the autoionization of water is equal to the $[H_3O^+]$, in this case 7.1 x $10^{-14} M$, and is negligible compared to the $[OH^-]$ produced by the strong base. The assumptions that the $[H_3O^+]$ concentration due to water is insignificant compared to that of the acid in acidic solutions and that the $[OH^-]$ contribution from water is insignificant compared to that of the base in basic solutions will be used throughout this discussion.

18-3 The pH and pOH Scales

The **pH scale** provides a convenient way of expressing the acidity or basicity of a solution. The **p** means **negative logarithm**; therefore,:

$$pH = -\log[H_3O^+] \text{ and } pOH = -\log[OH^-]$$

Similarly, $pK_a = -\log K_a$, $pCl = -\log[Cl^-]$, etc. When we take the logarithm of a number, only figures to the right of the decimal are significant. Numbers to the left of the decimal are related to the exponential term. Therefore, if $[H_3O^+] = 2.88 \times 10^{-5}$ M, pH $= -\log[2.88 \times 10^{-5}] = 4.541$.

If any of the four variables, $[OH^-]$, $[H_3O^+]$, pH and pOH are known, the other three can be calculated using the previous K_w relationship and the relationship that:

$$pH + pOH = 14.00 \text{ (at } 25°C\text{)}$$

In a neutral solution at 25°C, pH = pOH = 7.00. An acid solution has pH < 7.00 and a basic solution has pH > 7.00.

Example 18-3 Calculations Involving pH and pOH

Calculate the $[H_3O^+]$, the $[OH^-]$, the pH, and the pOH of a 0.20 M solution of hydroiodic acid, HI.

Plan
We know that hydroiodic acid is a strong acid and ionizes essentially completely. We set up a reaction summary and then apply the relationships above.

Solution

	HI	+ H$_2$O	→	H$_3$O$^+$	+	I$^-$
initial:	0.20 M			≈0 M		0 M
change:	-0.20 M			+0.20 M		+0.20 M
final:	0 M			0.20 M		0.20 M

The $[H_3O^+] = \boxed{0.20 \ M}$.

The pH is $-\log[H_3O^+] = -\log(0.20) = -(-0.70) = \boxed{0.70}$.

Finding the other two values can be approached two ways. K_w can be used first to determine $[OH^-]$ as follows:

$$K_w = [H_3O^+][OH^-] = 1.0 \times 10^{-14}$$

$$[OH^-] = \frac{1.0 \times 10^{-14}}{[H_3O^+]} = \frac{1.0 \times 10^{-14}}{0.20\,M} = \boxed{5.0 \times 10^{-14}\,M}$$

$$pOH = -\log[OH^-] = -\log(5.0 \times 10^{-14}) = -(-13.30) = 13.30$$

Alternatively, the pOH can be found first from pH + pOH = 14.00:

$$pOH = 14.00 - pH = 14.00 - 0.70 = \boxed{13.30}$$

$$[OH^-] = 10^{-pOH} = 10^{-13.30} = \boxed{5.0 \times 10^{-14}\,M}$$

18-4 Ionization Constants for Weak Monoprotic Acids and Bases

The distinction between strong and weak acids and bases is that the latter ionize only slightly in aqueous solution. The ionization of any weak monoprotic acid, HA, in dilute solutions can be represented as:

$$HA + H_2O \rightleftharpoons H_3O^+ + A^-$$

The ionization constant expression is: $K_a = \dfrac{[H_3O^+][A^-]}{[HA]}$ Appendix F in the text lists the ionization constants for many common acids. A higher K_a indicates a stronger acid. Table 18-1 includes a few examples of these. Because some chemists prefer to use pK_a ($-\log K_a$) values, these are also tabulated. Note that a stronger acid has a lower pKa.

Table 18-1 Ionization Constants and pK_a Values for Some Acids

Acid	Formula	K_a at 25°C	pK_a
acetic	CH_3COOH	1.8×10^{-5}	4.74
carbonic	H_2CO_3	4.2×10^{-7}	6.38
bicarbonate	HCO_3^-	4.8×10^{-11}	10.32
hypochlorous	$HClO$	3.5×10^{-8}	7.46
hydrocyanic	HCN	4.0×10^{-10}	9.40

Among these acids, acid strength increases as: $CH_3COOH > H_2CO_3 > HClO > HCN > HCO_3^-$.
Remember, from Brønsted-Lowry acid base theory that a stronger acid has a weaker conjugate base. Conjugate base strength increases as: $CO_3^{2-} > CN^- > ClO^- > HCO_3^- > CH_3COO^-$.

If we know the K_a and initial concentration of a weak acid in dilute aqueous solution, the equilibrium concentration of all species, the pH and pOH, and the percent ionization can be found. In working the equilibrium problems, a useful approximation that may sometimes be used is: any term containing x added to or subtracted from a number greater than 0.05 may be simplified by deleting the x if the exponent of 10 is -5 or less (-6, -7 etc.). This simplification is often valid and shortens problem solving time compared to using the quadratic formula.

Example 18-4 Calculation of Concentrations and pH from K_a

The K_a for HN_3 is 1.9×10^{-5}. What are the equilibrium concentrations of HN_3 and H_3O^+, the pH, and the percent ionization of a 0.20 M solution of HN_3?

Plan
We write the equation for the ionization of the weak acid and then prepare a reaction summary, with x the molarity of HN_3 that ionizes. We substitute into the K_a expression, checking to see if we can use the approximation, and solve for x. We can finally use the definition of pH and the formula for percent ionization to finish the problem.

Solution

	HN_3	+ H_2O	$\rightleftharpoons$	H_3O^+	+	N_3^-
initial:	0.20 M			≈0 M		0 M
change:	-x M			+x M		+x M
final:	(0.20-x) M			x M		x M

$$K_a = \frac{[H_3O^+][N_3^-]}{[HN_3]} = \frac{(x)(x)}{0.20-x} \approx \frac{x^2}{0.20} = 1.9 \times 10^{-5}$$

$$x^2 = 0.20(1.9 \times 10^{-5}) = 3.8 \times 10^{-6}$$

$$x = [H_3O^+] = \boxed{1.9 \times 10^{-3} \, M}$$

$$[HN_3] = 0.20 \, M - 1.9 \times 10^{-3} \, M = \boxed{0.20 \, M}$$

Since the value of $[HN_3]$ is unchanged with respect to appropriate significant figures, the assumption above is valid. Whenever the assumption that x is negligible is used, final answers should be checked in the manner above. If the x causes a change of more than 5% in the concentration of the acid, the problem should be reworked with the added or subtracted x present, using the quadratic formula.

The pH of the 0.20 solution of HN_3 is:

$$pH = -\log [H_3O^+] = -\log (1.9 \times 10^{-3}) = -(-2.72) = \boxed{2.72}$$

Finally, the percent ionization of an electrolyte is the amount that ionizes divided by the concentration initially present, multiplied by 100. In this case, x is the amount which ionizes and is equal to the $[H_3O^+]$:

$$\% \text{ ionization} = \frac{[HN_3]_{ionized}}{[HN_3]_{initial}} \times 100 = \frac{1.9 \times 10^{-3} M}{0.20 M} \times 100 = \boxed{0.95\%}$$

For a given weak acid at a given temperature, the percent ionization decreases as the initial concentration increases; however, concentrations of ionic species increase. Table 18-2 compares the percent ionization, $[H_3O^+]$, pH, and K_a values for 0.20 M solutions of three acids of different strengths. The calculations for the first two acids are shown in Examples 18-3 and 18-4. The third can be calculated in similar manner to Example 18-4.

Table 18-2 Comparison of Acid Strengths

Acid Solution	K_a	$[H_3O^+]$	pH	%ionization
0.20 M HI	no meaning	0.20 M	0.70	100%
0.20 M HN_3	1.9×10^{-5}	1.9×10^{-3}	2.72	0.95%
0.20 M HBrO	2.5×10^{-9}	2.2×10^{-5}	4.66	0.011%

If the molarity and pH of a monoprotic acid are known, the K_a can be determined as in the following example.

Example 18-5 Calculation of K_a from pH

A 0.30 M solution of a weak monoprotic acid, HA, was found to have a pH of 4.68. What is the ionization constant, K_a, of the acid?

Plan

We write the generalized equation for the ionization of the acid and the equilibrium constant expression. We can use the pH to find the $[H_3O^+]$ that is produced in the ionization. We then find the equilibrium concentrations using a reaction summary and substitute values into the K_a expression.

Solution

Because the pH is 4.68: $[H_3O^+] = 10^{-pH} = 10^{-4.68} = 2.1 \times 10^{-5} M$.

This is the amount that the acid must dissociate. It must also be equal to the $[A^-]$. The system can then be analyzed as in previous equilibrium cases.

	HA		+ H_2O ⇌	H_3O^+		+ A^-	
initial:	0.30	M		≈0	M	0	M
change:	-2.1×10^{-5}	M		$+2.1 \times 10^{-5}$	M	$+2.1 \times 10^{-5}$	M
final:	0.30	M		2.1×10^{-5}	M	2.1×10^{-5}	M

Now the K_a can be determined:

$$K_a = \frac{[H_3O^+][A^-]}{[HA]} = \frac{(2.1 \times 10^{-5})(2.1 \times 10^{-5})}{0.30} = \boxed{1.5 \times 10^{-9}}$$

The only common weak bases are aqueous ammonia, NH_3, and its derivatives, the amines. The basicity of this class of compounds arises from the presence of the lone pair of electrons of the nitrogen atom. They make these compounds both Brønsted-Lowry bases (proton acceptors) and Lewis bases (electron pair donors). The ionization of an amine is similar to that of ammonia.

$$NH_3 + H_2O \rightleftharpoons NH_4^+ + OH^- \quad \text{(ammonia)}$$

$$NH(C_2H_5)_2 + H_2O \rightleftharpoons NH(C_2H_5)_2H^+ + OH^- \quad \text{(diethylamine)}$$

Such solutions are basic since they contain an excess of OH^- ions over H_3O^+ ions. The ionization reaction and the base ionization constant of a weak base may be represented as:

$$B + H_2O \rightleftharpoons BH^+ + OH^- \qquad K_b = \frac{[BH^+][OH^-]}{[B]}$$

Table 18-3 lists the basic ionization constants for several weak bases and the pK_b ($-\log K_b$) values. Appendix G in the text contains a more extensive list. Equilibrium calculations are exactly analogous to those of the weak acids.

Table 18-3 Ionization Constants of Some Bases

Name	Formula	K_b at 25°C	pK_b
ammonia	NH_3	1.8×10^{-5}	4.74
methylamine	NH_2CH_3	5.0×10^{-4}	3.30
dimethylamine	$NH(CH_3)_2$	7.4×10^{-4}	3.13
trimethylamine	$N(CH_3)_3$	7.4×10^{-5}	4.13
aniline	$C_6H_5NH_2$	4.2×10^{-10}	9.38

Example 18-6 pH of a Weak Base Solution

Calculate the pH and percent ionization of 0.20 M aniline solution.

Plan

The ionization constant, K_b, is in Table 18-3 above. We proceed as in Example 18-4. In this case, x will be the $[OH^-]$ concentration. We can find the pH by finding the $[H_3O^+]$ from the K_w and $[OH^-]$ or by finding the pOH and using the pH + pOH = 14.00 relationship.

Solution

	$C_6H_5NH_2$	$+ H_2O$	$\rightleftharpoons$	$C_6H_5NH_3^+$	$+$	OH^-
initial:	0.20 M			≈0 M		0 M
change:	-x M			+x M		+x M
final:	(0.20-x) M			x M		x M

$$K_b = \frac{[C_6H_5NH_3^+][OH^-]}{[C_6H_5NH_2]} = \frac{(x)(x)}{0.20-x} \approx \frac{x^2}{0.20} = 4.2 \times 10^{-10}$$

$$x^2 = 0.20(4.2 \times 10^{-10}) = 8.4 \times 10^{-11} \text{ so } x = [OH^-] = 9.2 \times 10^{-6} M$$

$[C_6H_5NH_2] = 0.20\ M - 9.2 \times 10^{-6} M = 0.20\ M$ so the assumption that the subtracted x is negligible is valid. Now the pH can be found in the following manner.

$$pOH = -\log [OH^-] = -\log (9.2 \times 10^{-6}) = 5.04$$

$$pH = 14.00 - pOH = 14.00 - 5.04 = \boxed{8.96}$$

Finally, the percent ionization can be determined. The amount that ionizes is equal to x, the $[OH^-]$ or $[C_6H_5NH_3^+]$.

$$\%\ \text{ionization} = \frac{[C_6H_5NH_2]_{\text{ionized}}}{[C_6H_5NH_2]_{\text{initial}}} \times 100 = \frac{9.2 \times 10^{-6}\ M}{0.20\ M} \times 100 = \boxed{0.0046\%}$$

18-5 Acid-Base Indicators

Many common acid-base indicators are weak, complex organic acids, which can be represented by the general formula HIn. The important characteristic of an acid-base indicator is that the unionized acid molecule is one color and the anion of the acid is another color.

$$\underset{\text{Color 1}}{HIn} + H_2O \rightleftharpoons \underset{\text{Color 2}}{H_3O^+ + In^-}$$

$$K_{HIn} = \frac{[H_3O^+][In^-]}{[HIn]} \quad \text{or rearranging,} \quad \frac{K_{HIn}}{[H_3O^+]} = \frac{[In^-]}{[HIn]}$$

The relative amounts of [HIn] and [In$^-$] determine the color of the solution, with the one with a concentration at least a factor of ten higher than that of the other predominating. Usually, only a few drops of indicator are used; therefore, the [H$_3$O$^+$] is determined by the constituents of the bulk of the solution. As an acid-base reaction is performed, the ratio $K_{HIn}/[H_3O^+]$ changes. This concurrently causes a shift in the predominance of one color to the other color. The $K_{HIn}/[H_3O^+]$ change can occur quite sharply and then the color change will be distinct as well.

18-6 The Common Ion Effect and Buffer Solutions

The **common ion effect** is observed in solutions containing two or more compounds, both of which produce the same ion. Buffer solutions are an example of the common ion effect. **Buffer solutions** resist changes in pH caused by addition of either acid or base. The two major types of buffer solutions are:

1. a weak acid plus a soluble, ionic salt of the acid.
2. a weak base plus a soluble, ionic salt of the base.

For such acid/salt or base/salt solutions containing reasonable concentrations of each member of the conjugate pair, the presence of the relatively high concentration of anion (conjugate base) of the acid or cation (conjugate acid) of the base from ionization of the salt suppresses ionization of the weak acid or base. A solution of a weak acid and its salt is less acidic than a solution of the same concentration of the acid alone. Similarly, a solution of a weak base and its salt is less basic than a solution of the identical concentration of the weak base alone.

Example 18-7 Weak Acid/Salt of Weak Acid Buffers

Calculate the $[H_3O^+]$, the pH, and the percent ionization of the acid in a solution that is 0.20 M in HN_3 and 0.20 M in NaN_3.

Plan

We write the appropriate equations for both the HN_3 and the NaN_3 and the ionization constant expression for HN_3. We represent the equilibrium concentrations algebraically and substitute into the K_a expression.

Solution

$$NaN_3 \rightarrow Na^+ + N_3^-$$
$$0\,M \qquad 0.20\,M \quad 0.20\,M \qquad \text{(strong, soluble salt)}$$
$$\text{(100\% dissociated)}$$

	HN_3	$+ H_2O$	$\rightleftharpoons$	H_3O^+	$+$	N_3^-
initial:	0.20 M			$\approx 0\,M$		0.20 M
change:	$-x$ M			$+x\,M$		$+x\,M$
final:	$(0.20-x)\,M$			$x\,M$		$0.20+x\,M$

$$K_a = \frac{[H_3O^+][N_3^-]}{[HN_3]} = \frac{(x)(0.20+x)}{0.20-x} \approx \frac{(x)(0.20)}{0.20} = 1.9 \times 10^{-5}$$

$$x = [H_3O^+] = \boxed{1.9 \times 10^{-5}\,M} \quad [HN_3] = 0.20\,M - 1.9 \times 10^{-5}\,M = \boxed{0.20\,M}$$

Since the values of $[HN_3]$ and $[N_3^-]$ are unchanged with respect to appropriate significant figures, the assumption above that added and subtracted x values are negligible is valid. The pH of the 0.20M solution of HN_3 is:

$$pH = -\log[H_3O^+] = -\log(1.9 \times 10^{-5}) = -(-4.72) = \boxed{4.72}$$

Finally, the percent ionization can be found.

$$\% \text{ ionization} = \frac{[HN_3]_{\text{ionized}}}{[HN_3]_{\text{initial}}} \times 100 = \frac{1.9 \times 10^{-5} M}{0.20 M} \times 100 = \boxed{0.0095\%}$$

Table 18-5 compares the results found in Example 18-7 with those of Example 18-4 to illustrate the effect of the suppression of the ionization of the acid by the common ion, N_3^-.

Table 18-5 Comparison of $[H_3O^+]$, pH and Percent Ionization

Solution	$[H_3O^+]$	pH	% ionization
0.20 M HN$_3$	$1.9 \times 10^{-3} M$	2.72	0.95%
0.20 M HN$_3$ and 0.20 M NaN$_3$	1.9×10^{-5}	4.72	0.0095%

Example 18-8 Weak Acid/Salt of Weak Acid Buffers

A buffer is made by mixing 250.0 mL of 1.00 M HNO$_2$ with 500.0 mL 1.00 M Ba(NO$_2$)$_2$. Calculate the concentration of all species and the pH of the solution. For HNO$_2$, $K_a = 4.5 \times 10^{-4}$.

Plan
This buffer contains a weak acid and its salt. We can find the concentrations just after mixing, before reaction, using the dilution relationship ($M_1V_1 = M_2V_2$), with V_2 equal to the total volume of 750.0 mL. Then we can proceed as in Example 18-7.

Solution
After dilution but before reaction:

$$? M_{HNO_2} = \frac{(250.0 \text{ mL})(1.00 M)}{750.0 \text{ mL}} = 0.333 \ M \text{ HNO}_2$$

$$? M_{Ba(NO_2)_2} = \frac{(500.0 \text{ mL})(1.00 M)}{750.0 \text{ mL}} = 0.667 M \text{ Ba(NO}_2)_2$$

The reactions that occur and the equilibrium concentrations are:

$$\begin{array}{llll}
\text{Ba(NO}_2)_2 & \rightarrow \text{Ba}^{2+} & + \; 2\text{NO}_2^- & \\
0M & 0.667 \; M & 1.33 \; M & \text{(strong, soluble salt)} \\
& & & \text{(100\% dissociated)}
\end{array}$$

	HNO_2 + H_2O	$\rightleftharpoons$	H_3O^+	+	NO_2^-
initial:	0.333 M		≈ 0 M		1.33 M
change:	-x M		+x M		+x M
final:	(0.333 - x) M		x M		1.33 + x M

$$K_a = \frac{[H_3O^+][NO_2^-]}{[HNO_2]} = \frac{(x)(1.33+x)}{0.33-x} \approx \frac{(x)(1.33)}{0.333} = 4.5 \times 10^{-4}$$

$$x = \frac{(0.333)(4.5 \times 10^{-4})}{1.33} = [H_3O^+] = \boxed{1.1 \times 10^{-4} M}$$

$$pH = -\log[H_3O^+] = -\log(1.1 \times 10^{-4}) = -(-3.96) = \boxed{3.96}$$

$$[HNO_2] = 0.333 \; M - 1.1 \times 10^{-4} M = \boxed{0.333 \; M}$$

$$[OH^-] = \frac{K_w}{[H_3O^+]} = \frac{1.00 \times 10^{-14}}{1.1 \times 10^{-4}} = \boxed{9.1 \times 10^{-11} M}$$

$$[Ba^{2+}] = \boxed{0.667 \; M}$$

$$[NO_2^-] = 1.33 + x = 1.33 + 1.1 \times 10^{-4} = \boxed{1.33 \; M}$$

Example 18-9 Weak Base/Salt of Weak Base Buffers

What is the pH of a buffer solution that is 0.20 M NH_3 and 0.20 M NH_4Br? For NH_3, $K_b = 1.8 \times 10^{-5}$.

Plan

This is a buffer solution composed of a base and its conjugate acid. We write the appropriate equations, use the K_b, and algebraically solve. In this case, we find the OH^- directly and must convert from pOH to pH.

Solution

$$NH_4Br \rightarrow NH_4^+ + Br^- \quad \text{(strong, soluble salt)}$$
$$0M \qquad 0.20\ M \quad 0.20M \quad \text{(100\% dissociated)}$$

	NH_3	+ H_2O	$\rightleftharpoons$	NH_4^+	+	OH^-
initial:	0.20 M			0.20 M		$\approx 0\ M$
change:	$-x$ M			$+x$ M		$+x\ M$
final:	$(0.20 - x)\ M$			$0.20 + x\ M$		$x\ M$

$$K_b = \frac{[NH_4^+][OH^-]}{[NH_3]} = \frac{(0.20+x)(x)}{0.20-x} \approx \frac{(0.20)(x)}{0.20} = 1.8 \times 10^{-5}$$

$$x = [OH^-] = 1.8 \times 10^{-5}\ M$$

$$pOH = -\log[OH^-] = -\log(1.8 \times 10^{-5}) = -(-4.74) = 4.74$$

$$pH = 14.00 - pOH = 14.00 - 4.74 = \boxed{9.26}$$

18-7 Buffering Action

Buffer solutions resist changes in pH upon addition of either acid or base because they are able to react with either H_3O^+ or OH^- ions. In the buffer of Example 18-8, an added acid will react with the NO_2^- ion to produce unionized acid, while an added base reacts with HNO_2 to produce the conjugate NO_2^-. This minimizes the pH change that the addition causes. Buffer solutions composed of a weak base and its salt react similarly. The added acid can react with the base itself, while an added base will react with the conjugate acid (cation) of the base.

Example 18-10 Buffering Action

A 20.0 mL sample of 0.20 M HBr is added to 100.0 mL of the buffer solution from Example 18-9, which is 0.20 M in NH_3 and 0.20 M in NH_4Br. How much does the pH of the NH_3/NH_4Br solution change? $K_b = 1.8 \times 10^{-5}$

Plan
We find the diluted concentrations of HBr, NH_3, and NH_4Br by using the dilution formula with a total volume of 120.0 mL. HBr, the strong acid, will react with NH_3 to produce additional NH_4^+. We look at a reaction summary to find the new concentrations of NH_3 and NH_4^+ and then substitute these into the K_b expression, proceeding as in Example 18-9.

Solution
The molarities after addition, before reaction are:

$$M \text{ HBr} = \frac{(20.0 \text{ mL})(0.20 M)}{120.0 \text{ mL}} = 0.033 \, M \text{ HBr}$$

$$M \, NH_3 = M \, NH_4Br = \frac{(100.0 \text{ mL})(0.20 M)}{120.0 \text{ mL}} = 0.17 \, M \, NH_3 \text{ and } NH_4Br$$

The added HBr reacts completely with the NH_3:

	NH_3	+	HBr	→	NH_4Br
initial:	0.17 M		0.033 M		0.17 M
change:	-0.033 M		-0.033 M		+0.033 M
final:	0.14 M		0 M		0.20 M

(Br^- is a spectator ion.) These new concentrations can be substituted into the K_b expression as in Example 18-9, but now $[NH_4^+] = (0.20+x) \approx 0.20 \, M$ and $[NH_3] = (0.14-x) \approx 0.14 \, M$.

$$K_b = \frac{[NH_4^+][OH^-]}{[NH_3]} = \frac{(0.20+x)(x)}{0.14-x} \approx \frac{(0.20)(x)}{0.14} = 1.8 \times 10^{-5}$$

$$x = [OH^-] = \frac{(0.14)(1.8 \times 10^{-5})}{0.20} = 1.3 \times 10^{-5} \, M$$

$$pOH = -\log[OH^-] = -\log(1.3 \times 10^{-5}) = -(-4.89) = 4.89$$

$$pH = 14.00 - pOH = 14.00 - 4.89 = 9.11$$

The change in pH is 9.26 - 9.11 = $\boxed{0.15 \text{ pH units}}$.

If 20.0 mL of 0.20 M HBr is added to 100.0 mL of 0.20 M NH_3, the pH is decreased by 1.40 pH units, while if that amount of HBr is added to pure water the pH change is 5.52 pH units. (You can verify this with the appropriate series of calculations.) The buffering action considerably decreases the amount by which a given amount of acid (or of base) changes the pH of a specific volume of solution or solvent.

18-8 Preparation of Buffer Solutions

It is useful to be able to calculate how to prepare a buffer of a desired pH. An example of one common method follows.

Example 18-11 Preparation of a Buffer Solution

How many grams of KBrO must be added to 1.50 liters of 0.200 M HBrO to prepare a buffer at pH 8.50? Assume no volume change. $K_a = 2.5 \times 10^{-9}$.

Plan

We know the desired pH so can find the $[H_3O^+]$ needed which is equal to the molarity of HBrO that ionizes. We can let y be the concentration of the KBrO needed. We write summary equations and use the K_a as in previous examples to find the $[BrO^-]$ which equals the [KBrO]. Finally, we use the volume of solution needed and the molar mass of KBrO to find the mass required.

Solution

If the pH is 8.50, then $[H_3O^+] = 10^{-8.50} = 3.2 \times 10^{-9}$ M.
The pertinent reaction equations are:

$$KBrO \rightarrow K^+ + BrO^-$$
$$0\,M \qquad y\,M \quad y\,M$$

$$HBrO + H_2O \rightleftharpoons H_3O^+ + BrO^-$$
$$(0.200 - 3.2 \times 10^{-9})\,M \approx 0.200\,M \qquad\qquad 3.2 \times 10^{-9}\,M \quad 3.2 \times 10^{-9}\,M$$

The total concentration of BrO^- will be $(y + 3.2 \times 10^{-9})$ M. However, 3.2×10^{-9} is expected to be negligible compared to y as it is to 0.200 M. Using the expression for K_a yields:

$$K_a = \frac{[H_3O^+][BrO^-]}{[HBrO]} = \frac{(3.2 \times 10^{-9}\,M)(y)}{0.200\,M} = 2.5 \times 10^{-9}$$

$$y = [BrO^-] = [KBrO] = \frac{(2.5 \times 10^{-9})(0.200\ M)}{3.2 \times 10^{-9}\ M} = 0.16\ M\ KBrO$$

We can then find the mass of KBrO from the molarity, volume, and molar mass.

$$?\ g\ KBrO = \frac{0.16\ mol\ KBrO}{L} \times 1.50\ L \times \frac{135.8\ g\ KBrO}{1\ mol\ KBrO} = \boxed{33\ g\ KBrO}$$

Some scientists prefer to use the Henderson-Hasselbach equations for buffer calculations. The equations are:

$$pH = pK_a + \frac{\log[\text{conjugate base}]}{\log[\text{acid}]}\quad \text{(for acids)}\ \text{or}$$

$$pOH = pK_b + \frac{\log[\text{conjugate acid}]}{\log[\text{base}]}\quad \text{(for bases)}$$

18-9 Polyprotic Acids

Polyprotic acids are those containing two or more ionizable hydrogens that ionize in a stepwise manner. The ionization constants are represented by K_{a1}, K_{a2}, K_{a3}, etc. The steps in the ionization of a triprotic acid, H_3A, are:

$$H_3A + H_2O \rightleftharpoons H_3O^+ + H_2A^- \qquad K_{a1} = \frac{[H_3O^+][H_2A^-]}{[H_3A]}$$

$$H_2A^- + H_2O \rightleftharpoons H_3O^+ + HA^{2-} \qquad K_{a2} = \frac{[H_3O^+][HA^{2-}]}{[H_2A^-]}$$

$$HA^{2-} + H_2O \rightleftharpoons H_3O^+ + A^{3-} \qquad K_{a3} = \frac{[H_3O^+][A^{3-}]}{[HA^{2-}]}$$

Working with polyprotic acids is similar to the buffer situation, because initial ionizations suppress later ones. Successive constants usually decrease by a factor of 10^4 to 10^6. The $[H_3O^+]$ produced in the second and subsequent steps is negligible compared to that in the first step. It is important to remember that the $[H_3O^+]$ from the first step is used in K_{a2}, K_{a3}, etc., as well. The anion resulting from the second step has a concentration numerically equal to K_{a2} if K_{a1} is at least 10^3 times that of K_{a2}, because $[H_3O^+] \approx [H_2A^-]$.

Example 18-12 Solutions of a Weak Polyprotic Acid

Calculate the concentrations of all species present in 0.050 M H_3AsO_4. The necessary ionization constants for this acid are:

$$K_{a1} = 2.5 \times 10^{-4}, \quad K_{a2} = 5.6 \times 10^{-8}, \quad K_{a3} = 3.0 \times 10^{-13}.$$

Plan

We will treat each step individually, looking at the reaction summary and using the K_a for that step. Concentrations from a previous step must be used where appropriate in subsequent steps.

Solution

Let x = concentration of H_3AsO_4 that ionizes in the first step.

Step 1	H_3AsO_4	+ H_2O	$\rightleftharpoons$	H_3O^+	+	$H_2AsO_4^-$
initial:	0.050 M			≈ 0 M		0 M
change:	$-x$ M			$+x$ M		$+x$ M
final:	$(0.050-x)$ M			x M		x M

$$K_{a1} = \frac{[H_3O^+][H_2AsO_4^-]}{[H_3AsO_4]} = \frac{(x)(x)}{(0.050-x)}$$

Since K_{a1} is relatively large and $[H_3AsO_4]$ is fairly low, it may not be assumed that $x \ll 0.050$ M. Solving with the quadratic formula gives:

$$x = 3.4 \times 10^{-3} \text{ M} = [H_2AsO_4^-] = [H_3O^+] \text{ from the first step.}$$

Let y = concentration of $H_2AsO_4^-$ that ionizes in the second step.

Step 2	$H_2AsO_4^-$	+ H_2O	$\rightleftharpoons$	H_3O^+	+	$HAsO_4^{2-}$
initial:	$\approx 3.4 \times 10^{-3}$ M			3.4×10^{-3} M		0 M
change:	$-y$ M			$+y$ M		$+y$ M
final:	$(3.4 \times 10^{-3} - y)M$			$(3.4 \times 10^{-3}+y)$ M		y M

$$K_{a2} = \frac{[H_3O^+][HAsO_4^{2-}]}{[H_2AsO_4^-]} = \frac{(3.4 \times 10^{-3}+y)(y)}{3.4 \times 10^{-3}-y} \approx y = 5.6 \times 10^{-8}$$

The approximation that y is negligible compared to 3.4×10^{-3} is a valid one because $5.6 \times 10^{-8} \ll 3.4 \times 10^{-3}$. Therefore, $[HAsO_4^{2-}] = 5.6 \times 10^{-8}$ M.

Finally, let z = concentration of $HAsO_4^{2-}$ that ionizes in the third step.

Step 3

	$HAsO_4^{2-}$	+ H_2O ⇌	H_3O^+	+	AsO_4^{3-}
initial:	≈5.6×10^{-8} M		≈3.4×10^{-3} M		0 M
change:	-z M		+z M		+z M
final:	$(5.6 \times 10^{-8} - z)$ M		$(3.4 \times 10^{-3} + z)$ M		z M

$$K_{a3} = \frac{[H_3O^+][AsO_4^{3-}]}{[HAsO_4^{2-}]} = \frac{(3.4 \times 10^{-3}+z)(z)}{5.6 \times 10^{-8}-z} \approx \frac{(3.4 \times 10^{-3})(z)}{5.6 \times 10^{-8}} = 3.0 \times 10^{-13}$$

$$z = [AsO_4^{3-}] = \frac{(3.0 \times 10^{-13})(5.6 \times 10^{-8})}{3.4 \times 10^{-3}} = 4.9 \times 10^{-18}$$

A summary of the concentrations is:

$[H_3AsO_4] = 0.050\ M - x = 0.050\ M - 3.4 \times 10^{-3}\ M = \boxed{0.047\ M}$

$[H_2AsO_4^-] = 3.4 \times 10^{-3} - y = 3.4 \times 10^{-3}\ M - 5.6 \times 10^{-8}\ M = \boxed{3.4 \times 10^{-3}\ M}$

$[HAsO_4^{2-}] = 5.6 \times 10^{-8} - z = 5.6 \times 10^{-8}\ M - 4.9 \times 10^{-18}\ M = \boxed{5.6 \times 10^{-8}\ M}$

$[AsO_4^{3-}] = z = \boxed{4.9 \times 10^{-18}\ M}$

$[H_3O^+] = 3.4 \times 10^{-3} + y + z = 3.4 \times 10^{-3}\ M + 5.6 \times 10^{-8}\ M + 4.9 \times 10^{-18}\ M = \boxed{3.4 \times 10^{-3}\ M}$

$pH = -\log[H_3O^+] = -\log(3.4 \times 10^{-3}\ M) = -(-2.47) = \boxed{2.47}$

$[OH^-] = \dfrac{K_w}{[H_3O^+]} = \dfrac{1.0 \times 10^{-14}}{3.4 \times 10^{-3}} = \boxed{2.9 \times 10^{-12}\ M}$

Note that the pH is determined by the $[H_3O^+]$ produced in the first step. The production of this species, which is a product in subsequent ionizations, helps to suppress those ionizations by LeChatelier's Principle.

EXERCISES

Necessary K_a and K_b values are in Appendices F and G in the text. All solutions are aqueous.

Strong Acids and Strong Bases

1. What are the concentrations of ions and the pH of the following solutions?
   - (a) 1.2×10^{-3} M HBr
   - (b) 0.0046 M LiOH
   - (c) 0.0185 M Ba(OH)$_2$
   - (d) 0.028 M HNO$_3$

2. Calculate $[H_3O^+]$ and $[OH^-]$ for solutions of the following pH or pOH.
   - (a) pH = 3.825
   - (b) pOH = 1.33
   - (c) pH = 12.07
   - (d) pH = -0.220

3. Calculate $[H_3O^+]$, $[OH^-]$, pH and pOH of these solutions.
   - (a) 0.091 M HClO$_3$
   - (b) 0.45 M Ba(OH)$_2$
   - (c) 1.8×10^{-5} M KOH
   - (d) 3.00 M HCl

4. How many grams of NaOH must be added to water to make 400.0 mL of solution with a pH of 12.18?

Weak Acids and Weak Bases

5. Calculate the pH and the concentrations of all dissolved species in the following solutions.
   - (a) 0.75 M C$_6$H$_5$OH
   - (b) 0.66 M C$_6$H$_5$NH$_2$
   - (b) 1.5 M HCOOH
   - (d) 0.60 M C$_5$H$_5$N

6. Calculate the percent ionization of each acid or base at the molarities listed in Exercise 5.

7. How many grams of HF must be dissolved in water to make 750.0 mL of solution at pH 1.80?

8. How many grams of CH$_3$NH$_2$ must be dissolved in water to give 25.0 mL of solution at pH 12.28?

9. A 0.033 M solution of periodic acid, HIO$_4$, has a pH of 4.87. What is the K_a of this acid?

10. (a) What is the weakest weak acid listed in Appendix F? What are its K_a and pK_a?
    (b) What is the strongest weak base listed in Appendix G? What are its K_b and pK_b?
    (c) What is the weakest weak base listed in Appendix G? What are its K_b and pK_b?

11. A 0.36 M solution of the weak base ethylamine, $C_2H_5NH_2$, has a pH of 12.18. What are the [OH$^-$] in this solution and the K_b of this base?

Buffers

12. Calculate the pH and concentrations of all species in each of the following solutions.
    (a) 3.00 L solution containing 81.4 g of HOBr and 80.2 g LiOBr
    (b) 0.75 M HN_3 and 0.45 M KN_3
    (c) 0.020 M CH_3NH_2 and 0.080 M $CH_3NH_3NO_3$

13. What mass of KOCN must be added to 125.0 mL of 0.55 M CH_3NH_2 to make a buffer at pH 3.85? Assume no volume change.

14. What is the pH when 8.3 g of KOBr is added to 125.0 mL of 0.33 M HOBr. Assume no volume change.

15. What is the pH of a buffer made from 100.0 mL of 0.20 M NH_3 and 100.0 mL of 0.300 M NH_4NO_3? Assume volumes are additive.

16. What is the pH of a solution prepared by dissolving 10.0 g of NH_4I in 100.0 mL of 0.0500 M aqueous NH_3? Assume no volume change.

17. What will be the pH resulting from the addition of 50.0 mL of 0.40 M NaCN to 100.0 mL of 0.60 M HCN? Assume that the volumes are additive.

18. What is the pH of the solution which results when equal volumes of solutions of 0.60 M HN_3 and 0.20 M KN_3 are mixed? Assume that the volumes are additive.

19. Phosphate buffers are often used in biochemical work, because they are present in the natural environment of many biological components and they easily give a pH near the physiological pH desired. Two standard solutions of 0.50 M KH_2PO_4 and 0.50 M K_2HPO_4 are available. Assume that the volumes are additive and calculate the volume in mL of each needed to make 1.00 L of a solution at:
    (a) pH = 7.21   (b) pH = 7.50   (c) pH = 7.00

20. Calculate the pH of a buffer solution prepared by mixing 20.0 mL of 0.200 M BaF_2 and 25.0 mL of 0.250 M HF. Assume that the volumes are additive.

Buffering Action

21. Calculate the pH change resulting from addition of 0.200 mole of solid KOH to one liter of each of the following. Assume no volume change occurs. Compare results.

   (a) pure water
   (b) 0.75 M HNO_2
   (c) 0.75 M HNO_2 and 0.75 M KNO_2

22. Calculate the pH change resulting from addition of 0.200 mole of gaseous HBr to one liter of each of the following. Assume no volume change. Compare the results with each other and with those obtained in Exercise 21.

   (a) pure water
   (b) 0.75 M HNO_2
   (c) 0.75 M HNO_2 and 0.75 M KNO_2

23. Calculate the pH change resulting from addition of 0.025 mole of solid KOH to 500.0 mL of each of the following solutions. Assume no volume change. Compare results.

   (a) 0.200 M C_5H_5N
   (b) 0.200 M C_5H_5N and 0.200 M $C_5H_5NHNO_3$

24. Calculate the pH change resulting from addition of 0.025 mole of gaseous HBr to 500.0 mL of each of the following solutions. Assume no volume change. Compare results.

   (a) 0.200 M C_5H_5N
   (b) 0.200 M C_5H_5N and 0.200 M $C_5H_5NHNO_3$

25. How many grams of gaseous HCl must be added to 1.00 L of a buffer made of 0.50 M CH_3COOH and 0.30 M $NaCH_3COO$ to drop the pH by 1.00 pH unit? Assume no volume change.

26. A buffer is 0.36 M $HClO_2$ and 0.57 M $NaClO_2$. ($K_a = 1.1 \times 10^{-2}$ for $HClO_2$)

   (a) What is the initial pH of the buffer?
   (b) What are the final pH of the solution and the pH change if 0.080 mol HCl is added to 500.0 mL of this buffer solution?

Polyprotic Acids and Bases

27. Calculate the pH and the concentrations of all species present in the following solutions.
    (a) $0.10\,M\,(CH_2)_2(NH_2)_2$
    (b) $2.4\,M\,H_3AsO_4$
    (c) $2.8\,M\,H_2SeO_3$
    (d) $0.0100\,M\,(COOH)_2$

Miscellaneous Exercises

28. What are the $[H_3O^+]$, the $[OH^-]$ and the pH of the following solutions?
    (a) $2.5\,M\,(COOH)_2$
    (b) $0.0075\,M\,Ca(OH)_2$
    (c) $0.65\,M\,C_5H_5N$
    (d) $0.085\,M\,C_6H_5OH$

29. How many grams of NH_4Cl must be added to 350.0 mL of $2.33\,M\,NH_3$ solution to prepare a buffer at pH 9.10?

30. What are the $[H_3O^+]$, pH, and $[Se^{2-}]$ in a $2.3\,M\,H_2Se$ solution? $K_{a1} = 1.7 \times 10^{-4}$ and $K_{a2} = 1.0 \times 10^{-10}$.

31. What is the pH of a buffer solution that is $2.00\,M$ in HIO_3 and $2.25\,M$ in $NaIO_3$? The K_a of HIO_3 is 1.7×10^{-3}.

32. A $0.66\,M$ solution of hypoiodous acid, HIO, has a pH of 5.41. What is the K_a of this acid?

33. Calculate the initial pH, the final pH, and the pH change which occurs when 0.025 mol KOH is added to one liter of the following solutions. Assume no volume change.
    (a) $0.245\,M\,HCOOH$
    (b) $0.245\,M\,HCOOH$ and $0.320\,M\,NaCOOH$

34. How many grams of LiOCN must be added to 750.0 mL of a solution of $1.50\,M$ HOCN to make a buffer at pH 3.85? Assume no volume change.

35. What is the pH of a solution made by diluting 4.47 grams of $HClO_3$ with enough water to make a solution with a volume of 275.0 mL?

ANSWERS TO EXERCISES

1. (a) $1.2 \times 10^{-3}\ M\ H_3O^+$, $1.2 \times 10^{-3}\ M\ Br^-$, $8.3 \times 10^{-12}\ M\ OH^-$, $pH = 2.92$
   (b) $2.2 \times 10^{-12}\ M\ H_3O^+$, $0.0046\ M\ Li^+$, $0.0046\ M\ OH^-$, $pH = 11.66$
   (c) $0.0185\ M\ Ba^{2+}$, $0.037\ M\ OH^-$, $2.7 \times 10^{-13}\ M\ H_3O^+$, $pH = 12.57$
   (d) $0.028\ M\ H_3O^+$, $0.028\ M\ NO_3^-$, $3.2 \times 10^{-13}\ M\ OH^-$, $pH = 1.55$

2. (a) $[H_3O^+] = 1.50 \times 10^{-4}\ M$, $[OH^-] = 6.68 \times 10^{-11}\ M$
   (b) $[H_3O^+] = 2.1 \times 10^{-13}\ M$, $[OH^-] = 4.7 \times 10^{-2}\ M$
   (c) $[H_3O^+] = 8.5 \times 10^{-13}\ M$, $[OH^-] = 1.2 \times 10^{-2}\ M$
   (d) $[H_3O^+] = 1.66\ M$, $[OH^-] = 6.03 \times 10^{-15}\ M$

3. (a) $[H_3O^+] = 0.091\ M$, $[OH^-] = 1.1 \times 10^{-13}\ M$, $pH = 1.04$, $pOH = 12.96$
   (b) $[H_3O^+] = 1.1 \times 10^{-14}\ M$, $[OH^-] = 0.90\ M$, $pH = 13.95$, $pOH = 0.045$
   (c) $[H_3O^+] = 5.6 \times 10^{-10}\ M$, $[OH^-] = 1.8 \times 10^{-5}\ M$, $pH = 9.25$, $pOH = 4.74$
   (d) $[H_3O^+] = 3.00\ M$, $[OH^-] = 3.33 \times 10^{-15}\ M$, $pH = -0.477$, $pOH = 14.48$

4. $0.24\ g\ NaOH$

5. (a) $pH = 5.01$, $[C_6H_5OH] = 0.75\ M$, $[H_3O^+] = [C_6H_5O^-] = 0.9.8 \times 10^{-6}\ M$, $[OH^-] = 1.0 \times 10^{-9}\ M$
   (b) $pH = 9.22$, $[C_6H_5NH_2] = 0.66\ M$, $[OH^-] = [C_6H_5NH_3^+] = 1.7 \times 10^{-5}\ M$, $[H_3O^+] = 5.9 \times 10^{-10}\ M$
   (c) $pH = 1.78$, $[HCOOH] = 1.5\ M$, $[OH^-] = 6.2 \times 10^{-13}\ M$, $[H_3O^+] = [HCOO^-] = 0.016\ M$
   (d) $pH = 9.48$, $[C_5H_5N] = 0.60\ M$, $[H_3O^+] = 3.3 \times 10^{-10}\ M$, $[OH^-] = [C_5H_5NH^+] = 3.0 \times 10^{-5}\ M$

6. (a) 0.0013% (b) 0.0026% (c) 1.1% (d) 0.0050%

7. $5.1\ g\ HF$ 8. $0.55\ g\ CH_3NH_2$ 9. $K_a = 5.6 \times 10^{-9}$

10. (a) $BO_2(OH)_2^{2-}$ ($K_a = 1.6 \times 10^{-14}$; $pK_a = 13.80$)
    (b) $(CH_3)_2NH$ ($K_b = 7.4 \times 10^{-4}$; $pK_a = 3.13$)
    (c) $N_2H_5^+$ ($K_b = 8.9 \times 10^{-16}$; $pK_b = 15.05$)

11. $[OH^-] = 2.8 \times 10^{-10}$; $K_b = 6.4 \times 10^{-4}$

12. (a) pH = 8.61, $[HOBr] = 0.28 M$, $[Li^+] = [OBr^-] = 0.26 M$
    $[H_3O^+] = 2.5 \times 10^{-9} M$, $[OH^-] = 4.0 \times 10^{-6} M$

    (b) pH = 4.50, $[HN_3] = 0.75 M$, $[N_3^-] = [Na^+] = 0.45 M$
    $[H_3O^+] = 3.2 \times 10^{-5} M$, $[OH^-] = 3.2 \times 10^{-10} M$

    (c) pH = 10.08, $[CH_3NH_2] = 0.020 M$, $[H_3O^+] = 8.0 \times 10^{-11} M$
    $[CH_3NH_3^+] = [NO_3^-] = 0.080 M$, $[OH^-] = 1.2 \times 10^{-4} M$

13. 14 g KOCN   14. pH = 8.85   15. pH = 9.08

16. pH = 8.12   17. pH = 8.92   18. pH = 4.24

19. (a) 500 mL KH_2PO_4, 500 mL K_2HPO_4
    (b) 338 mL KH_2PO_4, 662 mL K_2HPO_4
    (c) 617 mL KH_2PO_4, 383 mL K_2HPO_4

20. pH = 3.25

21. (a) 13.30 (final) - 7.00 (initial) = 6.30 pH units
    (b) 2.92 (final) - 1.74 (initial) = 1.18 pH units
    (c) 3.58 (final) - 3.35 (initial) = 0.23 pH units
    The same amount of base added makes the largest change of the pH in the pure water and the smallest change in the pH of the buffer.

22. (a) 7.00 (initial) - 0.70 (final) = 6.30 pH units
    (b) 1.74 (initial) - 0.70 (final) = 1.04 pH units
    (c) 3.35 (initial) - 3.11 (final) = 0.24 pH units
    Acid and base raise and lower pH by about the same amount if added to the same solutions at the same concentration. There is a much smaller change in the pH of the buffer system than in the pH of the other two systems.

23. (a) 12.70 (final) - 9.24 (initial) = 3.46 pH units
    (b) 5.40 - 5.18 = 0.22 pH units
    The same amount of base added makes a much smaller change in the pH of the buffer.

4. (a) 9.24 (initial) − 5.65 (final) = 3.59 pH units
   (b) 5.18 (initial) − 4.95 (final) = 0.23 pH units
   The same amount of acid added makes a much smaller change in the pH of the buffer.

5. 9.0 g HCl

6. 2.16 (initial) − 1.86 (final) = 0.30 pH units

27. (a) pH = 11.46, $[(CH_2)_2(NH_2)_2]$ = 0.10 M, $[OH^-]$ = $[(CH_2)_2(NH_2)_2H^+]$ = 0.0029 M, $[(CH_2)_2(NH_2)_2H_2^{2+}]$ = 2.7 × 10^{-8} M, $[H_3O^+]$ = 3.5 × 10^{-12} M

   (b) pH = 1.61, $[H_3AsO_4]$ = 2.4 M, $[OH^-]$ = 4.2 × 10^{-13} M, $[HAsO_4^{2-}]$ = 5.6 × 10^{-8} M, $[H_2AsO_4^-]$ = $[H_3O^+]$ = 0.024 M, $[AsO_4^{3-}]$ = 6.9 × 10^{-19} M

   (c) pH = 1.06, $[H_2SeO_3]$ = 2.7 M, $[HSeO_3^-]$ = $[H_3O^+]$ = 8.7 × 10^{-2} M, $[SeO_3^{2-}]$ = 2.5 × 10^{-7} M, $[OH^-]$ = 1.2 × 10^{-13} M

   (d) pH = 2.07, $[(COOH)_2]$ = 0.0013 M, $[H_3O^+]$ = 0.0088 M, $[H(COO)_2^-]$ = 0.0086 M, $[(COO)_2^{2-}]$ = 6.4 × 10^{-5} M, $[OH^-]$ = 1.2 × 10^{-12} M

28. (a) pH = 0.45, $[H_3O^+]$ = 0.36 M (need quadratic), $[OH^-]$ = 2.8 × 10^{-14} M
    (b) pH = 12.18, $[H_3O^+]$ = 6.6 × 10^{-13} M, $[OH^-]$ = 0.015 M
    (c) pH = 9.49, $[H_3O^+]$ = 3.2 × 10^{-10} M, $[OH^-]$ = 3.1 × 10^{-5} M
    (d) pH = 5.48, $[H_3O^+]$ = 3.3 × 10^{-6} M, $[OH^-]$ = 3.0 × 10^{-9} M

29. 6.0 × 10^1 g NH_4Cl

30. pH = 1.70; $[H_3O^+]$ = 2.0 × 10^{-2}; $[Se^{2-}]$ = 1.0 × 10^{-10}

31. pH = 2.82

32. K_a = 2.3 × 10^{-11}

33. (a) 2.80 (final) − 2.18 (initial) = 0.62 pH units
    (b) 3.94 (final) − 3.86 (initial) = 0.08 pH units

34. 88 g LiOCN

35. pH = 0.717

Chapter 19: Ionic Equilibria II: Hydrolysis

19-1 Introduction

Solvolysis is the reaction of a solute with the solvent in which it is dissolved. When the solvent is water, the reaction is called **hydrolysis**. Many salts hydrolyze to produce acidic or basic solutions. This discussion will be restricted to normal salts—those containing neither free acid nor free base—whose acid-base behavior of salts follows clear patterns. Normal salts can be classified by the character of their cations and anions into four general types: (1) salts of strong bases and strong acids; (2) salts of strong bases and weak acids; (3) salts of weak bases and strong acids; and, (4) salts of weak bases and weak acids.

19-2 Salts of Strong Soluble Bases and Strong Acids

Salts derived from strong soluble bases and strong acids give neutral solutions. Consider KBr, the salt of KOH and HBr. KBr is soluble and ionizes completely in dilute aqueous solution, while water only slightly ionizes;

$$KBr \xrightarrow{100\%} K^+ + Br^-$$

$$2H_2O \rightleftharpoons H_3O^+ + OH^-$$

The cation of the salt, K^+, is too weak an acid to react with the anion of water, OH^-, while the anion, Br^- is too weak a base to react with the cation of water, H_3O^+. The solution is neutral because neither reacts to upset the H_3O^+/OH^- balance. There is no acid-base reaction of salts of a strong acid and a strong soluble base in water. Therefore, **salts of strong soluble acids and strong acids are always neutral.**

19-3 Salts of Strong Soluble Bases and Weak Acids

When KNO$_2$, a typical salt of a strong soluble base and a weak acid, is dissolved in water, the following occurs:

$$KClO \xrightarrow{100\%} K^+ + NO_2^-$$
$$+ \rightleftharpoons HNO_2 + H_2O$$
$$2H_2O \rightleftharpoons OH^- + H_3O^+$$

K$^+$ and OH$^-$ ions do not react with each other in dilute aqueous solution because KOH is a strong base. However, according to Brønsted-Lowry theory, the NO$_2^-$ ion is the conjugate base of HNO$_2$. The NO$_2^-$ ions react with H$_3$O$^+$ ions from the H$_2$O to produce undissociated molecules of the weak acid HNO$_2$. The reaction shifts the water equilibrium to the right and produces an excess of OH$^-$ ions and, therefore, a basic solution. The net reaction is represented as:

$$NO_2^- + H_2O \rightleftharpoons HNO_2 + OH^- \qquad K_b = \frac{[HNO_2][OH^-]}{[NO_2^-]}$$

Because HNO$_2$ is a weak acid:

$$HNO_2 + H_2O \rightleftharpoons H_3O^+ + NO_2^- \qquad K_a = \frac{[H_3O^+][NO_2^-]}{[HNO_2]} = 4.5 \times 10^{-4}$$

If the K_a and K_b expressions are multiplied together:

$$K_a \times K_b = \frac{[H_3O^+][NO_2^-]}{[HNO_2]} \times \frac{[HNO_2][OH^-]}{[NO_2^-]} = [H_3O^+][OH^-] = 1.0 \times 10^{-14}$$

Now K_b for NO$_2^-$ can be found from: $K_b = \dfrac{K_w}{K_a} = \dfrac{1.0 \times 10^{-14}}{4.5 \times 10^{-4}} = 2.2 \times 10^{-11}$

For **any** conjugate pairs: $\mathbf{K_a \times K_b = K_w}$

If the equilibrium constant of one conjugate is known, the other can be found. The inverse relationship between an acid and its conjugate base reinforces our knowledge that a stronger acid has a weaker conjugate base and vice versa. In any aqueous solution that contains a **soluble salt of a strong base and monoprotic weak acid**, the **anion** hydrolyzes to produce a **basic** solution. These hydrolysis constants are used in exactly the same manner as are the ionization constants of standard weak acids or bases.

Example 19-1 Calculation Based on Hydrolysis

Determine the pH, the concentrations of all dissolved species and the percent ionization in 0.50 M KOBr. K_a for HOBr = 2.5×10^{-9}.

Plan

We know that KOBr is a soluble ionic salt. The OBr^- ions hydrolyze to HOBr and OH^-. We can write the equilibration reaction and reaction summary. Then we can calculate the K_b from $K_a \times K_b = K_w$. We then proceed to solve in the same manner as for other weak bases in Chapter 18.

Solution

	OBr^-	+ H_2O	⇌	HOBr	+	OH^-
initial:	0.50 M			≈0 M		0 M
change:	-x M			+x M		+x M
final:	(0.50-x) M			x M		x M

$$K_b = \frac{[HOBr][OH^-]}{[OBr^-]} = \frac{(x)(x)}{0.50-x} \approx \frac{x^2}{0.50} = \frac{K_w}{K_a} = \frac{1.0 \times 10^{-14}}{2.5 \times 10^{-9}} = 4.0 \times 10^{-6}$$

$$x^2 = 0.50(4.0 \times 10^{-6}) = 2.0 \times 10^{-6}; \text{ so } x = [OH^-] = [HOBr] = \boxed{1.4 \times 10^{-3} M}$$

$[OBr^-] = 0.50 M - 1.4 \times 10^{-3} M = \boxed{0.50 M}$. The assumption that the subtracted x is negligible is valid. Now the pH can be found in the following manner.

$$pOH = -\log[OH^-] = -\log(1.4 \times 10^{-3}) = 2.85$$

$$pH = 14.00 - pOH = 14.00 - 2.85 = \boxed{11.15}$$

$$[H_3O^+] = 10^{-pH} = 10^{-11.15} = \boxed{7.1 \times 10^{-12}}$$

Finally, the percent hydrolysis can be determined. The amount of OBr^- that ionizes is equal to x, which is the $[OH^-]$ or the $[HOBr]$.

$$\% \text{ hydrolysis} = \frac{[OBr^-]_{ionized}}{[OBr^-]_{initial}} \times 100 = \frac{1.4 \times 10^{-3} M}{0.50 M} \times 100 = \boxed{0.28\%}$$

-4 Salts of Weak Bases and Strong Acids

e reactions that occur when NH_4NO_3, a soluble ionic salt of the weak base NH_3, which has a of 1.8×10^{-5}, and the strong acid HNO_3, is dissolved in water are shown below.

$$NH_4NO_3 \xrightarrow{100\%} NO_3^- + NH_4^+$$
$$+$$
$$2H_2O \rightleftharpoons H_3O^+ + OH^-$$
$$\rightleftharpoons NH_3 + H_2O$$

ecause HNO_3 is a strong acid, and, therefore, NO_3^- is a very weak base, H_3O^+ and NO_3^- ions o not react with each other. However, NH_4^+ ions react with OH^- ions to form H_2O and nionized NH_3 molecules. This reaction occurs because NH_4^+ is the cation (conjugate acid) of NH_3. This shifts the water equilibrium to the right and produces an acidic solution. The net eaction may be represented as:

$$NH_4^+ + H_2O \rightleftharpoons NH_3 + H_3O^+$$

The expression $K_w = K_a \times K_b$ is valid for **any** weak acid-base conjugate pair so:

$$K_a = \frac{[H_3O^+][NH_3]}{[NH_4^+]} = \frac{K_w}{K_b} = \frac{1.0 \times 10^{-14}}{1.8 \times 10^{-5}} = 5.6 \times 10^{-10}$$

The conjugate acid of any weak base will behave in a similar way, so the **salt of a strong acid and a weak monoprotic base is always basic**.

Example 19-2 Calculations Based on Hydrolysis

Determine the pH and percent hydrolysis in $0.060\ M\ CH_3NH_3I$. The K_b for $CH_3NH_2 = 5.0 \times 10^{-4}$.

Plan
CH_3NH_3I is the soluble, ionic salt of the weak base, CH_3NH_2 and the strong acid HI. We know that the I^- will not hydrolyze. We write the reaction summary for the conjugate acid, $CH_3NH_3^+$, calculate K_a of this conjugate acid, and then proceed as in the previous example for a weak acid.

Solution

$$\begin{array}{lcccc}
 & CH_3NH_3^+ & + H_2O & \rightleftharpoons H_3O^+ & + CH_3NH_2 \\
\text{initial:} & 0.060 \ M & & \approx 0 \ M & 0 \ M \\
\text{change:} & -x \ M & & +x \ M & +x \ M \\
\text{final:} & (0.060-x) \ M & & x \ M & x \ M
\end{array}$$

$$K_a = \frac{[H_3O^+][CH_3NH_3]}{[CH_3NH_3^+]} = \frac{(x)(x)}{0.060-x} \approx \frac{x^2}{0.060} = \frac{K_w}{K_b} = \frac{1.0 \times 10^{-14}}{5.0 \times 10^{-4}} = 2.0 \times 10^{-11}$$

$$x^2 = 0.060(2.0 \times 10^{-11}) = 1.2 \times 10^{-12}$$

$$x = [H_3O^+] = [CH_3NH_2] = \boxed{1.1 \times 10^{-6} \ M}$$

$[CH_3NH_3^+] = 0.060 \ M - 1.1 \times 10^{-6} \ M = 0.060 \ M$, so the assumption that the subtracted x is negligible is valid. Now the pH can be found:

$$pH = -\log[H_3O^+] = -\log(1.1 \times 10^{-6}) = -(-5.96) = \boxed{5.96}$$

Finally, the percent hydrolysis can be determined.

$$\% \text{ hydrolysis} = \frac{[CH_3NH_3^+]_{ionized}}{[CH_3NH_3^+]_{initial}} \times 100 = \frac{1.1 \times 10^{-6} \ M}{0.0.060 \ M} \times 100 = \boxed{0.0018\%}$$

19-5 Salts of Weak Bases and Weak Acids

The reactions that occur when $C_6H_5NH_3F$, a soluble ionic salt, is dissolved in water are:

$$\begin{array}{c}
C_6H_5NH_3F \xrightarrow{100\%} C_6H_5NH_3^+ \ + \ F^- \ + \\
+ \\
2H_2O \rightleftharpoons \quad OH^- \quad + \quad H_3O^+ \\
\quad \quad \quad \quad \quad \quad \updownarrow \quad \quad \quad \quad \updownarrow \\
\quad \quad \quad \quad C_6H_5NH_2 + H_2O \quad HF + H_2O
\end{array}$$

The $C_6H_5NH_3^+$ and the F^- both react with the ions of H_2O to produce unionized molecules of the weak base, $C_6H_5NH_2$, and the weak acid, HF. The net reactions and hydrolysis constants are:

$$C_6H_5NH_3^+ + H_2O \rightleftharpoons H_3O^+ + C_6H_5NH_2 \quad (C_6H_5NH_2, K_b = 4.2 \times 10^{-10})$$

$$K_a = \frac{[H_3O^+][C_6H_5NH_2]}{[C_6H_5NH_3^+]} = \frac{K_w}{K_b} = \frac{1.0 \times 10^{-14}}{4.2 \times 10^{-10}} = 2.4 \times 10^{-5}$$

$$F^- + H_2O \rightleftharpoons HF + OH^- \quad (HF, K_a = 7.2 \times 10^{-4})$$

$$K_b = \frac{[HF][OH^-]}{[F^-]} = \frac{K_w}{K_a} = \frac{1.0 \times 10^{-14}}{7.2 \times 10^{-4}} = 1.4 \times 10^{-11}$$

Because, **in this case**, $C_6H_5NH_3^+$ hydrolyzes more than F^- (K_a for $C_6H_5NH_3^+$ > K_b for F^-), an excess of H_3O^+ is produced and the solution is acidic. The net reaction is:

$$C_6H_5NH_3^+ + F^- + 2H_2O \rightleftharpoons C_6H_5NH_2 + HF + H_3O^+ + OH^-$$

Finding the value of the constant and determining the actual pH of these solutions is beyond the scope of this course. Because **both** the cations and anions of salts of weak acids and weak bases hydrolyze, the acidity or basicity of the solution depends upon which ion hydrolyzes to the greater extent. It is apparent that most salt solutions are **not** neutral. The only salts that produce neutral solutions are:

1. salts of strong acids and strong soluble bases
2. salts of weak acids and weak bases in the restricted case that:
   $K_{a(parent\ acid)} = K_{b(parent\ base)}$.

A summary of salts of weak acids and weak bases is below:

If	Then, solution is:
$K_{a(cation)} > K_{b(anion)}$ $K_{b(parent\ base)} < K_{a(parent\ acid)}$	acidic
$K_{a(cation)} < K_{b(anion)}$ $K_{b(parent\ base)} > K_{a(parent\ acid)}$	basic
$K_{a(cation)} = K_{b(anion)}$ $K_{b(parent\ base)} = K_{a(parent\ acid)}$	neutral

19-6 Hydrolysis of Small Highly Charged Cations

Salts containing small, highly charged cations also hydrolyze to produce acidic solutions. Such cations are almost exclusively the metal ions that form insoluble hydroxides. Although not usually represented as such, the metal ions are hydrated when dissolved, usually with four to six water molecules bonded to the metal ion through coordinate covalent bonds. (See Figures 19-1 and 19-2 in the text.) For example, the ion represented as Fe^{3+}(aq) is actually $[Fe(OH_2)_6]^{3+}$(aq). The donation of electron pairs by the water molecule to the metal causes a weakening of the H- bonds in H_2O as the electrons are shifted toward the metal. As a consequence, it is easier for coordinated H_2O molecules to act as acids and to produce H_3O^+ ions than for uncoordinated H_2O molecules to ionize. Such hydrolysis reactions can be represented as:

$$[Fe(OH_2)_6]^{3+} + H_2O \rightleftharpoons [Fe(OH_2)_5(OH)]^{2+} + H_3O^+$$

or more simply as: $Fe^{3+} + 2H_2O \rightleftharpoons Fe(OH)^{2+} + H_3O^+$

The hydrolysis constant for the reaction has the usual form:

$$K_a = \frac{[[Fe(OH_2)_5(OH)]^{2+}][H_3O^+]}{[[Fe(OH_2)_6]^{3+}]} = 4.0 \times 10^{-13}$$

and is often written in simplified form as:

$$K_a = \frac{[Fe(OH)^{2+}][H_3O^+]}{[Fe^{3+}]} = 4.0 \times 10^{-13}$$

The strength of metal-oxygen bonds in hydrated ions, and therefore, the degree of hydrolysis, **generally** increases as the positive charge on the cation increases and as its size decreases. This trend and some values of the hydrolysis constant are shown in Table 19-2 in the text. Hydrolysis reactions of a small highly-charged metal ion, M^{n+}, may be generalized as:

$$M^{n+}(aq) + H_2O(\ell) \rightleftharpoons M(OH)^{(n-1)+}(aq) + H_3O^+(aq)$$

$$K_a = \frac{[M(OH)^{(n-1)+}][H_3O^+]}{[M^{n+}]}$$

If the hydrolysis constant, K_a is known, the pH of a given solution can be calculated, or if the pH of a given solution is known, K_a, can be calculated. The calculations involved are similar to those for a Brønsted-Lowry acid.

Example 19-3 Hydrolysis of a Highly Charged Cation

Calculate the pH and concentrations of all dissolved species in 0.15 M $CoBr_2$. The hydrolysis constant for $[Co(OH_2)_6]^{2+} = 5.0 \times 10^{-10}$.

Plan
$CoBr_2$ is a soluble ionic compound and a strong electrolyte. We write its dissociation reaction and reaction summary to determine ion concentrations. We recognize that the Br^- ions do not hydrolyze because Br^- is the conjugate base of the strong acid HBr. However, Table 19-2 indicates that the Co^{2+} ions coordinate with six H_2O molecules and hydrolyze as discussed above. We write a second reaction summary for this hydrolysis with x, the amount that hydrolyzes. We set the ionization expression equal to K_a and solve as for other weak acids.

Solution
The strong electrolyte ionizes as below.

	$CoBr_2$	$\rightarrow$	Co^{2+}	+	$2Br^-$
initial:	0.15 M		0 M		0 M
change:	-0.15 M		+0.15 M		+2(0.15) M
final:	0 M		0.15 M		0.30 M

The Co^{2+} ions coordinate with 6 water molecules to form $Co(OH_2)_6^{2+}$, which then hydrolyzes as follows:

	$Co(OH_2)_6^{2+}$	+ H_2O	$\rightleftharpoons$	H_3O^+	+	$[Co(OH_2)_5(OH)]^+$
initial:	0.15 M			≈0 M		0 M
change:	-x M			+x M		+x M
final:	(0.15-x) M			x M		x M

$$K_a = \frac{[H_3O^+][[Co(OH_2)_5(OH)]^+]}{[Co(OH_2)_6^{2+}]} = \frac{(x)(x)}{0.15-x} \approx \frac{x^2}{0.15} = 5.0 \times 10^{-10}$$

$$x^2 = (0.15)(5.0 \times 10^{-10}) = 7.5 \times 10^{-11}$$

$$x = [H_3O^+] = [Co(OH_2)_5(OH)]^+ = \boxed{8.7 \times 10^{-6} \, M}$$

$$[OH^-] = \frac{K_w}{[H_3O^+]} = \frac{1.0 \times 10^{-14}}{8.7 \times 10^{-6} \, M} = \boxed{1.1 \times 10^{-9} \, M}$$

$$[Co(OH_2)_6^{2+}] = 0.15 \, M - x = 0.15 \, M - 8.7 \times 10^{-6} \, M = \boxed{0.15 \, M}$$

$$[Br^-] = \boxed{0.30 \, M}$$

$$pH = -\log[H_3O^+] = -\log(8.7 \times 10^{-6}) = -(-5.06) = \boxed{5.06}$$

19-7 Strong Acid-Strong Base Titration Curves

We can visually assess pH changes as an acid is added to a base by preparing titration curves. A titration curve is a plot of pH versus volume or other specified quantity of an added titrant. Titration curves are useful for determining concentrations and pK_a values of an acid or base. Titration curves also indicate what indicator could be used for a titration.

For example, 25.0 mL of a 0.200 M solution of HBr can be titrated by addition of 0.200 M KOH. The salt formed by this strong acid, strong base reaction is soluble and neutral for reasons discussed in Section 19-1. We will present reaction summaries in this example in terms of millimoles. As an alternate approach we could find the diluted molarities of both the acid and the base and use those diluted molarities in the reaction summary.

1. Before we add any KOH solution, the HBr is 100% ionized:

   $HBr + H_2O \rightarrow H_3O^+ + Br^-$

   $[H_3O^+] = 0.200\,M$ and $pH = -\log[H_3O^+] = 0.699$

2. After we have added 10.0 mL of KOH, the total volume is 35.0 mL. Remember that volume in milliliters times molarity equals millimoles. The reaction summary is:

	KOH +	HBr	→	KBr	+ H$_2$O
Initial	2.00 mmol	5.00 mmol		0.00 mmol	
Change	-2.00 mmol	-2.00 mmol		+2.00 mmol	
Final	0.00 mmol	3.00 mmol		2.00 mmol	

   $[HBr]_{excess} = [H_3O^+] = \dfrac{3.00\,\text{mmol}}{35.0\,\text{mL}} = 0.0857\,M$

   $pH = -\log[H_3O^+] = -\log(0.0857\,M) = -(-1.067) = 1.067$

   We can calculate all points up to the equivalence point in the same manner. See the following Table 19-1 for a tabulation.

3. At the equivalence point, we have added 25.0 mL of 0.200 M KOH.

	KOH +	HBr	→	KBr	+ H$_2$O
Initial	5.00 mmol	5.00 mmol		0.00 mmol	
Change	-5.00 mmol	-5.00 mmol		+5.00 mmol	
Final	0.00 mmol	0.00 mmol		5.00 mmol	

The resulting solution contains only KBr, a soluble salt of a strong acid and strong soluble base. As we discussed in Section 19-2, all such salts are neutral. Only the autoionization of water occurs; therefore, $[H_3O^+] = [OH^-]$ and pH = 7.00.

4. Beyond the equivalence point, the KOH is in excess and determines the pH of the solution. After we have added 26.0 mL:

	KOH +	HBr	$\rightarrow$	KBr	+	H_2O
Initial	5.20 mmol	5.00 mmol		0.00 mmol		
Change	-5.00 mmol	-5.00 mmol		+5.00 mmol		
Final	0.20 mmol	0.00 mmol		5.00 mmol		

Because KOH is a strong soluble base, $M_{KOH} = [OH^-] = 0.0039\ M$

pOH = -log (0.0039 M) = 2.41, pH = 14.00 - 2.41 or pH = 11.59

We can calculate all points beyond the equivalence point in a similar manner. The points at which the pH of the solution have been calculated are shown in Table 19-1, with enough other additional points calculated to more clearly show the shape of the titration curve.

Table 19-1 Titration Data for 25.0 mL of 0.200 M KBr with 0.200 M KOH

mL 0.200 M KOH added	mmol KOH added	mmol excess acid or base	pH
0.00	0.00	5.00 H_3O^+	0.699
5.00	1.00	4.00 H_3O^+	0.875
10.0	2.00	3.00 H_3O^+	1.067
12.5	2.50	2.50 H_3O^+	1.176
20.0	4.00	1.00 H_3O^+	1.653
24.0	4.80	0.20 H_3O^+	2.39
24.9	4.98	0.02 H_3O^+	3.4
25.0	5.00	0.00	7.00
25.1	5.02	0.02 OH^-	10.6
26.0	5.20	0.20 OH^-	11.59
27.0	5.40	0.40 OH^-	11.89
30.0	6.00	1.00 OH^-	12.260

Figure 19-1 shows the plot of this data. Note that the "vertical" portion of the curve is quite l[...] and extends from about pH = 3.4 to pH = 10.6. Indicators that would change color within this range would be suitable for determining the equivalence point. All the indicators in Table 19.3 [...] the text except methyl orange would be appropriate. Note that this curve looks very similar to that of Figure 19-3 in the text. This is because all strong acid-strong soluble base reactions producing a soluble salt have the net reaction: $H_3O^+ + OH^- \rightarrow 2H_2O$

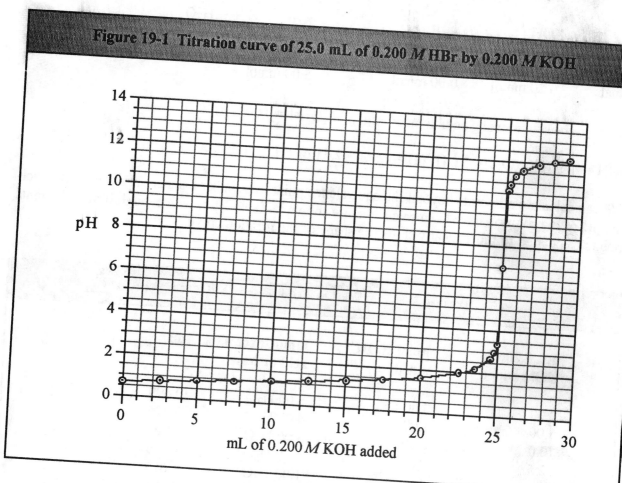

Figure 19-1 Titration curve of 25.0 mL of 0.200 M HBr by 0.200 M KOH

19-8 Weak Base–Strong Acid Titration Curves

Section 19-7 of the text addresses weak acid/strong base titration curves and the similarity and differences to those of strong acids with a strong base. Here we will consider the titration of 25.0 mL of 0.200 M NH_3 (a weak base with K_b = 1.8 x 10^{-5}) with 0.200 M HNO_3, a strong acid. This system is more complicated than that seen in the previous titration because, after some HNO_3 has been added, a buffer solution consisting of aqueous NH_4NO_3 and NH_3 exists. At the equivalence point, only the salt, NH_4NO_3, is present in solution and the NH_4^+ ion hydrolyzes to produce an acidic solution. Beyond the equivalence point, the excess HNO_3 determines the pH.

Chapter 19 Ionic Equilibria II: Hydrolysis

1. We find the initial pH, before any we have added any HNO_3, as in previous examples. (See Section 18-4 in this book if necessary.)

	NH_3	$+ H_2O \rightleftharpoons$	NH_4^+	$+$	OH^-
initial:	0.200 M		≈ 0 M		≈ 0 M
change:	-x M		+x M		+x M
final:	(0.200-x) M		x M		x M

$$K_b = \frac{[NH_4^+][OH^-]}{[NH_3]} = \frac{(x)(x)}{0.200-x} \approx \frac{x^2}{0.200} = 1.8 \times 10^{-5}$$

$x^2 = 0.200(1.8 \times 10^{-5}) = 3.6 \times 10^{-6}$ so $x = [OH^-] = 1.9 \times 10^{-3} M$

$[NH_3] = 0.200\ M - 1.9 \times 10^{-3} M = 0.198\ M$. The result is within 5% of the initial molarity, so the assumption that the subtracted **x** is negligible is valid. Now we can find the pH in the following manner.

$$pOH = -\log [OH^-] = -\log (1.9 \times 10^{-3}) = 2.72$$

$$pH = 14.00 - pOH = 14.00 - 2.72 = \boxed{11.28}$$

2. After we have added 10.0 mL (2.00 mmol) of HNO_3, the total volume is 35.0 mL and reaction summary is:

	$HNO_3\ +$	NH_3	$\rightarrow$	NH_4^+	$+$	NO_3^-
Initial	2.00 mmol	5.00 mmol		0.00 mmol		
Change	-2.00 mmol	-2.00 mmol		+2.00 mmol		
Final	0.00 mmol	3.00 mmol		2.00 mmol		

The solution is now a buffer solution and the new pH can be calculated. The new molarities are:

$$[NH_3] = \frac{3.00\ mmol}{35.0\ mL} = 0.0857\ M \text{ and } [NH_4^+] = \frac{2.00\ mmol}{35.0\ mL} = 0.0571\ M$$

The reaction summary and solution are:

	NH_3	$+ H_2O \rightleftharpoons$	NH_4^+	$+$	OH^-
initial:	0.0857 M		0.0571 M		≈ 0 M
change:	-x M		+x M		+x M
final:	(0.0857-x) M		0.0571+ x M		x M

$$K_b = \frac{[NH_4^+][OH^-]}{[NH_3]} = \frac{(0.0571+x)(x)}{0.0857-x} \approx \frac{(0.0571)(x)}{0.0857} = 1.8 \times 10^{-5}$$

$$x = [OH^-] = \frac{0.0857}{0.0571}(1.8 \times 10^{-5}) = 2.7 \times 10^{-5} M$$

$$pOH = -\log[OH^-] = -\log(2.7 \times 10^{-5}) = 4.57$$

$$pH = 14.00 - pOH = 14.00 - 4.57 = 9.43$$

Similar calculations at points before the equivalence point yield the values in Table 19-2.

3. At the equivalence point: mmol HNO_3 = mmol NH_3 = 5.00 mmol. This means that we must have added 25.0 mL of HNO_3 to react totally to form NH_4NO_3. The total volume will be 50.0 mL. Only the NH_4^+ will hydrolyze and the molarity of the NH_4^+ is:

$$[NH_4^+] = \frac{5.00 \text{ mmol}}{50.0 \text{ mL}} = 0.100 M$$

The NH_4^+ hydrolyzes as discussed in Section 19-5.

	NH_4^+	$+ H_2O$	$\rightleftharpoons$	NH_3	$+$	H_3O^+
initial:	0.100 M			≈0 M		≈0 M
change:	-x M			+x M		+x M
final:	(0.100-x) M			x M		x M

$$K_b = \frac{[NH_3][H_3O^+]}{[NH_4^+]} = \frac{(x)(x)}{0.100-x} \approx \frac{x^2}{0.100} = \frac{K_w}{K_b} = \frac{1.0 \times 10^{-14}}{1.8 \times 10^{-5}} = 5.6 \times 10^{-10}$$

$$x^2 = 0.100(5.6 \times 10^{-10}) = 5.6 \times 10^{-11} \text{ so } x = [H_3O^+] = 7.5 \times 10^{-6} M$$

$$pH = -\log[H_3O^+] = -\log(7.5 \times 10^{-6}) = -(-5.12) = 5.12$$

4. Beyond the equivalence point, the excess nitric acid governs the pH of the solution in the same manner as if a strong base were being titrated with a strong acid. We show the data for this titration in Table 19-2 and the titration curve plotted from the data in Figure 19-2.

Note that "vertical" portion of the curve is less dramatic than in the strong acid-strong base titration. An increment of acid near the equivalence point makes less of a change in the pH due to the buffering effect of the weak base and its conjugate acid. The best indicator from Table 19-3 in the text for this titration would be methyl red since its color change is from pH 4.4 to 6.2.

Table 19-2 Titration Data for 25.0 mL of 0.200 M NH₃ with 0.200 M HNO₃

mL 0.200 M HNO₃ added	mmol HNO₃ added	mmol excess acid or base	pH
0.00	0.00	5.00 NH₃	11.28
5.00	1.00	4.00 NH₃	9.85
10.0	2.00	3.00 NH₃	9.43
12.5	2.50	2.50 NH₃	9.28
20.0	4.00	1.00 NH₃	8.65
24.0	4.80	0.20 NH₃	7.88
24.9	4.98	0.02 NH₃	6.9
25.0	5.00	0.00 *	5.13
25.1	5.02	0.02 HNO₃	3.4
26.0	5.20	0.20 HNO₃	2.41
27.0	5.40	0.40 HNO₃	2.11
30.0	6.00	1.00 HNO₃	1.74

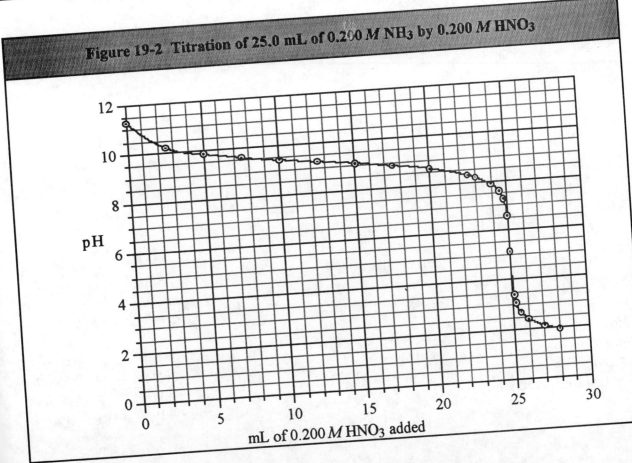

Figure 19-2 Titration of 25.0 mL of 0.200 M NH₃ by 0.200 M HNO₃

19-9 Weak Acid–Strong Base Titration Curves

A titration curve for the titration of a weak acid by adding a strong base is similar to the inverted image of the curve in Figure 19-2, beginning at low pH, having an equivalence point in the basic region (pH > 7) due to the hydrolysis of the salt produced and ending at high pH. Figure 19-4 in the text shows a typical weak acid-strong base curve for $CH_3COOH/NaOH$.

19-10 Weak Acid–Weak Base Titration Curves

Such titrations are impossible to perform using visual indicators because color changes are not sufficiently distinct. At the equivalence point, both the cation and the anion of the salt hydrolyze and, as a result, the pH change is gradual throughout the titration.

EXERCISES

Consult Appendices F and G and Table 19-2 in the text to obtain ionization and hydrolysis constants as necessary.

Hydrolysis of Salts

1. Indicate whether each salt below hydrolyzes, and, for those that do, whether they give acidic, basic or neutral solutions.

   (a) $CsClO_3$
   (b) $(CH_3)_2NH_2Br$
   (c) KHO_2
   (d) $Ba(CN)_2$
   (e) $CH_3NH_3NO_3$
   (f) $CaCl_2$
   (g) NH_4CH_3COO
   (h) $(CH_3)_3NHOBr$
   (i) $C_6H_5NH_3F$
   (j) $Sr(OCl)_2$
   (k) $RbClO_4$
   (l) $LiBr$

2. Write equations for the hydrolysis (if any) of each salt in Exercise 1. Also write the mass action expression and evaluate the hydrolysis constant (except for weak acid/weak base salts) from tabulated acid and base ionization constants.

3. Arrange the following salts in order of increasing basicity.

   (a) $KOBr$ (b) KCN (c) K_2S (d) KNO_3 (e) $KOCN$ (f) KN_3

4. Arrange the following salts in order of increasing acidity.

   (a) NH_2OH_2Cl (b) $(CH_3)_3NHCl$ (c) C_5H_5NHCl (d) $(CH_3)_2NH_2Cl$

5. Calculate the pH of each of the solutions below.

   (a) $0.76\ M\ LiOBr$
   (b) $0.20\ M\ CH_3NH_3ClO_4$
   (c) $0.17\ M\ NH_4I$
   (d) $1.75\ M\ Ba(IO)_2$ (K_a for $HIO = 2.3 \times 10^{-11}$)
   (e) $0.050\ M\ C_5H_5NHClO_4$
   (e) $0.25\ M\ CH_3NH_3Cl$

6. $Ca(ClO)_2$ is often used in swimming pools as an algicide and disinfectant. However, use of this chemical does affect the pH of the pool water. What would be the resultant pH if 6.0 pounds of $Ca(ClO)_2$ were added to 50,000 gallons of water? (Swimming pools are usually buffered to minimize this effect.)

7. In areas of acid rain, one of the attempts to raise the pH of a body of water has been to neutralize the acids present with soda ash, Na_2CO_3. What would the pH be if 6.0 pounds of Na_2CO_3 were added to 50,000 gallons of pure water?

Hydrolysis of Small, Highly-Charged Metal Cations

8. Write equations, showing coordinated the water molecules, for the first step in the hydrolysis of:
   (a) $[Co(OH_2)_6]^{2+}$ in aqueous $CoCl_2$
   (b) $[Cu(OH_2)_4]^{2+}$ in aqueous $CuCl_2$
   (c) $[Bi(OH_2)_6]^{3+}$ in aqueous $Bi(NO_3)_3$

9. Calculate the pH and the concentrations of all dissolved species in the following solutions.
   (a) $0.50\ M\ BeCl_2$
   (b) $0.10\ M\ CuCl_2$
   (c) $0.60\ M\ ZnCl_2$
   (d) $1.0\ M\ Hg(NO_3)_2$

Titration Curves and Indicators

10. What is the pH at the equivalence point of each of the following titrations?
    (a) 30.0 mL of $0.400\ M\ HNO_3$ with $0.400\ M\ NH_2OH$
    (b) 35.0 mL of $0.150\ M\ NH_3$ with $0.150\ M\ HBr$
    (c) 40.0 mL of $0.30\ M\ C_5H_5N$ with $0.15\ M\ HI$
    (d) 50.0 mL of $0.20\ M\ HCN$ with $0.50\ M\ KOH$

11. Methyl red has a $K_b = 2 \times 10^{-9}$. It is red in its acidic range and yellow in the basic range. What color is it at (a) pH = 3.00 (b) pH = 7.00 and (c) pH = 11.00?

12. Bromocresol green has a $K_a = 2.2 \times 10^{-5}$. It is yellow in its acidic range and blue in its basic range. What color is it at (a) pH = 2.00 (b) pH = 4.66 and (c) pH = 7.00?

13. What is the pH at the following points in some titrations?
    (a) 15.0 mL of $0.175\ M\ LiOH$ added to 15.0 mL of $0.300\ M\ HI$
    (b) 15.0 mL of $0.150\ M\ NaOH$ to 35.0 mL of $0.150\ M\ HClO_3$
    (c) 22.0 mL of $0.0500\ M\ RbOH$ to 25.0 mL of $0.050\ M\ HClO_4$
    (d) 50.0 mL of $0.750\ M\ HCl$ to 75.0 mL of $0.750\ M\ CH_3NH_2$
    (e) 50.0 mL of $0.25M\ HNO_3$ to 50.0 mL of $0.25\ M\ C_6H_5NH_2$
    (f) 22.0 mL of $0.220\ M\ NaOH$ to 28.0 mL of $0.220\ M\ HIO$ (K_a HIO $= 2.3 \times 10^{-11}$)
    (g) 40.0 mL of $0.200\ M\ Ba(OH)_2$ to 80.0 mL of $0.200\ M\ HClO_2$ (K_a $HClO_2 = 1.1 \times 10^{-2}$)

Miscellaneous Exercises

14. What is the pH of 1.50 M NH_2OH_2Br?

15. Write the hydrolysis reaction for $CH_3NH_3NO_2$ and tell if the solution is acidic, basic, or neutral.

16. What is the pH after 35.0 mL of 0.200 M LiOH has been added to 35.0 mL of 0.150 M $HClO_3$?

17. What is the pH at the equivalence point in the titration in Exercise 16.

18. What is the pH at the equivalence point in the titration of 40.0 mL of 0.20 M HN_3 solution by 0.40 M LiOH?

19. What is the pH of a 0.20 M $Ca(C_6H_5O)_2$ solution?

20. What is the pH after 14.5 mL of 0.220 M HNO_3 has been added to 2.75 mL of 0.220 M $Ba(OH)_2$?

21. A 0.220 M solution of $RbC_3H_5O_3$ has a pH of 8.60. What is the K_a of $HC_3H_5O_3$?

22. What is the pH after 14.0 mL of 0.400 M RbOH is added to 25.0 mL of 0.400 M HCl?

23. Assume that we use 0.0200 M NaOH to titrate 30.0 mL samples of the following acids and that volumes are additive. Determine the $[H_3O^+]$, $[OH^-]$, pH, and pOH after the following volumes of base have been added. Plot the titration curves and determine appropriate indicators for this titration from those in Table 19-3 in the text.

    (i) 0.0200 M $HClO_4$ (ii) HN_3

    Use the following number of milliliters of base added.

    (a) none   (b) 5.00   (c) 10.00   (d) 15.00 (50%)
    (e) 20.00   (f) 25.00   (g) 29.00   (h) 29.50
    (i) 29.90   (j) 30.00 (100%)   (k) 30.10   (l) 30.50
    (m) 31.00   (n) 35.00   (o) 40.00   (p) 45.00 (50% excess)

24. Assume that we use 0.0200 M HCl to titrate 30.0 mL samples of the following bases and that volumes are additive. Determine the $[H_3O^+]$, $[OH^-]$, pH, and pOH after the volumes of acid indicated in Exercise 23 have been added. Plot the titration curves and determine appropriate indicators for this titration from those in Table 19-3 in the text.

    (i) 0.0200 M NaOH   (ii) 0.0200 M CH_3NH_2

ANSWERS TO EXERCISES

1. (a) no hydrolysis
   (b) $(CH_3)_2NH_2^+$ hydrolyzes, acidic
   (c) HO_2^- hydrolyzes, basic
   (d) CN^- hydrolyzes, basic
   (e) $(CH_3)_3NH^+$ hydrolyzes, acidic
   (f) no hydrolysis
   (g) NH_4^+ and CH_3COO^- hydrolyze, neutral ($K_{a(parent)} = K_{b(parent)}$)
   (h) $(CH_3)_3NH^+$ and OBr^- hydrolyze, basic ($K_{a(parent)} < K_{b(parent)}$)
   (i) $C_6H_5NH_3^+$ and F^- hydrolyze, acidic ($K_{a(parent)} > K_{b(parent)}$)
   (j) OCl^- hydrolyzes, basic
   (k) no hydrolysis
   (l) no hydrolysis

2. (b) $(CH_3)_2NH_2^+ + H_2O \rightleftharpoons H_3O^+ + (CH_3)_2NH$

   $$K_a = \frac{[H_3O^+][(CH_3)_2NH]}{[(CH_3)_2NH_2^+]} = \frac{1.0 \times 10^{-14}}{7.4 \times 10^{-4}} = 1.4 \times 10^{-11}$$

   (c) $HO_2^- + H_2O \rightleftharpoons H_2O_2 + OH^-$ $\quad K_b = \frac{[H_2O_2][OH^-]}{[HO_2^-]} = \frac{1.0 \times 10^{-14}}{2.4 \times 10^{-12}} = 4.2 \times 10^{-3}$

   (d) $CN^- + H_2O \rightleftharpoons HCN + OH^-$ $\quad K_b = \frac{[HCN][OH^-]}{[CN^-]} = \frac{1.0 \times 10^{-14}}{4.0 \times 10^{-10}} = 2.5 \times 10^{-5}$

   (e) $CH_3NH_3^+ + H_2O \rightleftharpoons CH_3NH_2 + H_3O^+$

   $$K_a = \frac{[H_3O^+][CH_3NH_2]}{[CH_3NH_3^+]} = \frac{1.0 \times 10^{-14}}{5.0 \times 10^{-4}} = 2.0 \times 10^{-11}$$

   (g) $NH_4^+ + CH_3COO^- + 2H_2O \rightleftharpoons NH_3 + CH_3COOH + H_3O^+ + OH^-$

   (h) $(CH_3)_3NH^+ + OBr^- + 2H_2O \rightleftharpoons (CH_3)_3N + HOBr + H_3O^+ + OH^-$

   (i) $C_6H_5NH_3^+ + F^- + 2H_2O \rightleftharpoons C_6H_5NH_2 + HF + OH^- + H_3O^+$

   (j) $OCl^- + H_2O \rightleftharpoons HOCl + OH^-$ $\quad K_b = \frac{[HOCl][OH^-]}{[OCl^-]} = \frac{1.0 \times 10^{-14}}{3.5 \times 10^{-8}} = 2.9 \times 10^{-7}$

3. $KNO_3 < KOCN < KN_3 < KOBr < KCN < K_2S$ (strongest base)

4. $(CH_3)_2NH_2Cl < (CH_3)_3NHCl < C_5H_5NHCl < NH_2OH_2Cl$ (strongest acid)

Chapter 19 Ionic Equilibria II: Hydrolysis

5. (a) pH = 11.23 (b) pH = 5.70 (c) pH = 5.01
   (d) pH = 12.59 (e) pH = 3.24 (f) pH = 5.65

6. pH = 8.87   7. pH = 9.96

8. (a) $[Co(OH_2)_6]^{2+} + H_2O \rightleftharpoons [Co(OH_2)_5(OH)]^+ + H_3O^+$
   (b) $[Cu(OH_2)_4]^{2+} + H_2O \rightleftharpoons [Cu(OH_2)_3(OH)]^+ + H_3O^+$
   (c) $[Bi(OH_2)_6]^{3+} + H_2O \rightleftharpoons [Bi(OH_2)_5(OH)]^{2+} + H_3O^+$

9. (a) pH = 2.65, $[[Be(OH_2)_4]^{2+}] = 0.50\,M$, $[OH^-] = 4.5 \times 10^{-12}\,M$
       $[Cl^-] = 1.0\,M$, $[H_3O^+] = [[Be(OH_2)_3(OH)]^+] = 2.2 \times 10^{-3}\,M$
   (b) pH = 4.50, $[[Cu(OH_2)_4]^{2+}] = 0.10\,M$, $[OH^-] = 3.2 \times 10^{-10}\,M$
       $[Cl^-] = 0.20\,M$, $[H_3O^+] = [[CuOH_2)_3(OH)]^+] = 3.2 \times 10^{-5}\,M$
   (c) pH = 4.91, $[[Zn(OH_2)_6]^{2+}] = 0.60\,M$, $[OH^-] = 8.3 \times 10^{-10}\,M$
       $[Cl^-] = 1.2\,M$, $[H_3O^+] = [[Zn(OH_2)_5(OH)]^+] = 1.2 \times 10^{-5}\,M$
   (d) pH = 3.04, $[[Hg(OH_2)_6]^{2+}] = 1.0\,M$, $[OH^-] = 1.1 \times 10^{-11}\,M$
       $[NO_3^-] = 2.0\,M$, $[H_3O^+] = [[Fe(OH_2)_5(OH)]^+] = 9.1 \times 10^{-4}\,M$

10. (a) pH = 3.26 (b) pH = 5.19 (c) pH = 3.09 (d) pH = 11.25

11. (a) red (b) red (c) yellow   12. (a) yellow (b) pH = pK_a; green (c) blue

13. (a) pH = 1.21 (b) pH = 1.22 (c) pH = 2.49 (d) pH = 10.40
    (e) pH = 2.77 (f) pH = 11.21 (g) 7.54

14. pH = 2.82

15. $CH_3NH_3^+ + NO_2^- + 2H_2O \rightleftharpoons CH_3NH_2 + NO_2^- + H_3O^+ + OH^-$
    The solution is basic because K_b of $CH_3NH_2 = 5.0 \times 10^{-4}$, while K_a of $HNO_2 = 4.5 \times 10^{-4}$

16. pH = 12.40   17. pH = 7.00   18. 8.92   19. pH = 11.74   20. pH = 0.946

21. $K_a = 1.4 \times 10^{-4}$ 22. 0.948

23. (i)

# mL of NaOH added	$[H_3O^+]$	$[OH^-]$	pH	pOH
0	2.0×10^{-2}	5.0×10^{-13}	1.70	12.30
5.00	1.4×10^{-2}	7.0×10^{-13}	1.85	12.15
10.00	1.0×10^{-2}	1.0×10^{-12}	2.00	12.00
15.00	6.7×10^{-3}	1.5×10^{-12}	2.18	11.82
20.00	4.0×10^{-3}	2.5×10^{-12}	2.40	11.60
25.00	1.8×10^{-3}	5.6×10^{-12}	2.74	11.26
29.00	3.4×10^{-4}	2.9×10^{-11}	3.47	10.53
29.50	1.7×10^{-4}	5.9×10^{-11}	3.77	10.23
29.90	3.3×10^{-5}	3.0×10^{-10}	4.48	9.52
30.00	1.0×10^{-7}	1.0×10^{-7}	7.00	7.00
30.10	3.0×10^{-10}	3.3×10^{-5}	9.52	4.48
30.50	6.0×10^{-11}	1.6×10^{-4}	10.22	3.78
31.00	3.0×10^{-11}	3.3×10^{-4}	10.52	3.48
35.00	6.5×10^{-12}	1.5×10^{-3}	11.19	2.81
40.00	3.5×10^{-12}	2.9×10^{-3}	11.46	2.54
45.00	2.5×10^{-12}	4.0×10^{-3}	11.60	2.40

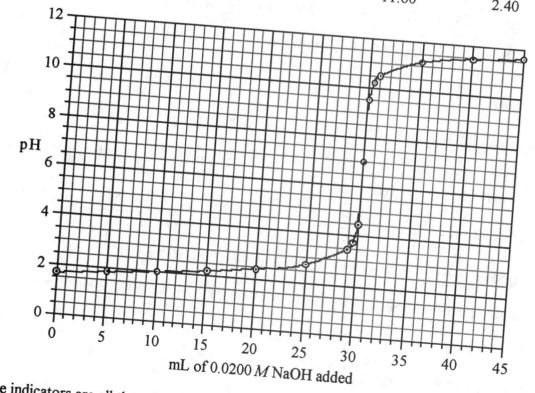

Suitable indicators are all those in Table 19-3 except methyl orange. Phenolphtalein is marginal for this titration because it finishes changing color at pH 10.0, just past the equivalence point.

23. (ii)

# mL of NaOH added	$[H_3O^+]$	$[OH^-]$	pH	pOH
0	6.2×10^{-4}	1.6×10^{-11}	3.21	10.79
5.00	9.5×10^{-5}	1.0×10^{-10}	4.02	9.98
10.00	3.8×10^{-5}	2.6×10^{-10}	4.42	9.58
15.00	1.9×10^{-5}	5.2×10^{-10}	4.72	9.28
20.00	9.5×10^{-6}	1.0×10^{-9}	5.02	8.98
25.00	3.8×10^{-6}	2.6×10^{-9}	5.42	8.58
29.00	6.6×10^{-7}	1.5×10^{-8}	6.18	7.82
29.50	3.2×10^{-7}	3.1×10^{-8}	6.49	7.51
29.90	6.4×10^{-8}	1.6×10^{-7}	7.20	6.80
30.00	4.4×10^{-9}	2.2×10^{-6}	8.36	5.64
30.10	3.0×10^{-10}	3.3×10^{-5}	9.52	4.48
30.50	6.0×10^{-11}	1.6×10^{-4}	10.22	3.78
31.00	3.0×10^{-11}	3.3×10^{-4}	10.52	3.48
35.00	6.5×10^{-12}	1.5×10^{-3}	11.19	2.81
40.00	3.5×10^{-12}	2.9×10^{-3}	11.46	2.54
45.00	2.5×10^{-12}	4.0×10^{-3}	11.60	2.40

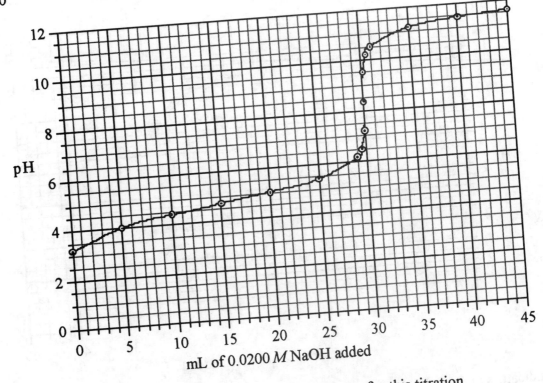

Bromothymol blue and neutral red could be used as indicators for this titration.

24. (i)

# mL of HCl added	[H₃O⁺]	[OH⁻]	pH	pOH
0	5.0 x 10⁻¹³	2.0 x 10⁻²	12.30	1.70
5.00	7.0 x 10⁻¹³	1.4 x 10⁻²	12.15	1.85
10.00	1.0 x 10⁻¹²	1.0 x 10⁻²	12.00	2.00
15.00	1.5 x 10⁻¹²	6.7 x 10⁻³	11.82	2.18
20.00	2.5 x 10⁻¹²	4.0 x 10⁻³	11.60	2.40
25.00	5.6 x 10⁻¹²	1.8 x 10⁻³	11.26	2.74
29.00	2.9 x 10⁻¹¹	3.4 x 10⁻⁴	10.53	3.47
29.50	5.9 x 10⁻¹¹	1.7 x 10⁻⁴	10.23	3.77
29.90	3.0 x 10⁻¹⁰	3.3 x 10⁻⁵	9.52	4.48
30.00	1.0 x 10⁻⁷	1.0 x 10⁻⁷	7.00	7.00
30.10	3.3 x 10⁻⁵	3.0 x 10⁻¹⁰	4.48	9.52
30.50	1.6 x 10⁻⁴	6.0 x 10⁻¹¹	3.78	10.22
31.00	3.3 x 10⁻⁴	3.0 x 10⁻¹¹	3.48	10.52
35.00	1.5 x 10⁻³	6.5 x 10⁻¹²	2.81	11.19
40.00	2.9 x 10⁻³	3.5 x 10⁻¹²	2.54	11.46
45.00	4.0 x 10⁻³	2.5 x 10⁻¹²	2.40	11.60

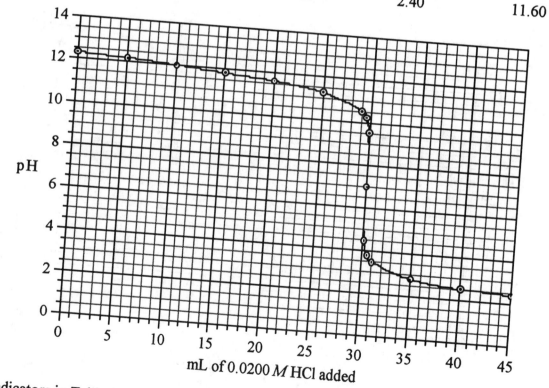

All indicators in Table 19-3 except methyl orange could be used.

24. (ii)

# mL of HCl added	[H$_3$O$^+$]	[OH$^-$]	pH	pOH
0	3.1 x 10^{-12}	3.2 x 10^{-3}	11.51	2.49
5.00	4.0 x 10^{-12}	2.5 x 10^{-3}	11.40	2.60
10.00	1.0 x 10^{-11}	1.0 x 10^{-3}	11.00	3.00
15.00	2.0 x 10^{-11}	5.0 x 10^{-4}	10.70	3.30
20.00	4.0 x 10^{-11}	2.5 x 10^{-4}	10.40	3.60
25.00	1.0 x 10^{-10}	1.0 x 10^{-4}	10.00	4.00
29.00	5.8 x 10^{-10}	1.7 x 10^{-5}	9.24	4.76
29.50	1.2 x 10^{-9}	8.5 x 10^{-6}	8.93	5.07
29.90	6.0 x 10^{-9}	1.7 x 10^{-6}	8.22	5.78
30.00	4.5 x 10^{-7}	2.2 x 10^{-8}	6.35	7.65
30.10	3.3 x 10^{-5}	3.0 x 10^{-10}	4.48	9.52
30.50	1.6 x 10^{-4}	6.0 x 10^{-11}	3.78	10.22
31.00	3.3 x 10^{-4}	3.0 x 10^{-11}	3.48	10.52
35.00	1.5 x 10^{-3}	6.5 x 10^{-12}	2.81	11.19
40.00	2.9 x 10^{-3}	3.5 x 10^{-12}	2.54	11.46
45.00	4.0 x 10^{-3}	2.5 x 10^{-12}	2.40	11.60

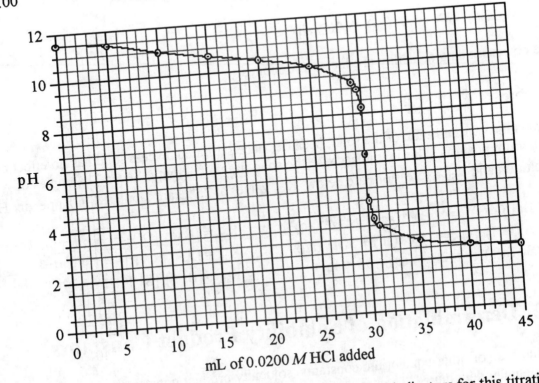

Methyl red, bromothymol blue, and neutral red could be used as indicators for this titration.

Ionic Equilibria III: The Solubility Product Principle

20-1 Introduction

Most "insoluble" substances are actually soluble to a very small extent (less than 10^{-2} M). Ionic compounds dissociate into ions as they dissolve in water to produce a saturated solution. The equilibrium constant for the dissolution process is called a **solubility product constant, K_{sp}**. The general dissolution reaction may be represented as:

$$M_nX_y(s) \rightleftharpoons nM^{y+}(aq) + yX^{n-}(aq)$$

The corresponding solubility product expression is:

$$K_{sp} = [M^{y+}]^n[X^{n-}]^y$$

It omits the concentration of the solid, because, once saturated, excess solid has no effect on the equilibrium position. This is true because the solid has unit activity. Several examples of dissolution and K_{sp} expressions are below. K_{sp} values at 25°C are tabulated in Appendix H.

$BaCO_3(s) \rightleftharpoons Ba^{2+}(aq) + CO_3^{2-}(aq)$ $K_{sp} = [Ba^{2+}][CO_3^{2-}] = 8.1 \times 10^{-9}$

$Al(OH)_3(s) \rightleftharpoons Al^{3+}(aq) + 3OH^-(aq)$ $K_{sp} = [Al^{3+}][OH^-]^3 = 1.9 \times 10^{-33}$

$Bi_2S_3(s) \rightleftharpoons 2Bi^{3+}(aq) + 3S^{2-}(aq)$ $K_{sp} = [Bi^{3+}]^2[S^{2-}]^3 = 1.6 \times 10^{-72}$

20-2 Determination of Solubility Product Constants

As in the case of other equilibrium constants, solubility product constants must be experimentally determined. They can be determined by using solubility data as seen in the next two examples.

Example 20-1 Solubility Product Constant

The molar solubility of mercury(I) thiocyanate is $1.98 \times 10^{-7} M$ at 25°C. This means that 1.98×10^{-7} mole of $Hg_2(SCN)_2$ dissolves to give one liter of saturated solution. Calculate K_{sp} for $Hg_2(SCN)_2$.

Plan

We know the molar solubility of this salt and recognize that this is a strong electrolyte that will be totally dissociated in solution. We write the appropriate equation, determine the concentrations of the ions by considering the stoichiometry and finally substitute those concentrations into the K_{sp} expression.

Solution

The dissolution process can be represented as follows:

$$Hg_2(SCN)_2(s) \rightleftharpoons Hg_2^{2+}(aq) + 2SCN^-(aq)$$

solubility: $1.98 \times 10^{-7} M \Rightarrow 1.98 \times 10^{-7} M \quad 2(1.98 \times 10^{-7} M)$

$$K_{sp} = [Hg_2^{2+}][SCN^-]^2 = [1.98 \times 10^{-7}][2(1.98 \times 10^{-7})]^2 = \boxed{3.10 \times 10^{-20}}$$

Example 20-2 Solubility Product Constant

The solubility of $Ba_3(PO_4)_2$ is 3.94×10^{-5} g/100 mL of water at 25°C. Evaluate the K_{sp} of $Ba_3(PO_4)_2$ at this temperature.

Plan

First, we must convert the concentration to molar solubility. We then write the dissolution equation and proceed as in the previous example.

Solution

$$? M = \frac{3.94 \times 10^{-5} \text{ g } Ba_3(PO_4)_2}{100 \text{ mL}} \times \frac{1000 \text{ mL}}{1 \text{ L}} \times \frac{1 \text{ mol } Ba_3(PO_4)_2}{601.9 \text{ g } Ba_3(PO_4)_2}$$

$$= 6.55 \times 10^{-7} M \; Ba_3(PO_4)_2$$

Now the equilibrium concentrations of the ions can be determined.

solubility: $Ba_3(PO_4)_2(s)$ $\rightleftharpoons$ $3Ba^{2+}(aq) + 2PO_4^{3-}(aq)$
$6.55 \times 10^{-7} M$ $\Rightarrow$ $3(6.55 \times 10^{-7})M$ $2(6.55 \times 10^{-7})M$

$K_{sp} = [Ba^{2+}]^3[PO_4^{3-}]^2 = [3(6.55 \times 10^{-7})]^3[2(6.55 \times 10^{-7})]^2 = \boxed{1.30 \times 10^{-29}}$

20-3 Uses of Solubility Product Constants

Molar solubilities and concentrations of constituent ions in saturated solutions may be calculated from the K_{sp} values.

Example 20-3 Molar Solubility

At 25°C, the solubility product constant for $CaCO_3$ is 4.8×10^{-9}. Calculate the molar solubility of $CaCO_3$ and the Ca^{2+} and CO_3^{2-} concentrations in a saturated aqueous solution. How many grams of $CaCO_3$ must dissolve to give 500.0 mL of saturated solution?

Plan

We are given the value of the K_{sp}. We write the appropriate dissolution equation in terms of x the molar solubility. We substitute into the K_{sp} expression to solve for x, then use dimensional analysis to solve for the mass of $CaCO_3$ required to make the desired solution.

Solution

The dissolution equation is:

solubility: $CaCO_3(s)$ $\rightleftharpoons$ $Ca^{2+}(aq) + CO_3^{2-}(aq)$
xM $\Rightarrow$ xM xM

$K_{sp} = [Ca^{2+}][CO_3^{2-}] = (x)(x) = x^2 = 4.8 \times 10^{-9}$

$x = \boxed{6.9 \times 10^{-5} M}$ = molar solubility of $CaCO_3$ in water

Also, $x = [Ca^{2+}] = [CO_3^{2-}] = \boxed{6.9 \times 10^{-5} M}$

Finally, the molar solubility can be used to determine how many grams of $CaCO_3$ dissolve to give 500.0 mL of saturated solution.

$$? \text{ g CaCO}_3 = \frac{6.9 \times 10^{-5} \text{ mol CaCO}_3}{1000 \text{ mL}} \times 500.0 \text{ mL} \times \frac{100.1 \text{ g CaCO}_3}{1 \text{ mol CaCO}_3} = \boxed{3.5 \times 10^{-3} \text{ g CaCO}_3}$$

Example 20-4 Molar Solubility

What is the molar solubility of $Sr_3(AsO_4)_2$ and the concentration of Sr^{2+} and AsO_4^{3-} in a saturated solution at 25°C? The K_{sp} is 1.3×10^{-18}.

Plan
We are given the value of the K_{sp}. We write the appropriate dissolution equation in terms of x the molar solubility. We substitute into the K_{sp} expression to solve for x and then use stoichiometry to solve for the ion concentrations.

Solution
The dissolution equation is:

$$\begin{array}{cccc} & Sr_3(AsO_4)_2(s) & \rightleftharpoons & 3Sr^{2+}(aq) + 2AsO_4^{3-}(aq) \\ \text{solubility:} & xM & \Rightarrow & 3xM \quad\quad 2xM \end{array}$$

$$K_{sp} = [Sr^{2+}]^3[AsO_4^{3-}]^2 = (3x)^3(2x)^2 = 108x^5 = 1.3 \times 10^{-18}$$

$x^5 = (1.2 \times 10^{-20})$, so $x = (1.2 \times 10^{-20})^{1/5}$ (can use inv y^x on calculator to evaluate)

$x = \boxed{1.0 \times 10^{-4} M}$ = molar solubility of $Sr_3(AsO_4)_2$

$[Sr^{2+}] = 3x = 3(1.0 \times 10^{-4} M) = \boxed{3.0 \times 10^{-4} M}$

$[AsO_4^{3-}] = 2x = 2(1.0 \times 10^{-4} M) = \boxed{2.0 \times 10^{-4} M}$

20-4 The Common Ion Effect

The common ion effect applies to solubility equilibria just as it does to other ionic equilibria. By LeChatelier's Principle, the equilibrium shifts to the left when a common ion is added. Presence of a common ion from another compound, barring additional reactions caused by that ion, causes a decrease in the solubility of the compound of interest.

Example 20-5 Molar Solubility and the Common Ion Effect

Answer the question of Example 20-3, using a solution of 0.20 M Na_2CO_3 as the solvent instead of pure water.

Plan

We recognize that Na_2CO_3 is a soluble ionic salt that completely dissociates into ions, whereas, $CaCO_3$ is a slightly soluble salt. CO_3^{2-} is the common ion. We write the appropriate chemical equations, determine the concentrations of the ions and substitute into the K_{sp} expression. Once we have determined the concentrations, we can find the mass of $CaCO_3$ required as in Example 20-3.

Solution

The dissolution equations are:

dissociation:
$$Na_2CO_3(s) \rightleftharpoons 2Na^+(aq) + CO_3^{2-}(aq)$$
$$0.20\ M \Rightarrow 2(0.20)\ M \quad\quad 0.20\ M$$

solubility:
$$CaCO_3(s) \rightleftharpoons Ca^{2+}(aq) + CO_3^{2-}(aq)$$
$$x\ M \Rightarrow x\ M \quad\quad x\ M$$

The total concentration of CO_3^{2-} in the saturated solution is $(0.20 + x)\ M$. Because the molar solubility x was very small in water, it should be even smaller in this solution with a common ion; therefore, we assume $(0.20 + x)\ M \approx 0.20M$.

$$K_{sp} = [Ca^{2+}][CO_3^{2-}] = (x)(0.20) = 4.8 \times 10^{-9}$$

$x = \boxed{2.4 \times 10^{-8}\ M}$ = molar solubility of $CaCO_3$ in Na_2CO_3 solution = $[Ca^{2+}]$

$[CO_3^{2-}] = (0.20 + x)\ M = (0.20 + 2.4 \times 10^{-8}) = \boxed{0.20\ M}$

$$?\ g\ CaCO_3 = \frac{2.4 \times 10^{-8}\ mol\ CaCO_3}{1000\ mL} \times 500.0\ mL \times \frac{100.1\ g\ CaCO_3}{1\ mol\ CaCO_3} = \boxed{1.2 \times 10^{-6}\ g\ CaCO_3}$$

Note that the molar solubility and the mass of $CaCO_3$ required for a saturated solution are much less than before.

Solubility product constants can also be used to determine whether or not a precipitate will form when two solutions are mixed or a compound is added to a solution. The ion product, Q_{sp} is compared to the K_{sp} to aid in this determination.

If $Q_{sp} > K_{sp}$, a precipitate will form.
If $Q_{sp} = K_{sp}$, the solution is saturated.
If $Q_{sp} < K_{sp}$, more solid could dissolve (unsaturated).

Example 20-6 Predicting Precipitation

Will SnI_2 precipitate if 50.0 mL of 0.010 M KI and 100.0 mL of 0.015 M $Sn(NO_3)_2$ solutions are mixed? The K_{sp} for SnI_2 is 1.0×10^{-4}.

Plan
We are mixing two solutions of soluble ionic salts. We need to find the molarity of each at the instant of mixing. We then find the concentration of each ion in the new solution using the dilution formula. The ions Sn^{2+} and I^- could form an insoluble precipitate. We calculate Q_{sp} for this combination and compare it to the K_{sp}.

Solution
After mixing, but before any reaction occurs, the total volume is 150.0 mL and the concentrations are:

$$[Sn(NO_3)_2] = \frac{(100.0 \text{ mL})(0.015 M)}{150.0 \text{ mL}} = 1.0 \times 10^{-2} M$$

$$[KI] = \frac{(50.0 \text{ mL})(0.010 M)}{150.0 \text{ mL}} = 3.3 \times 10^{-3} M$$

Both the $Sn(NO_3)_2$ and KI are soluble and completely dissociated in dilute solutions.

$Sn(NO_3)_2(aq) \rightarrow Sn^{2+}(aq) \quad + \quad 2NO_3^-(aq)$
$1.0 \times 10^{-2} M \Rightarrow 1.0 \times 10^{-2} M \quad\quad 2(1.0 \times 10^{-2}) M$

$KI \rightarrow K^+(aq) \quad + \quad I^-(aq)$
$3.3 \times 10^{-3} M \Rightarrow 3.3 \times 10^{-3} M \quad\quad 3.3 \times 10^{-3} M$

Precipitation will occur if $Q_{sp} > K_{sp}$ or if the solution is "more than saturated".

$$Q_{sp} = [Sn^{2+}]_{initial}[I^-]^2_{initial} = (1.0 \times 10^{-2})(3.3 \times 10^{-3})^2 = 1.1 \times 10^{-7}$$

Since $Q_{sp} < K_{sp}$, no precipitation will occur.

Example 20-7 Common Ion Effect

How many grams of solid KI must be added to 50.0 mL of 0.10 M $Sn(NO_3)_2$ solution to just initiate precipitation of SnI_2. For SnI_2, $K_{sp} = 1.0 \times 10^{-4}$.

Plan

We know that $Sn(NO_3)_2$ is soluble and completely ionized. We write the appropriate equation to determine the initial Sn^{2+} concentration. We know that precipitation will occur when $Q_{sp} > K_{sp}$. We substitute the initial Sn^{2+} concentration into the Q_{sp} expression and solve for the concentration of I^- needed. Finally, we use dimensional analysis to determine the mass of KI required to cause precipitation.

Solution

$$Sn(NO_3)_2(aq) \rightarrow Sn^{2+}(aq) + 2NO_3^-(aq)$$
$$0.10\ M \Rightarrow 0.10\ M \quad\quad 2(0.10)\ M$$

Precipitation will begin when $Q_{sp} > K_{sp}$ or when:

$$[Sn^{2+}][I^-]^2 > 1.0 \times 10^{-4}$$

$$[I^-]^2 > \frac{1.0 \times 10^{-4}}{[Sn^{2+}]} = 1.0 \times 10^{-3}$$

$$[I^-] = \sqrt{1.0 \times 10^{-3}} = 3.2 \times 10^{-2}$$

$[I^-] = [KI] > 3.2 \times 10^{-2}\ M$ (At $[I^-] = 3.2 \times 10^{-2}$ M, the solution will be just saturated)

The mass of [KI] needed to initiate precipitation is:

$$?\ g\ KI > 50.0\ mL \times \frac{3.2 \times 10^{-2}\ mol\ KI}{1000\ mL} \times \frac{166.0\ g\ KI}{1\ mol\ KI} = 0.27\ g\ KI$$

Precipitation is initiated with $\boxed{\text{slightly greater than 0.27 g KI}}$.

Example 20-8 Concentration of a Common Ion

What concentration of PO_4^{3-} is required to precipitate 99.999% of the nickel from a solution that is $1.00 \times 10^{-3} M$ $Ni(NO_3)_2$? The K_{sp} of $Ni_3(PO_4)_2$ is 4.7×10^{-32}.

Plan

We use the percentage to determine the amount of nickel(II) that can remain in the solution. We then substitute that molarity into the K_{sp} expression to determine the molarity of PO_4^{3-} in equilibrium with that concentration.

Solution

If 99.999% of the nickel is to be removed, then only 0.001% (1.0×10^{-5} as a fraction) can remain in solution or :

$$[Ni^{2+}]_{remaining} = (1.0 \times 10^{-3} M) \times (1.0 \times 10^{-5}) = 1.0 \times 10^{-8} M$$

$$K_{sp} = [Ni^{2+}]^3[PO_4^{3-}]^2 = 4.7 \times 10^{-32}$$

$$[PO_4^{3-}]^2 = \frac{4.7 \times 10^{-32}}{[Ni^{2+}]^3} = \frac{4.7 \times 10^{-32}}{(1.0 \times 10^{-8})^3} = 4.7 \times 10^{-8}$$

$$[PO_4^{3-}] = \sqrt{4.7 \times 10^{-8}} = \boxed{2.2 \times 10^{-4} M}$$

20-5 Fractional Precipitation

Often chemically similar ions have differing solubility properties. The process by which one ion is separated from a solution of chemically similar ions by precipitation is called **fractional precipitation**.

Example 20-9 Concentration Needed to Initiate Precipitation

We add small incremental amounts of solid NaOH to an acidic solution with $1.0 \times 10^{-3} M$ concentrations of each of the three soluble salts $Bi(NO_3)_3$, $Cr(NO_3)_3$, and $Co(NO_3)_3$. Determine the order in which the metal hydroxides precipitate and the concentration of OH^- necessary to begin precipitation of each.

K_{sp} of $Bi(OH)_3 = 3.2 \times 10^{-40}$; K_{sp} of $Cr(OH)_3 = 6.7 \times 10^{-31}$; K_{sp} of $Co(OH)_3 = 4.0 \times 10^{-45}$

Plan

In each case the initial metal ion concentration is 1.0×10^{-3} M. We know that each metal hydroxide begins to precipitate when its ion product exceeds its K_{sp}. As in previous examples, we will calculate the concentration of [OH⁻] needed to initiate precipitation of each and compare those concentrations.

Solution

Precipitation begins when $Q_{sp} > K_{sp}$. The [OH⁻] needed to cause Bi(OH)₃ to begin to precipitate is:

$$[Bi^{3+}][OH^-]^3 > 3.2 \times 10^{-40} \text{ therefore, } [OH^-]^3 > \frac{3.2 \times 10^{-40}}{[Bi^{3+}]}$$

$$[OH^-]^3 > \frac{3.2 \times 10^{-40}}{1.0 \times 10^{-3}} \text{ and } [OH^-]^3 = 3.2 \times 10^{-37}$$

$$[OH^-] > \sqrt[3]{3.2 \times 10^{-37}} \text{ so } [OH^-] > \boxed{6.8 \times 10^{-13} \, M}$$

In order to begin precipitation of Bi(OH)₃, [OH⁻] > 6.8×10^{-13} M.

Similarly, for Cr(OH)₃ precipitation, $Q_{sp} > K_{sp}$ and:

$$[Cr^{3+}][OH^-]^3 > 6.7 \times 10^{-31} \text{ therefore, } [OH^-]^3 > \frac{6.7 \times 10^{-31}}{[Cr^{3+}]}$$

$$[OH^-]^3 > \frac{6.7 \times 10^{-31}}{1.0 \times 10^{-3}} \text{ and } [OH^-]^3 > 6.7 \times 10^{-28}$$

$$[OH^-] > \sqrt[3]{6.7 \times 10^{-28}} \text{ so } [OH^-] > \boxed{8.8 \times 10^{-10} \, M}$$

In order to begin precipitation of Cr(OH)₃, [OH⁻] > 8.8×10^{-10} M.

And for Co(OH)₃ precipitation, $Q_{sp} > K_{sp}$ and:

$$[Co^{3+}][OH^-]^3 > 4.0 \times 10^{-45} \text{ therefore } [OH^-]^3 > \frac{4.0 \times 10^{-45}}{[Co^{3+}]}$$

$$[OH^-]^3 > \frac{4.0 \times 10^{-45}}{1.0 \times 10^{-3}} \text{ and } [OH^-]^3 > 4.0 \times 10^{-42}$$

$[OH^-] > \sqrt[3]{4.0 \times 10^{-42}}$ so $[OH^-] > \boxed{1.6 \times 10^{-14} M}$

In order to begin precipitation of $Co(OH)_3$, $[OH^-] > 1.6 \times 10^{-14} M$.

$Co(OH)_3$ requires the lowest concentration of hydroxide ion to begin precipitating while $Cr(OH)_3$ requires the most. The order of precipitation is:
$\boxed{Co(OH)_3 \text{ before } Bi(OH)_3 \text{ before } Cr(OH)_3}$.

Example 20-10 Fractional Precipitation

Refer to Example 20-9 and determine the percentage of Co^{3+} still in solution at the point at which just enough NaOH has been added to remove as much Co^{3+} as possible without precipitating any $Bi(OH)_3$. Also determine the percentages of Co^{3+} and Bi^{3+} still in solution at the point at which the maximum amount of Bi^{3+} has been removed without any precipitation of $Cr(OH)_3$.

Plan

We know the molarity of hydroxide ion needed to initiate each precipitation. We use these values of $[OH^-]$ with the appropriate K_{sp} expression, in turn, to find the concentration of each metal ion that remains in solution. We express these concentrations as the percent of the concentration unprecipitated.

Solution

Precipitation of $Co(OH)_3$ begins when $[OH^-]$ exceeds $1.6 \times 10^{-14} M$. Only $Co(OH)_3$ precipitates until $[OH^-]$ just exceeds $6.8 \times 10^{-13} M$, at which point the $Bi(OH)_3$ begins to precipitate. The $[Co^{3+}]$ remaining in solution at this point is calculated below.

$$K_{sp} = [Co^{3+}][OH^-]^3 = [Co^{3+}](6.8 \times 10^{-13})^3 = 4.0 \times 10^{-45}$$

$$[Co^{3+}] = \frac{4.0 \times 10^{-45}}{(6.8 \times 10^{-13})^3} = \boxed{1.3 \times 10^{-8} M}$$

The percent $[Co^{3+}]$ still in solution is:

$$\%[Co^{3+}] = \frac{[Co^{3+}]_{solution}}{[Co^{3+}]_{original}} \times 100 = \frac{1.3 \times 10^{-8} M}{1.0 \times 10^{-3} M} \times 100 = \boxed{0.0013\%}$$

In order for precipitation of $Cr(OH)_3$ to begin, $[OH^-]$ must just exceed $8.8 \times 10^{-10}\,M$. The $[Bi^{3+}]$ still in solution at this point can be calculated.

$$K_{sp} = [Bi^{3+}][OH^-]^3 = [Bi^{3+}](8.8 \times 10^{-10})^3 = 3.2 \times 10^{-37}$$

$$[Bi^{3+}] = \frac{3.2 \times 10^{-37}}{(8.8 \times 10^{-10})^3} = \boxed{4.7 \times 10^{-10}\,M}$$

The percent $[Bi^{3+}]$ remaining is:

$$\%[Bi^{3+}] = \frac{[Bi^{3+}]_{solution}}{[Bi^{3+}]_{original}} \times 100 = \frac{4.7 \times 10^{-10}\,M}{1.0 \times 10^{-3}\,M} \times 100 = \boxed{4.7 \times 10^{-5}\%}$$

The $[Co^{3+}]$, still in solution, just before the $Cr(OH)_3$ begins to precipitate can be found in the same manner.

$$K_{sp} = [Co^{3+}][OH^-]^3 = [Co^{3+}](8.8 \times 10^{-10})^3 = 4.0 \times 10^{-45}$$

$$[Co^{3+}] = \frac{4.0 \times 10^{-45}}{(8.8 \times 10^{-10})^3} = \boxed{5.9 \times 10^{-18}\,M}$$

The percent $[Co^{3+}]$ still in solution is:

$$\frac{[Co^{3+}]_{solution}}{[Co^{3+}]_{original}} \times 100 = \frac{5.9 \times 10^{-18}\,M}{1.0 \times 10^{-3}\,M} \times 100 = \boxed{5.9 \times 10^{-13}\%}$$

It can be seen that 99.99987 % of the Co^{3+} precipitates before $Bi(OH)_3$ begins to precipitate, and 99.99991% of the Bi^{3+} and essentially 100% of the Co^{3+} precipitate before precipitation of the $Cr(OH)_3$ begins. For this system of ions, selective precipitation using hydroxide concentration, which can be monitored with a pH meter, provides an extremely effective means of separating the different metals.

20-6 Simultaneous Equilibria Involving Slightly Soluble Compounds

Many times acid/base equilibria are used to adjust concentrations of precipitating ions. This is especially effective for ions such as fluoride, cyanide and sulfide ions, which are conjugate bases of weak acids, and for hydroxide ions, which can be adjusted by buffering to specific pH's.

Example 20-11 Preventing Precipitation with Simultaneous Equilibria

If we wish to prepare a solution that is 0.010 M in $Cd(NO_3)_2$ and 0.010 M in aqueous NH_3, what concentration of NH_4NO_3 must we add to prevent precipitation of $Cd(OH)_2$? K_{sp} for $Cd(OH)_2 = 1.2 \times 10^{-14}$ and K_b for $NH_3 = 1.8 \times 10^{-5}$.

Plan

Because we know the K_{sp} for $Cd(OH)_2$ and the $[Cd^{2+}]$ in solution (from the complete dissociation of $Cd(NO_3)_2$) are known, the minimum $[OH^-]$ that would cause precipitation of $Cd(OH)_2$ can be calculated. We can use that minimum concentration in the K_b expression and solve for the $[NH_4^+]$ needed to maintain the $[OH^-]$ at that level as we did for buffer problems in Chapter 18.

Solution

First, we determine the maximum $[OH^-]$ allowable. This occurs when the solution is saturated, which means that the ion product is equal to K_{sp}.

$$K_{sp} = [Cd^{2+}][OH^-]^2 = (0.010)[OH^-]^2 = 1.2 \times 10^{-14}$$

$$[OH^-]^2 = \frac{1.2 \times 10^{-14}}{0.010} = 1.2 \times 10^{-12}$$

$$[OH^-] = \sqrt{1.2 \times 10^{-12}} = 1.1 \times 10^{-6} M$$

Now we can calculate the concentration of NH_4NO_3 necessary to buffer the 0.010 M NH_3 solution to give $[OH^-] = 1.1 \times 10^{-6} M$. NH_4NO_3 ionizes completely and suppresses the ionization of NH_3 by the common ion effect. Let x be the concentration of NH_4NO_3 needed.

$$NH_4NO_3 \xrightarrow{100\%} NH_4^+ + NO_3^-$$
$$x M \qquad\qquad x M \quad\; x M$$

Ionization of NH_3 must produce no more than $1.1 \times 10^{-6} M$ $[OH^-]$.

$$NH_3 + H_2O \rightleftharpoons NH_4^+ + OH^-$$
$$(0.010 - 1.1 \times 10^{-6}) M \qquad 1.1 \times 10^{-6} M \quad 1.1 \times 10^{-6} M$$

NH_4^+ is produced by both NH_4NO_3 and aqueous NH_3. It is obvious that for NH_3, $(0.010 - 1.1 \times 10^{-6}) M \approx 0.010 M$. Assume that the total $[NH_4^+]$, which equals $(x + 1.1 \times 10^{-6}) M \approx x M$. The K_b expression is now used to compute x, the $[NH_4^+]$.

$$K_b = \frac{[NH_4^+][OH^-]}{[NH_3]} = \frac{(x)(1.1 \times 10^{-6})}{0.010} = 1.8 \times 10^{-5}$$

$$x = [NH_4^+] = \frac{(1.8 \times 10^{-5})(0.010)}{1.1 \times 10^{-6}} = \boxed{0.16\ M}$$

The solution must be made $0.16\ M$ in $[NH_4^+]$ to prevent the precipitation.

($0.16\ M \gg 1.1 \times 10^{-6}$, so the assumption above is valid.)

20-7 Dissolving Precipitates

Precipitates can be dissolved by three main methods: (1) converting the ions in a saturated solution of a precipitate to a weak (predominantly unionized) electrolyte, which causes the ion product, Q_{sp}, to be less than K_{sp}; (2) converting one of the species by oxidation or reduction, thus decreasing the ion product; and, (3) formation of a soluble complex compound from one ion, effectively removing it from the solution and again, decreasing the ion product.

Complex ion formation occurs when certain electron donating species replace water molecules coordinated to aqueous metal ions. For example, Fe^{3+}(aq) ions, or more accurately $[Fe(OH_2)_6]^{3+}$ ions react with cyanide ions to form the hexacyanoferrate(III) complex ion.

$$[Fe(OH_2)_6]^{3+} + 6CN^- \rightleftharpoons [Fe(CN)_6]^{3-} + 6H_2O$$

Many complex ions are very stable, i.e., they dissociate only very slightly. Their stabilities are indicated by their dissociation constants, K_d, which are equilibrium constants for dissociation reactions (the reverse of the equation above). For example, the overall dissociation reaction for $[Fe(CN)_6]^{3-}$ ions can be represented:

$$[Fe(CN)_6]^{3-} + 6H_2O \rightleftharpoons [Fe(OH_2)_6]^{3+} + 6CN^-$$

or more simply, $[Fe(CN)_6]^{3+} \rightleftharpoons Fe^{3+} + 6CN^-$

$$K_d = \frac{[[Fe(OH_2)_6]^{3+}][CN^-]^6}{[[Fe(CN)_6]^{3-}]} \text{ or } \frac{[Fe^{3+}][CN^-]^6}{[[Fe(CN)_6]^{3-}]}$$

Dissociation constants for complex ions are tabulated in Appendix I in the text.

Example 20-12 Complex Ion Calculations

Calculate the concentrations of the various dissolved species in 0.25 M [Co(NH$_3$)$_6$](OH)$_3$ solution. K_d for [Co(NH$_3$)$_6$]$^{3+}$ is 2.2×10^{-34}.

Plan
We write the equation for the complete dissociation of this compound into its constituent ions, which gives us the complex ion of interest. We then write the equation for the dissociation of the complex ion, express the concentrations algebraically, and substitute into the dissociation expression.

Solution
The soluble compound is essentially completely dissociated into its constituent ions:

$$[Co(NH_3)_6](OH)_3 \xrightarrow{100\%} [Co(NH_3)_6]^{3+} + 3OH^-$$
$$0.25\ M \Rightarrow 0.25\ M \quad\quad 3(0.25)\ M$$

The complex ion, in turn, only slightly dissociates. Let x be the concentration of free Co^{3+} produced by that dissociation.

$$[Co(NH_3)_6]^{3+} \rightleftharpoons Co^{3+} + 6NH_3$$
$$(0.25-x)\ M \Rightarrow x\ M \quad\quad 6x\ M$$

$$K_d = \frac{[Co^{3+}][NH_3]^6}{[[Co(NH_3)_6]^{3+}]} = \frac{(x)(6x)^6}{(0.25-x)} = 2.2 \times 10^{-34}$$

Assume that $(0.25-x) \approx 0.25$ since K_d is very small, then:

$$\frac{(x)(6x)^6}{0.25} = \frac{46656\ x^7}{0.25} = 2.2 \times 10^{-34};\ x^7 = 1.2 \times 10^{-39}$$

$$x = [Co^{3+}] = (1.2 \times 10^{-39})^{1/7} = \boxed{2.8 \times 10^{-6}\ M}$$

$$[NH_3] = 6x = 6(2.8 \times 10^{-6}\ M) = \boxed{1.7 \times 10^{-5}\ M}$$

$$[[Co(NH_3)_6]^{3+}] = 0.25 - x = 0.25\ M - 2.8 \times 10^{-6} = \boxed{0.25\ M}$$

$$[OH^-] = \boxed{0.75\ M}$$

In Example 20-12, the presence of the high concentration of OH⁻ ion from the dissociation of the complex compound into its constituent ions prevents any significant ionization of NH_3 as a base. You can perform calculations to confirm this fact.

Example 20-13 Complex Ion Calculations

Calculate the mass of gaseous NH_3 that must be added to one liter of pure water to dissolve 0.0010 mole of $Co(OH)_3$. Assume no volume change occurs. K_{sp} for $Co(OH)_3 = 4.0 \times 10^{-45}$ and K_d for $[Co(NH_3)_6]^{3+} = 2.2 \times 10^{-34}$.

Plan

We write the appropriate chemical equations and equilibrium constants for the two reversible reactions. By considering stoichiometry, we can determine the amount of NH_3 required to convert all the $Co(OH)_3$ to the complex ion. We then use the dissociation equilibrium to determine the molarity of NH_3 that will be in equilibrium with that ion. The sum of those two numbers will be the molarity of NH_3 required for the dissolution. We can then use dimensional analysis to find the mass of NH_3 required.

Solution

The amount of $Co(OH)_3$ that is to be dissolved has an apparent molarity of:

$$? M\ Co(OH)_3 = \frac{0.0010\ \text{mol Co(OH)}_3}{1\ L} = 0.0010\ M$$

The overall reaction may be represented as:

reaction: $Co(OH)_3(s) + 6NH_3(aq) \rightleftharpoons [Co(NH_3)_6]^{3+}(aq) + 3OH^-(aq)$
            0.0010 M   6(0.0010) M $\Rightarrow$ 0.0010 M   3(0.0010) M

The reaction ratio indicates that $6(0.0010) = 0.0060\ M\ NH_3$ must combine with the 0.0010 M $Co(OH)_3$ to dissolve all the $Co(OH)_3$. Additional NH_3 must also be in solution in equilibrium with the complex ion. The overall reaction consists of two equilibria. The first is the equilibrium for the dissolution of $Co(OH)_3$:

$$Co(OH)_3(s) \rightleftharpoons Co^{3+}(aq) + 3OH^-(aq)$$

$$K_{sp} = [Co^{3+}][OH^-]^3 = 4.0 \times 10^{-45}$$

From the overall reaction, the [OH$^-$] must be 3(0.0010) = 0.0030 M if all the Co(OH)$_3$ dissolves. The [OH$^-$] from the ionization from water is negligible compared to this number. Additionally, this concentration of OH$^-$ will suppress the ionization of the weak base NH$_3$; therefore, negligible amounts of OH$^-$ would be produced from that source. The maximum [Co^{3+}] value is:

$$[Co^{3+}] = \frac{K_{sp}}{[OH^-]^3} = \frac{4.0 \times 10^{-45}}{(0.0030)^3} = 1.5 \times 10^{-37} \, M$$

Now the fact that [Co^{3+}] = 1.5 × 10^{-37} M may be used in the dissociation equilibrium of [Co(NH$_3$)$_6$]$^{3+}$ to calculate the equilibrium concentration of NH$_3$ needed in the solution.

$$[Co(NH_3)_6]^{3+} \rightleftharpoons Co^{3+} + 6NH_3$$
$$(0.0010 - 1.5 \times 10^{-37}) \, M \quad\Rightarrow\quad 1.5 \times 10^{-37} \, M \quad [NH_3]$$

Since (0.0010 - 1.5 × 10^{-37}) $M \approx$ 0.0010 M, the dissociation equilibrium expression is:

$$K_d = \frac{[Co^{3+}][NH_3]^6}{[[Co(NH_3)_6]^{3+}]} = \frac{(1.5 \times 10^{-37})[NH_3]^6}{0.0010} = 2.2 \times 10^{-34}$$

Rearranging yields:

$$[NH_3]^6 = \frac{(2.2 \times 10^{-34})(0.0010)}{(1.5 \times 10^{-37})} = 1.5$$

$$[NH_3] = \sqrt[6]{1.5} = 1.1 \, M \text{ at equilibrium}$$

[NH$_3$]$_{total}$ = [NH$_3$]$_{complexed}$ + [NH$_3$]$_{equilibrium}$

[NH$_3$]$_{total}$ = 0.0060 M + 1.1 $M \approx$ 1.1 M

The mass of NH$_3$ that must be used to dissolve 0.0010 mole of Co(OH)$_3$ in one liter of water can finally be found.

$$? \text{ g NH}_3 = \frac{1.1 \text{ mol NH}_3}{1 \text{ L}} \times 1.0 \text{ L} \times \frac{17.0 \text{ g NH}_3}{1 \text{ mol NH}_3} = \boxed{19 \text{ g NH}_3}$$

EXERCISES

You should consult the Appendixes in the text for the necessary constants.

Solubility Product Constants

1. Write solubility product expressions for the following slightly soluble compounds.
   - (a) $Ca_3(AsO_4)_2$
   - (b) Ni_2S_3
   - (c) $Mg_3(PO_4)_2$
   - (d) $Fe_4[(Fe(CN)_6)]_3$
   - (e) AuI_3
   - (f) $Be_3Al_2(SiO_3)_6$
   - (g) $CsAl(SO_4)_2 \cdot 12H_2O$

2. From the water solubilities given at 25°C, calculate solubility product constants for the following compounds.
   - (a) CeF_3, 1.45×10^{-3} g/100 mL
   - (b) $Cu(SCN)_2$, 4.0×10^{-4} g/100 mL
   - (c) SrF_2, 0.0129 g/100 mL
   - (d) Ag_3PO_4, 6.7×10^{-4} g/100 mL
   - (e) $AuBr_3$, 2.7×10^{-8} g/100 mL

3. Calculate the molar solubilities of the following solid compounds in pure water from their K_{sp} values.
   - (a) Cu_2S
   - (b) Ag_3AsO_4
   - (c) Bi_2S_3
   - (d) MgF_2
   - (e) $AuCl_3$
   - (f) $Ag_4[Fe(CN)_6]$

4. Calculate the masses of the compounds in Exercise 3 in grams per 100 milliliters of saturated solution.

5. A 7.00 g sample of SnI_2 was washed with 100 mL of water. Assuming that the water became saturated, what mass and what percent of the SnI_2 dissolved?

The Common Ion Effect

6. Calculate the molar solubilities of each of the solid compounds in the solutions indicated.
   - (a) Cu_2S in 0.025 M $CuClO_4$
   - (b) $Ca_3(AsO_4)_2$ in 0.030 M $CaCl_2$
   - (c) MgF_2 in 0.015 M NaF
   - (d) Bi_2S_3 in 1.0×10^{-6} M $Bi(NO_3)_3$

(e) $Sr_3(PO_4)_2$ in 0.0025 M $Sr(NO_3)_2$

(f) $Sr_3(PO_4)_2$ in 0.10 M Na_3PO_4

7. Calculate the number of grams of solute that will dissolve to give one liter of saturated solution of the substances listed in Exercise 6.

8. Calculate the molar solubility of $MgNH_4PO_4$ in 2.0 M NH_4NO_3. Neglect hydrolysis.

9. Calculate the concentration of KF that is necessary to initiate precipitation of BaF_2 from 0.11 M $Ba(NO_3)_2$ solution.

10. How many moles and how many grams of solid $CaCl_2$ are needed to begin precipitating AgCl from 250 mL of 0.065M $AgNO_3$ solution?

11. Will precipitation occur if the following pairs of aqueous solutions are mixed? Ignore hydrolysis.

    (a) 15.0 mL of 0.28M $Pb(NO_3)_2$; 55.0 mL of 1.4×10^{-3} M KBr

    (b) 150.0 mL of 0.0010 M $Ca(NO_3)_2$; 750.0 mL of 3.4 M NaF

    (c) 200.0 mL of 1.0×10^{-5} M $SrCl_2$; 100.0 mL of 2.0×10^{-6} M Na_3PO_4

    (d) 20.0 mL of 3.0×10^{-3} M $Pb(NO_3)_2$; 60.0 mL of 1.1×10^{-3} M HI

    (e) 30.0 mL of 0.0200 M $Ca(NO_3)_2$; 370.0 mL of 0.00186 M KOH

12. How many moles of solid NaI must be added to remove 99.9999% of the Hg_2^{2+} as Hg_2I_2 from 650.0 mL of 0.616 M $Hg_2(NO_3)_2$?

13. How many grams of NaOH must be added to 2.00 L of 0.0400 M $Cr(NO_3)_3$ solution to initiate precipitation of $Cr(OH)_3$?

14. A solution is 1.0×10^{-4} M in $Hg_2(NO_3)_2$. If 2.0×10^{-5} mole of solid $BaCl_2$ is added to 500 mL of this solution:

    (a) How many moles and how many grams of precipitate form?

    (b) What percent of the Hg_2^{2+} precipitates?

Fractional Precipitation

15. A solution is 0.10 M in each of the salts, NaCl, NaBr, and NaI. It is necessary to separate the halide ions by fractional precipitation by adding solid $Pb(NO_3)_2$.

    (a) What will be the order of precipitation?

    (b) What is the $[I^-]$ and the percent of I^- precipitated just before any $PbBr_2$ begins to precipitate?

(c) What is the [I⁻] and the percent I⁻ precipitated just before $PbCl_2$ begins to precipitate?

(d) What is the [Br⁻] and the percent of Br⁻ precipitated just before $PbCl_2$ begins to precipitate?

(e) Is it effective to try separating the halide ions as lead salts?

16. A solution is 0.20 M in KCN, 0.20 M in KCl, and 0.20 M in KI. Solid $CuNO_3$ is added. Assume no volume change.

   (a) What is the order of precipitation?

   (b) What is the [CN⁻] and the percent CN⁻ still in solution when CuI begins to precipitate?

   (c) What is the [I⁻] and the percent I⁻ still in solution when the CuCl begins to precipitate?

   (d) Is it effective to try separating these ions as copper(I) salts?

17. A solution is 0.040 M $CaCl_2$ and 0.040 M $CaBr_2$. $Au(NO_3)_3$ is added slowly. What [Br⁻] and what percent [Br⁻] remains in solution when $AuCl_3$ begins to precipitate?

Simultaneous Equilibria and Complex Ion Formation

18. Calculate the [OH⁻], pOH, and pH of a saturated $Zn(OH)_2$ solution.

19. Calculate the pH of a saturated solution of BaF_2. Consider hydrolysis of the F⁻ ion.

20. How many moles of NaCl would be needed to dissolve 1.0×10^{-4} mole of HgS to give 1.0 liter of aqueous solution by formation of the complex ion $HgCl_4^{2-}$? Is this possible?

21. What is the molar solubility of $Cd(OH)_2$ in a buffer that is 0.65 M NH_3 and 0.90 M NH_4Cl?

22. Will $Fe(OH)_2$ precipitate if 150 mL of 1.0×10^{-3} M $Fe(NO_3)_2$ is mixed with 200.0 mL of 4.0×10^{-3} M NH_3?

23. What molarity and how many moles of NH_4Cl would need to be added to the solution to prevent precipitation of $Fe(OH)_2$ in Exercise 22? Assume no volume change when the NH_4Cl is added.

24. How many moles of KCN must be added to dissolve 0.020 mole of AgCl in 1.0 liter of water buffered to pH 5.50. Assume no volume change. The $Ag(CN)_2^-$ complex ion will be formed.

Miscellaneous Exercises

25. A solution is 0.20 M $FeCl_3$ and 0.20 M $CoCl_3$. NaOH is slowly added to the solution.

    (a) What molarity of $[OH^-]$ is needed to initiate precipitation of each $Fe(OH)_3$ and $Co(OH)_3$?

    (b) When $Fe(OH)_3$ begins to precipitate, what is the $[Co^{3+}]$ left in the solution?

    (c) Refer to (b). What is the percent $[Co^{3+}]$ remaining in the solution?

26. What is the solubility of $Cd(OH)_2$ in 0.40 M NH_3?

27. What molarity of $[NH_4^+]$ would be required to prevent precipitation of $Cu(OH)_2$ from a solution of 3.6 x 10^{-10} M $Cu(NO_3)_2$ and 0.31 M NH_3?

28. The molar solubility of $Hg(SCN)_2$ in 0.030 M $Hg(NO_3)_2$ is 4.8 x 10^{-10} M. What is the K_{sp} of $Hg(SCN)_2$?

29. What is the solubility of $Bi(OH)_3$ in a solution buffered to pH 5.00?

30. The solubility of $Cu_3(PO_4)_2$ in water is 6.24 x 10^{-7} g/100 mL. What is the K_{sp} of $Cu_3(PO_4)_2$?

31. To what pH must a 2.0 M $Ni(NO_3)_3$ solution be buffered to prevent precipitation of $Ni(OH)_2$?

32. It is necessary to dissolve 0.010 mol of CuI in one liter of solution buffered to pH 6.50. How many moles of KCN must be added to this solution to achieve this dissolution? The complex ion $[Cu(CN)_2]^-$ will be formed. Also consider the hydrolysis of the CN^- ion, because cyanide is the conjugate base of the very weak acid HCN.

33. What is the pH of a saturated solution of $Mg(OH)_2$?

34. Will precipitation of $Ca(OH)_2$ occur if 0.012 mol of $CaCl_2$ is added to 500 mL of a buffer that is 0.22 M NH_3 and 0.33 M NH_4NO_3?

35. What is the molar solubility of CaF_2 in 2.0 M HF?

36. Will precipitation of $Cr(OH)_3$ occur if 0.00200 mol $Cr(NO_3)_3$ is added to 500 mL of a solution that is buffered to pH 4.30?

ANSWERS TO EXERCISES

1. (a) $[Ca^{2+}]_3[AsO_4^{3-}]_2$ (b) $[Ni^{3+}]_2[S^{2-}]_3$ (c) $[Mg^{2+}]_3[PO_4^{3-}]_2$
   (d) $[Fe^{3+}]_4[Fe(CN)_6]_3$ (e) $[Au^{3+}][I^-]_3$
   (f) $[Be^{2+}]_3[Al^{3+}]_2[SiO_3^{2-}]_6$ (g) $[Cs^+][Al^{3+}][SO_4^{2-}]_2$

2. (a) 7.91×10^{-16} (b) 4.3×10^{-14} (c) 4.33×10^{-9}
   (d) 1.8×10^{-18} (e) 4.3×10^{-36}

3. (a) $7.4 \times 10^{-17} M$ (b) $4.5 \times 10^{-6} M$ (c) $1.7 \times 10^{-15} M$
   (d) $1.2 \times 10^{-3} M$ (e) $1.2 \times 10^{-26} M$ (f) $2.3 \times 10^{-9} M$

4. (a) 1.2×10^{-15} g (b) 2.1×10^{-4} g (c) 8.8×10^{-13} g
   (d) 7.3×10^{-3} g (e) 3.6×10^{-25} g (f) 1.5×10^{-7} g

5. 1.1 g, 16%

6. (a) $2.6 \times 10^{-45} M$ (b) $7.9 \times 10^{-8} M$ (c) $2.8 \times 10^{-5} M$
   (d) $3.9 \times 10^{-21} M$ (e) $2.5 \times 10^{-12} M$ (f) $2.2 \times 10^{-10} M$

7. (a) 4.1×10^{-43} g (b) 3.1×10^{-5} g (c) 1.8×10^{-3} g
   (d) 2.0×10^{-18} g (e) 1.1×10^{-9} g (f) 1.0×10^{-7} g

8. $1.1 \times 10^{-6} M$ 9. $[KF] > 3.9 \times 10^{-3} M$

10. $CaCl_2 = 3.5 \times 10^{-10}$ mol; 3.9×10^{-8} g $CaCl_2$

11. (a) no precipitate (b) yes, CaF_2 (c) yes, $Sr_3(PO_4)_2$
    (d) no precipitate (e) no precipitate

12. The stoichiometric amount (0.800 mol) plus 5.6×10^{-12} mole which remains in solution.

13. 2.0×10^{-8} g NaOH

14. 2.0×10^{-5} mol Hg_2Cl_2, 0.0094 g, 99.99%

15. (a) PbI_2, then $PbBr_2$, then $PbCl_2$
    (b) $[I^-] = 3.7 \times 10^{-3} M$, 96.3% I^- precipitates
    (c) $[I^-] = 2.3 \times 10^{-3} M$, 97.7% I^- precipitates
    (d) $[Br^-] = 6.1 \times 10^{-2} M$, 39% Br^- precipitates
    (e) The method would not be effective, because separation is very poor.

(a) CuCN, then CuI, then CuCl
(b) $[CN^-] = 1.2 \times 10^{-9}\,M$, only $6.2 \times 10^{-7}\,\%$ in solution
(c) $[I^-] = 5.4 \times 10^{-6}\,M$, only $2.7 \times 10^{-3}\,\%$ in solution
(d) The separation would be very effective under these conditions because K_{sp} values are very different.

17. $[Br^-] = 1.9 \times 10^{-5}\,M$, 2.4% left in solution

18. $[OH^-] = 2.2 \times 10^{-6}\,M$, pOH = 5.65, pH = 8.35   19. pH = 7.51

20. 2.3×10^7 mole NaCl, impossible--too concentrated

21. $7.1 \times 10^{-5}\,M\,Cd(OH)_2$ 22. yes

23. $[NH_4^+] = 0.0096\,M$, 3.4×10^{-3} mole NH_4Cl 24. 0.16 mole KCN

25. (a) for $Fe(OH)_3$, $[OH^-] > 6.8 \times 10^{-13}\,M$; for $Co(OH)_3$, $[OH^-] > 2.7 \times 10^{-15}$
    (b) $[Co^{3+}] < 1.3 \times 10^{-8}\,M$ (c) $6.4 \times 10^{-6}\%$ Co^{3+} left in solution

26. $1.6 \times 10^{-9}\,M$ 27. $[NH_4^+] = 0.26\,M$ 28. $K_{sp} = 2.8 \times 10^{-20}$

29. $3.2 \times 10^{-13}\,M$ 30. $K_{sp} = 1.28 \times 10^{-37}$ 31. pH 6.07 or less

32. $[CN^-]_{total} = 0.020\,M$(complexed) $+ 4.5 \times 10^{-5}\,M$ (free) $+ 0.036\,M$ (in HCN) = 0.056 mol; so slightly greater than 0.056 mol KCN are required.

33. pH = 10.52   34. no precipitation   35. $2.7 \times 10^{-8}\,M\,CaF_2$

36. no precipitate

Chapter 21 Electrochemistry

21-1 Introduction

All **electrochemical** reactions involve electron transfer and are, therefore, **oxidation-reduction** reactions. The reacting system is contained in a cell with the reduction and the oxidation reactions physically separated. Electrons are transferred at an electrode surface as the reaction occurs. Conduction occurs either by metallic conduction using wires or by ionic or electrolytic conduction with the charge being carried by ions in solution or pure liquid.

There are two kinds of electrochemical cells: (1) electrolytic; and, (2) voltaic or galvanic cells. In **electrolytic** cells, nonspontaneous redox reactions are forced to occur by passage of electrical current from an external source in a process called electrolysis. In **voltaic** cells, electrical current is produced by spontaneous redox reactions. In both kinds of cells, the **cathode** is the electrode at which reduction occurs and the **anode** is the electrode at which oxidation occurs.

21-2 Electrolytic cells

Essential features of electrolytic cells can be deduced by experimental observation. These include the identification of the oxidation and reduction half-reactions, the overall cell reaction, the anode and cathode, the positive and negative electrodes, and the direction of flow of electrons. For example, we can make a simplified but complete diagram of the cell used for the electrolysis of molten (melted) potassium bromide, KBr, (melting point 730°C), from the following experimental observations. We use inert (nonreactive) electrodes such as platinum for this process.

1. Molten potassium forms at one electrode and floats on top of the molten KBr.
2. Reddish-brown bromine, Br_2, (B.P. 59°C) is produced at the other electrode.

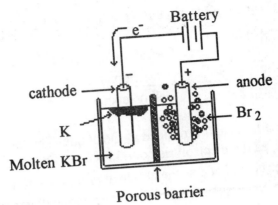

The half-reactions involve the reduction of K^+ to K at the cathode and the oxidation of Br^- to Br_2 at the anode. In the complete cell reaction, the electrons must be balanced as in previous oxidation-reduction equations. Review Chapter Four if necessary.

$$2(K^+ + e^- \rightarrow K) \quad \text{(reduction at cathode)}$$
$$2Br^- \rightarrow Br_2 + 2e^- \quad \text{(oxidation at anode)}$$
$$2K^+(\ell) + 2Br^-(\ell) \rightarrow 2K(\ell) + Br_2(g) \quad \text{(cell reaction)}$$

Since electrons are **forced** to flow by application of an external source of direct current, they move away from the positive electrode (to which they would flow spontaneously) toward the negative electrode through the wire. Therefore, the anode is the positive electrode and the cathode is the negative electrode. This is true in all electrolytic cells.

In a second example, the following observations have been made during the electrolysis of aqueous sulfuric acid, H_2SO_4, using inert electrodes.

1. Gaseous hydrogen, H_2, is produced at one electrode and the solution becomes less acidic around this electrode.
2. Gaseous oxygen, O_2, is produced at the other electrode and the solution becomes more acidic around this electrode.

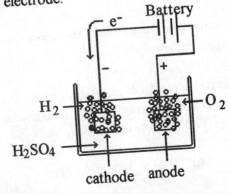

The cathode half-reaction must be the reduction of H^+ from the acid to H_2 while the anode half-reaction must be the oxidation of H_2O to O_2 and H^+.

$$2(2H^+ + 2e^- \rightarrow H_2) \quad \text{(reduction at cathode)}$$
$$2H_2O \rightarrow O_2 + 4H^+ + 4e^- \quad \text{(oxidation at anode)}$$
$$2H_2O(\ell) \rightarrow 2H_2(g) + O_2(g) \quad \text{(cell reaction)}$$

The H^+ is consumed at the cathode and produced at the anode. The net reaction is the decomposition of water and, as the electrolysis continues, the concentration of sulfuric acid increases. The fact that water is oxidized at the anode in preference to the sulfate ion, SO_4^{2-}, or hydrogen sulfate ion, HSO_4^-, while H^+ is reduced at the cathode illustrates one of the basic principles of electrochemistry: the most easily oxidized species present is oxidized and the most easily reduced species is reduced.

21-3 Faraday's Law of Electrolysis

Faraday's Law of Electrolysis states that the amount of substance undergoing oxidation or reduction at each electrode during electrolysis is directly proportional to the amount of electricity that passes through the cell. One **faraday** is the amount of electricity that reduces one equivalent weight of oxidizing agent at the cathode or oxidizes one equivalent weight of reducing agent at the anode. An **equivalent weight** of an oxidizing agent or a reducing agent is the mass of that substance that provides or consumes one mole of electrons in a reaction. A faraday corresponds to 96,485 coulombs (C) or the charge on one mole (6.022×10^{23}) electrons.

$$1 \text{ Faraday} = 96,485 \text{ C} = 1 \text{ mol } e^- = 6.022 \times 10^{23} \text{ } e^-$$

Electrical current is usually expressed in units of amperes (A). One ampere corresponds to the passage of one coulomb of charge per second.

$$\text{one ampere} = \text{one coulomb/second} \quad \text{or} \quad 1 \text{ A} = 1 \text{ C/s}$$

Example 21-1 Electrolysis

Metallic aluminum can be obtained by electrolysis of aluminum fluoride, AlF_3, dissolved in molten cryolite, K_3AlF_6. What mass of aluminum is obtained at the cathode if a 5.00 ampere current is applied for 4.00 hours? The appropriate reaction is:

$$Al^{3+}(aq) + 3e^- \rightarrow Al(s)$$

Plan

The half-reaction tells us that three moles of electrons are required to produce one mole of aluminum. We know that each mole of electrons corresponds to 96,485 coulombs and that the product of current and time is coulombs. We use dimensional analysis to finish the problem:

$$\text{current, time} \rightarrow \text{coulombs} \rightarrow \text{mol e}^- \rightarrow \text{mol Al} \rightarrow \text{mass Al}$$

Solution

$$? \text{ C} = 4.00 \text{ hr} \times \frac{60 \text{ min}}{1 \text{ hr}} \times \frac{60 \text{ s}}{1 \text{ min}} \times \frac{5.00 \text{ C}}{\text{s}} = 7.20 \times 10^4 \text{ C}$$

$$? \text{ g Al} = 7.20 \times 10^4 \text{ C} \times \frac{1 \text{ mol e}^-}{96485 \text{ C}} \times \frac{1 \text{ mol Al}}{3 \text{ mol e}^-} \times \frac{27.0 \text{ g Al}}{1 \text{ mol Al}} = \boxed{6.72 \text{ g Al}}$$

Example 21-2 Electrolysis

Refer to Example 21-1. How many grams and what volume of fluorine, F_2, (measured at STP) could be liberated at the anode during that time? The reaction is:

$$2F^-(aq) \rightarrow F_2(g) + 2e^-$$

Plan

From Example 21-1, we know that 7.20×10^4 coulombs pass through the cell in 4.00 hours. We can use the stoichiometry of the reaction, the molar mass of F_2 and the knowledge that one mole of any gas at STP occupies 22.4 L to solve the problem.

$$\text{coulombs} \rightarrow \text{mol e}^- \rightarrow \text{mol } F_2 \rightarrow \text{mass of } F_2 \text{ or volume of } F_2 \text{ at STP}$$

Solution

$$? \text{ g } F_2 = 7.20 \times 10^4 \text{ C} \times \frac{1 \text{ mol e}^-}{96485 \text{ C}} \times \frac{1 \text{ mol } F_2}{2 \text{ mol e}^-} \times \frac{38.0 \text{ g } F_2}{1 \text{ mol } F_2} = \boxed{14.2 \text{ g } F_2}$$

$$? \text{ L } F_2 = 7.20 \times 10^4 \text{ C} \times \frac{1 \text{ mol e}^-}{96485 \text{ C}} \times \frac{1 \text{ mol } F_2}{2 \text{ mol e}^-} \times \frac{22.4 \text{ L}}{1 \text{ mol } F_2} = \boxed{8.36 \text{ L } F_2}$$

Example 21-3 Electrolysis

Refer to Example 21-1. How many hours would the electrolysis need to continue to produce 75.0 grams of Al with a current of 15.0 amperes?

Plan

We wish to find the time required and we recognize that this problem is the reverse of Example 21-1. We need to make the conversions:

Solution

$$\text{mass Al} \rightarrow \text{mol Al} \rightarrow \text{mol e}^- \rightarrow \text{coulombs} \rightarrow \text{time}$$

$$? \, C = 75.0 \, g \, Al \times \frac{1 \, mol \, Al}{27.0 \, g \, Al} \times \frac{3 \, mol \, e^-}{1 \, mol \, Al} \times \frac{96485 \, C}{1 \, mol \, e^-} = 8.04 \times 10^5 \, C$$

$$? \, hr = 8.04 \times 10^5 \, C \times \frac{1 \, s}{15.0 \, C} \times \frac{1 \, min}{60 \, s} \times \frac{1 \, hr}{60 \, min} = \boxed{14.9 \, hr}$$

Example 21-4 Electrolysis

A 2.00 ampere current is passed through an electrolytic cell containing an aqueous divalent (2+) metal salt for 4.75 hours. During that time, 10.40 grams of the reduced metal are deposited at the cathode. Identify the metal.

Plan

To identify the metal, we must find the atomic weight (molar mass) in g/mol and compare it with masses on the periodic chart. We know the mass of the metal, therefore, we need to find the number of moles of metal present. We can find the number of moles of electrons from the product of the time and the current and from Faraday's constant. We know that 2 moles of electrons are required to reduce a metal from the 2+ state to the 0 state, so we can find the number of moles of metal. The pathway is:

$$\text{time, charge} \rightarrow \text{coulombs} \rightarrow \text{mol e}^- \rightarrow \text{mol metal}$$

Solution

$$? \, mol \, metal = 4.75 \, hr \times \frac{60 \, min}{1 \, hr} \times \frac{60 \, s}{1 \, min} \times \frac{2.00 \, C}{s} \times \frac{1 \, mol \, e^-}{96485 \, C} \times \frac{1 \, mol \, metal}{2 \, mol \, e^-}$$

$$= 0.177 \, mol \, metal$$

$$? \text{ g/mol} = \frac{10.40 \text{ g}}{0.177 \text{ mol}} = 58.8 \text{ g/mol}$$

Since the molar mass is 58.8 g/mol, the metal must be $\boxed{\text{Ni}}$.

Example 21-5 Oxidation State of a Metal

Determine the oxidation state of rhodium, Rh, which is used as a brilliant plating for jewelry settings, from the following electrolysis data. A 1.92 gram sample of metallic rhodium is deposited during the electrolysis of an aqueous solution of a rhodium salt under a 1.50 ampere current for 1.00 hour.

Plan

We know that metals always have positive oxidation numbers in salts. We can find the magnitude of the charge if we determine the number of moles of electrons required to reduce 1 mole of rhodium. We can find the number of moles of rhodium by using the mass and molar mass of the metal. We can find the number of moles of electrons by using the charge, current, and Faraday's constant as in previous examples. The reaction can be represented as below, where n is the oxidation number of the rhodium ion:

$$\text{Rh}^{n+}(aq) + n e^- \rightarrow \text{Rh}(s)$$

Solution

$$? \text{ mol Rh} = 1.92 \text{ g Rh} \times \frac{1 \text{ mol Rh}}{102.9 \text{ g Rh}} = 0.0187 \text{ mol Rh}$$

$$? \text{ mol } e^- = 1.00 \text{ hr} \times \frac{3600 \text{ s}}{1 \text{ hr}} \times \frac{1.50 \text{ C}}{\text{s}} \times \frac{1 \text{ mol } e^-}{96485 \text{ C}} = 0.0560 \text{ mol } e^-$$

Finally, the desired ratio can be computed:

$$\frac{0.0560 \text{ mol } e^-}{0.0187 \text{ mol Rh}} = 2.99 \text{ so the ion is } \boxed{\text{Rh}^{3+}} \text{ and the oxidation number is } \boxed{+3}.$$

21-4 Voltaic Cells (Galvanic Cells)

In voltaic (galvanic) cells, the two halves of a spontaneous redox reaction are physically separated, requiring electron transfer to occur through an external circuit, often a wire. A salt bridge usually completes the circuit. Voltaic cells can also be diagrammed on the basis of experimental observation. Cell potential is reported in **volts**. One volt (V) is a joule/coulomb. For example, a voltaic cell can be made in the following manner. One electrode consists of a st of magnesium metal immersed in 1.00 M aqueous magnesium nitrate, $Mg(NO_3)_2(aq)$, solution. The other electrode consists of a strip of silver metal immersed in 1.00 M aqueous silver nitrate $AgNO_3$. The cell is completed with a 5% agar-KNO_3 salt bridge. We construct a diagram of the cell from the following experimental observations.

1. The initial cell potential read from the voltmeter is 3.16 V.
2. The magnesium strip loses mass and the Mg^{2+} concentration increases in the solution surrounding the magnesium strip.
3. The silver strip gains mass and the concentration of Ag^+ decreases in the solution surrounding the silver strip.

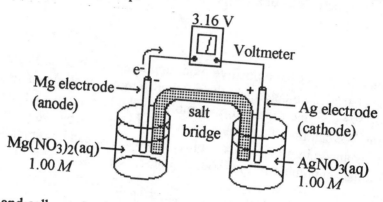

The half-reactions and cell reactions are as follows:

$Mg \rightarrow Mg^{2+} + 2e^-$ (oxidation at anode)
$2(Ag^+ + e^- \rightarrow Ag)$ (reduction at cathode)
$Mg(s) + 2Ag^+(aq) \rightarrow Mg^{2+}(aq) + 2Ag(s)$ (cell reaction)

The cell reaction is spontaneous and the flow of electrons is from the anode to the cathode. The anode is negative and the cathode positive (true in all **voltaic** cells). Although the electrodes have opposite signs in electrolytic cells, in both cases, oxidation is at the anode while reduction occurs at the cathode.

21-5 The Standard Hydrogen Electrode (SHE)

Standard electrodes consist of oxidized and reduced forms of a substance in contact with each other at unit activities. Unit activities are one molar concentrations of all dissolved species and one atmosphere partial pressure of all gases. Any solids or liquids must be pure.

is useful to be able to predict spontaneous cell reactions and voltages of cells. To facilitate this, half-cells are compared to a **standard hydrogen electrode, SHE**. The SHE is shown in Figure 21-9 in the text. It consists of a strip of inert platinum, immersed in 1.00 M H^+ solution. Hydrogen gas, H_2, is bubbled through a glass envelope and over the platinum at a pressure of one atmosphere.

Either of two reactions can occur at the SHE, depending upon whether it functions as an anode or cathode. This in turn depends upon the electrode to which it is connected. The **standard electrode potential, E°**, of the SHE is arbitrarily assigned a value of exactly zero volts (whether for oxidation or reduction). As we saw in thermodynamics, the ° means at standard conditions.

(as cathode) $2H^+ + 2e^- \rightarrow H_2$ $E° =$ exactly 0.0 V
(as anode) $H_2 \rightarrow 2H^+ + 2e^-$ $E° =$ exactly 0.0 V

The potentials of other electrodes can be measured relative to the potential of the SHE (or to other reference electrodes, which have been compared to the SHE and which are less cumbersome to use). Since the potential of the SHE is zero volts, the measured potential of a cell containing the SHE is taken as the potential of the other electrode.

For example, we can construct a voltaic cell that consists of a strip of aluminum that is immersed in 1.00 M aqueous aluminum nitrate, $Al(NO_3)_3$, connected to a SHE by a wire and a 5% agar-saturated KCl salt bridge. We can construct a diagram for the cell from the following observations.

1. The strip of aluminum loses mass and the concentration of Al^{3+} increases in solution around the aluminum.
2. The partial pressure of H_2 increases and the $[H^+]$ decreases at the SHE.
3. The initial voltage (potential) of the cell is 1.66 V.

Since the cell potential is 1.66 V, and because the contribution due to the SHE is defined as 0 V, this is also the potential of the standard aluminum electrode (for oxidation). The half-reactions and net cell reaction must be the following:

(oxidation at anode) $2(Al \rightarrow Al^{3+} + 3e^-)$ 1.66 V
(reduction at cathode) $3(2H^+ + 2e^- \rightarrow H_2)$ 0.00 V
(net cell reaction) $2Al(s) + 6H^+(aq) \rightarrow 2Al^{3+}(aq) + 3H_2(g)$ 1.66 V

The potential of this standard cell is called E°cell. For this combination of these two standard electrodes, E°cell = 1.66 V.

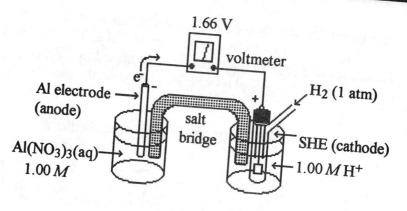

21-6 Standard Electrode Potentials

By repeating the process above, comparing standard cells to the SHE or to other electrodes, we can generate a table of standard electrode potentials. By international convention, we tabulate all standard potentials, $E°$, as reduction potentials. A **reduction potential** indicates the relative tendencies of electrodes to behave as cathodes versus the SHE. We assign positive values to electrodes that act as cathodes versus the SHE, and negative values to those that act as anodes versus the SHE. The more positive the $E°$ value of a half-reaction is, the greater is the tendency for the reaction to proceed in the forward direction. Reversing a half-reaction changes the sign, but not the magnitude, of $E°$. For the aluminum standard electrode detailed in Section 21-5, the standard reduction potential is:

$$Al^{3+} + 3e^- \rightarrow Al \qquad E° = -1.66 \text{ V}$$

Table 21-2 in the text gives some standard reduction potentials, while Appendix J is a more extensive tabulation.

Standard reduction potentials are directly related to the electromotive or activity series discussed in Chapter Four. The species listed on the left side of such tabulations can act as oxidizing agents and those on the right can act as reducing agents. The strongest oxidizing agents have the most positive $E°$ values and are at the lower left, while the weakest oxidizing agents are at the upper left. The strongest reducing agents are at the upper right and have the most negative standard reduction potentials and, therefore, the most positive oxidation potentials. The weakest reducing agents are at the lower right.

-7 Uses of Standard Reduction Potentials

...dox reactions that have positive $E°_{cell}$ values are spontaneous under standard conditions, while ...se with negative $E°_{cell}$ values are nonspontaneous. Therefore, standard reduction potentials ... be used to determine the spontaneous reaction that occurs when two half cells are combined. ...is enables us to predict whether a given redox process will be spontaneous or not. The general ocedure is:

1. Choose the appropriate half-reactions from a table of standard reduction potentials.
2. Write the half-reaction with the more positive (or less negative) $E°$ value for **reduction** first, along with its $E°$ value.
3. Write the other half-reaction as an **oxidation** and write its oxidation potential. To do this, reverse the tabulated reduction half-reaction and change the sign of $E°$.
4. Balance the electron transfer. Do **not** multiply the potential because $E°$ is a measure of the tendency for a process to occur and is independent of how many times it occurs.
5. Add the half-reactions and $E°$ values and eliminate common terms. When this procedure is followed, $E°_{cell}$ will be positive, indicating the spontaneity of the forward reaction.

Example 21-6 Predicting the Direction of a Reaction

Can metallic aluminum reduce Mg^{2+} to metallic magnesium and be oxidized to Al^{3+} ions, or will the reverse process be spontaneous when all species are present at unit activities?

Plan

We refer to Appendix J for the appropriate $E°$ values and then determine the spontaneous reaction by following the procedure above. The aluminum reaction has the less negative reduction potential and occurs as written. The magnesium reaction has the more negative reduction potential and must be reversed to occur as the oxidation. We balance the electron transfer and then add the two half-reactions and their potentials.

Solution

$$
\begin{array}{ll}
 & E° \\
2(Al^{3+}(aq) + 3e^- \rightarrow Al(s)) & -1.66 \text{ V} \quad \text{(cathode)} \\
3(Mg(s) \rightarrow Mg^{2+}(aq) + 2e^-) & -(-2.37 \text{ V}) \quad \text{(anode)} \\
\hline
2Al^{3+}(aq) + 3Mg(s) \rightarrow 2Al(s) + 3Mg^{2+}(aq) & +0.71 \text{ V} = E°_{cell}
\end{array}
$$

Since this combination produces a positive $E°_{cell}$, the forward reaction as written is spontaneous and Mg reduces the Al^{3+} to Al and is oxidized to Mg^{2+}.

Example 21-7 Predicting the Direction of a Reaction

Can permanganate ions, MnO_4^-, oxidize Cl^- to Cl_2 and be reduced to Mn^{2+} in acidic solution when all species are at unit activity?

Plan

We again follow the procedure as outlined above. In this case the permanganate half-reaction has the more positive cell potential and will be written first as the reduction. The chloride half-reaction will be reversed and be the oxidation. We balance electron transfer and add the equations and the $E°$ values to determine the spontaneous reaction and its cell potential. (For this and subsequent examples, we will omit states of matter in the half-reactions for brevity.)

Solution

$$2(MnO_4^- + 8H^+ + 5e^- \rightarrow Mn^{2+} + 4H_2O)$$
$$5(2Cl^- \rightarrow Cl_2 + 2e^-)$$

$E°$
+1.51 V (cathode)
−(+1.36) V (anode)

$$2MnO_4^-(aq) + 16H^+(aq) + 10Cl^-(aq) \rightarrow 2Mn^{2+}(aq) + 5Cl_2(g) + 8 H_2O(\ell) \quad 0.15 \text{ V} = E°_{cell}$$

The answer is yes because $E°_{cell}$ is positive for the oxidation of Cl^- to Cl_2 by MnO_4^- in acidic solution.

Example 21-8 Determining a Standard Reduction Potential

When we set up a voltaic cell, the following net reaction is observed at standard conditions. If $E°_{cell}$ is 1.40 V, what is $E°_{red}$ for $VO_2^+ + 2H^+ + e^- \rightarrow VO^{2+} + H_2O$?

$$2VO_2^+(aq) + 4H^+(aq) + Cd(s) \rightarrow 2VO^{2+}(aq) + 2 H_2O(\ell) + Cd^{2+}(aq)$$

Plan

We follow the procedure as outlined above. In this case, the vanadium reaction we want is the reduction half-reaction. The known cadmium reaction will be the oxidation. We balance electron transfer and add the equations. From the known $E°_{ox}$ of cadmium and the $E°_{cell}$ we can find the missing $E°_{red}$ value for the vanadium reaction.

Solution
2(VO$_2^+$ + 2H$^+$ + e$^-$ → VO^{2+} + H$_2$O)   x   V   (cathode)
Cd → Cd^{2+} + 2e$^-$ -(-0.403)) V   (anode)
_____
2VO$_2^+$(aq) + 4H$^+$(aq) + Cd(s) → 2VO^{2+}(aq) + 2 H$_2$O(ℓ) + Cd^{2+}(aq)

x + 0.403 V = 1.40 V E°$_{cell}$

E°$_{red}$ for the vanadium half-reaction is 1.40 V - 0.403 V = $\boxed{1.00 \text{ V}}$.

Example 21-9 Predicting the Direction of a Reaction

When a Cr^{3+}/Cr$_2$O$_7^{2-}$ half-cell and a Pb^{2+}/Pb half-cell are connected, what is the spontaneous reaction and the E° of the cell when all species are present at unit activity?

Plan
We proceed as in previous examples. The chromium half-cell has the more positive E° and will be the cathode. The lead reaction will occur at the anode. We balance electron transfer and add the equations and E° values to determine the spontaneous reaction and its cell potential.

Solution   E°
Cr$_2$O$_7^{2-}$ + 14H$^+$ + 6e$^-$ → 2Cr^{3+} + 7H$_2$O   +1.33   V
3(Pb → Pb^{2+} + 2e$^-$)   -(-0.126) V
_____
$\boxed{\text{Cr}_2\text{O}_7^{2-}(aq) + 14\text{H}^+(aq) + 3\text{Pb}(s) \rightarrow 2\text{Cr}^{3+}(aq) + 7\text{H}_2\text{O}(\ell) + 3\text{Pb}^{2+}(aq)}$ +1.46 V = E°$_{cell}$

21-8 The Nernst Equation

The tabulated standard reduction potentials, E°, apply only when all species are present at unit activity. Under other conditions, the electrode potential, E, can be calculated from the Nernst equation:

$$E = E° - \frac{RT}{nF}\ln Q, \text{ which at 25°C reduces to } E = E° - \frac{0.0592}{n}\log Q$$

E   =   the electrode potential under nonstandard conditions

E° = the standard (tabulated) electrode potential
R = the ideal gas constant (8.314 J/mol·K)
T = the temperature in Kelvin
n = the number of electrons transferred in the reaction
Q = the reaction quotient
F = Faraday's constant, 96,487 C/mol e⁻

Example 21-10 The Nernst Equation

Calculate the potential for the Al^{3+}/Al electrode as a cathode when $[Al^{3+}] = 0.0050\ M$ at 25°C.

Plan

We know that the reduction reaction occurs at the cathode. From the table of standard reduction potentials, $E° = -1.66$ V for: $Al^{3+}(aq) + 3e^- \rightarrow Al(s)$. We use the balanced half-reactions and the given concentration to calculate the value of Q. We use this data and the Nernst equation with n = 3, the number of electrons transferred in the half-reaction, to calculate E.

Solution

$$Q = \frac{1}{[Al^{3+}]} = \frac{1}{0.0050} = 2.0 \times 10^2$$

$$E = E° - \frac{0.0592}{n} \log Q = -1.66\ V - \frac{0.0592}{3} \log(2.0 \times 10^2)$$

$$E = -1.66\ V - (0.0197)(2.30) = -1.66\ V - (0.0453\ V) = \boxed{-1.71\ V}$$

Example 21-11 The Nernst Equation

Calculate the potential for the hydrogen electrode as a cathode when the $[H^+]$ is 5.00 M and the partial pressure of H_2 is 0.250 atmospheres at 25°C.

Plan

We use the same approach as in the previous example. In this case, the half-reaction is: $2H^+ + 2e^- \rightarrow H_2$; n = 2 and E° = 0.0000 V. We use the concentration of H^+ and the partial pressure of H_2 in the expression for Q.

Solution

$$Q = \frac{P_{H_2}}{[H^+]^2} = \frac{0.250}{(5.00)^2} = 0.0100 \text{ and } \log Q = -2.000$$

$$E = E° - \frac{0.0592}{n} \log Q = 0.0000 \text{ V} - \frac{0.0592}{2}(-2.000)$$

$$E = 0.0000 + 0.0592 = \boxed{+0.0592 \text{ V}}$$

Example 21-12 The Nernst Equation

Refer to Examples 21-10 and 21-11. If these two half-cells are connected to make a voltaic cell at 25°C, what will be the cell potential?

Plan

Since the H^+/H_2 half-cell has the positive potential written as a reduction while the Al^{3+}/Al electrode has a negative potential written as a reduction, the former must be the cathode and the latter the anode. The spontaneous reaction and cell potential are determined in the same way as if E° values were used under standard conditions.

Solution

	E_{cell}
$3(2H^+ + 2e^- \rightarrow H_2)$	+0.0592 V
$2(Al \rightarrow Al^{3+} + 3e^-)$	-(-1.71) V
$6H^+(aq) + 2Al(s) \rightarrow 3H_2(g) + 2Al^{3+}(aq)$	$\boxed{+1.77 \text{ V} = E_{cell}}$

Note that this is the initial E_{cell}. As the reaction progresses, concentrations change; therefore, E_{cell} also changes until $E_{cell} = 0$ and the system reaches equilibrium.

Example 21-13 The Nernst Equation

Find $E°_{cell}$ for the net reaction in Example 21-12, then find Q of the net reaction for the non-standard conditions detailed in Examples 21-10 and 21-11. Finally, calculate E_{cell} using the Nernst equation applied to the net equation. Compare the result to the answer given for Example 21-12.

Plan

We compute E° of the cell in the same manner as in previous examples. We find Q from the net equation and substitute in the appropriate pressure or concentration. We then use the Nernst equation with n = 6, the number of electrons transferred in the overall cell reaction.

Solution

$$3(2H^+ + 2e^- \rightarrow H_2) \qquad\qquad E°$$
$$2(Al \rightarrow Al^{3+} + 3e^-) \qquad\qquad +0.0000 \text{ V}$$
$$\qquad\qquad\qquad\qquad\qquad\qquad\qquad -(-1.66) \text{ V}$$

$$6H^+(aq) + 2Al(s) \rightarrow 3H_2(g) + 2Al^{3+}(aq) \quad \boxed{+1.66 \quad V = E°}$$

$$Q = \frac{P_{H_2}^3 [Al^{3+}]^2}{[H^+]^6} = \frac{(0.250)^3 (0.0050)^2}{(5.00)^6} = \boxed{2.5 \times 10^{-11}} \text{ and } \log Q = -10.60$$

$$E = E° - \frac{0.0592}{n} \log Q = 1.61 \text{ V} - \frac{0.0592}{6}(-10.60) = 1.66 \text{ V} + 0.105 = \boxed{1.76 \text{ V}}$$

The results are the same within rounding differences.

Examples 21-12 and 21-13 illustrate that non-standard cell potentials can be determined by finding the cell potentials of the half-reactions separately and adding them, or by applying the Nernst equation to the standard cell potential of the net reaction under standard conditions. Occasionally, the second approach will give an apparent negative cell potential. If this occurs, then the reaction under non-standard conditions is spontaneous in the reverse direction from that of the system under standard conditions. We then reverse the reaction, changing the sign of E_{cell}, to show the true spontaneous reaction.

21-9 The Relationship of E°cell to ΔG° and K

In the study of thermodynamics, we saw that the standard Gibbs free energy change, ΔG°, and the equilibrium constant, K, for a reaction are related by the equation below.

$$\Delta G° = -RT \ln K = -2.303 \, RT \log K$$

R = 8.314 J/mol·K
T = 298K for normal standard state conditions

The standard cell potential, $E°_{cell}$, for a redox reaction is related to the standard free energy, $\Delta G°$, by the following equation.

$$\Delta G° = -nFE°_{cell}$$

n = number of electrons transferred in the reaction
F = Faraday's constant = 96,485 C/mol e⁻

These relationships can be set equal to give:

$$\Delta G° = -RT \ln K = -2.303\, RT \log K = -nFE°_{cell}$$

Under nonstandard state conditions, ΔG (not $\Delta G°$) is related to E_{cell} (not $E°_{cell}$) by the equation:

$$\Delta G = -nFE_{cell}$$

The relationships to spontaneity are summarized below.

Forward Reaction	ΔG	E_{cell}	K
spontaneous	−	+	>1
equilibrium	0	0	1
non-spontaneous	+	−	<1

The value of an equilibrium constant changes only with temperature, and does not change with concentrations or partial pressures. Therefore, it is related only to $\Delta G°$ and $E°_{cell}$ at a specific temperature.

Example 21-14 Calculation of K from E°

Calculate the equilibrium constant, K, at 25°C from the standard reduction potentials for the reaction below.

$$Sn(s) + 2H^+(aq) + S(s) \rightarrow Sn^{2+}(aq) + H_2S(g)$$

The necessary half-reactions and standard reduction potentials are:

$S(s) + 2H^+(aq) + 2e^- \rightarrow H_2S(g)$ $E° = +0.14$ V
$Sn^{2+}(aq) + 2e^- \rightarrow Sn(s)$ $E° = -0.14$ V

Plan

To generate the equation of interest, the Sn^{2+}/Sn half-reaction must be reversed, i.e., changed to an oxidation, and the sign of its $E°$ changed. We compute $E°$ as in previous examples and then use the formula $-nFE° = -RT \ln K$. We remember that a volt is a joule/coulomb to simplify dimensional analysis.

Solution

	$E°$	
$S + 2H^+ + 2e^- \rightarrow H_2S$	+0.14	V
$Sn \rightarrow Sn^{2+} + 2e^-$	-(-0.14)	V
$S(s) + Sn(s) + 2H^+(aq) \rightarrow Sn^{2+}(aq) + H_2S(g)$	+0.28	$V = E°_{cell}$

$-nFE°_{cell} = -RT \ln K$, therefore, $\ln K = \dfrac{nFE°_{cell}}{RT}$

$$\ln K = \frac{(2 \text{ mol } e^-)\left[96485 \dfrac{C}{\text{mol } e^-}\right](+0.28 \dfrac{J}{C})}{\left[8.314 \dfrac{J}{\text{mol·K}}\right](298K)} = 22$$

$K = e^{22} = \boxed{4 \times 10^9}$ (thermodynamic equilibrium constant)

Example 21-15 Calculation of $\Delta G°$

Calculate $\Delta G°$ for the reaction of Example 21-14.

Plan

We know $E°_{cell}$ from the previous example and, therefore, can simply use the formula: $\Delta G° = -nFE°_{cell}$. In this case, $n = 2$, the number of moles of electrons that are transferred in the net reaction.

Solution

$$\Delta G° = -nFE°_{cell} = -(2 \text{ mol } e^-)\left[96485 \dfrac{C}{\text{mol } e^-}\right](+0.28 \text{ J/C}) \times \dfrac{1 \text{ kJ}}{1000 \text{ J}} = \boxed{-54 \text{ kJ}}$$

The negative sign for $\Delta G°$ indicates that the reaction is spontaneous under standard conditions at 25°C. This is consistent with a large value of K and a positive $E°_{cell}$ value.

Example 21-16 Calculation of ΔG, $\Delta G°$, and K

Refer to Examples 21-10 through 21-13. Calculate ΔG, $\Delta G°$, and K at 25°C from the information given. The reaction of interest is:

$$6H^+ (5.00\,M) + 2Al(s) \rightarrow 2Al^{3+} (0.0050\,M) + H_2 (0.250\,atm)$$

Plan

We see that the concentrations in the reaction above are those from Examples 21-10 and 21-11. We determined E_{cell} in Exercise 21-12 and found $E°_{cell}$ in Example 21-13. We use this data in the appropriate formulas, $\Delta G = -nFE_{cell}$ or $\Delta G° = -nFE°_{cell}$. In this case, n, number of electrons transferred in the net reaction, is 6. We find K from $\Delta G°$ and the formula above.

Solution

$$\Delta G = -(6\,mol\,e^-)\left[96485\,\frac{C}{mol\,e^-}\right](1.71\,V) \times \frac{1\,kJ}{1000\,J} = \boxed{-9.90 \times 10^2\,kJ}$$

$$\Delta G° = -(6\,mol\,e^-)\left[96485\,\frac{C}{mol\,e^-}\right](+1.66\,V) \times \frac{1\,kJ}{1000\,J} = \boxed{-961\,kJ}$$

$$\log K = \frac{\Delta G°}{-2.303RT} = \frac{-961\,kJ}{(-2.303)\left[8.314 \times 10^{-3}\,\frac{kJ}{mol \cdot K}\right](298K)} = 168$$

$K = \boxed{10^{168}}$ (the equilibrium lies far to the right)

Example 21-17 Calculation of K

Given the following information, estimate K_{sp} for $Zn(OH)_2$ at 25°C.

$$Zn(OH)_2(s) + 2e^- \rightarrow Zn(s) + 2OH^-(aq) \quad E° = -1.245\,V$$
$$Zn^{2+}(aq) + 2e^- \rightarrow Zn(s) \quad E° = -0.763\,V$$

Plan

We know that the solubility equilibrium and solubility product constant expression for $Zn(OH)_2$ are:

$$Zn(OH)_2(s) \rightarrow Zn^{2+}(aq) + 2OH^-(aq) \quad K_{sp} = [Zn^{2+}][OH^-]^2$$

We see that the equation for the dissolution equilibrium can be obtained by reversing the Zn^{2+}/Zn half-reaction (making it an oxidation), changing the sign of $E°$ and adding it to the $E°$ from the first half-reaction given above. Once we determine $E°_{cell}$, we can use the relationship that $-nFE° = -RT \ln K$ to find K_{sp}. In this case, $n = 2$ moles of electrons transferred in the reaction.

Solution

$$Zn(OH)_2(s) + 2e^- \rightarrow Zn(s) + 2OH^-(aq) \qquad E° = -1.245 \text{ V}$$
$$Zn(s) \rightarrow Zn^{2+}(aq) + 2e^- \qquad E° = -(-0.763 \text{ V})$$
$$\overline{Zn(OH)_2(s) \rightarrow Zn^{2+}(aq) + 2OH^-(aq) \qquad E°_{cell} = -0.482 \text{ V}}$$

The K_{sp} for this process can now be calculated. Note that $K_{sp} = K$, the thermodynamic equilibrium constant for this reaction.

$$-nFE° = -RT \ln; \text{ therefore } \ln K = \frac{-nFE°}{-RT}$$

$$\ln K = \frac{(2 \text{ mol } e^-)\left[96485 \frac{C}{\text{mol } e^-}\right](-0.482 \frac{J}{C})}{\left[8.314 \frac{J}{\text{mol} \cdot K}\right](298K)} = -37.5$$

$$K_{sp} = e^{-37.5} = \boxed{5 \times 10^{-17}}$$

The value calculated with this method compares well with the value for the K_{sp} of $Zn(OH)_2$ listed in Appendix H, 4.5×10^{-17}.

EXERCISES

Electrolytic Cells

1. We electrolyze a sample of aqueous silver nitrate, using silver electrodes. Diagram the cell completely as in the illustrative examples and write balanced equations for half-reactions and the overall reaction from the observations below. When the cell is in operation:

    (a) metallic silver plates out on one electrode

    (b) gaseous oxygen, O_2, bubbles off at the other electrode and the solution becomes acidic about that electrode.

2. Repeat Exercise 1 for the electrolysis of aqueous calcium bromide at 98°C using inert electrodes. The observations are:

    (a) reddish-brown gaseous bromine, Br_2, (a liquid at room temperature) bubbles off at one electrode

    (b) gaseous hydrogen is produced at the other electrode and the solution becomes basic around that electrode.

Faraday's Law of Electrolysis

3. How many grams of the following elements could be produced at the cathode by the passage of a 2.50 ampere current for 4.00 hours through a liquid or solution that contains the indicated species?

    (a) Fe from Fe^{3+}
    (b) K from K^+
    (c) Ag from $AgNO_3$
    (d) Ni from Ni^{2+}
    (e) Zn from Zn^{2+}
    (f) Zr from Zr^{4+}

4. What volume (at STP) of the following gases could be produced at an electrode of an electrolytic cell by the passage of a 12.0 ampere current for 1.75 hours? Assume no other half-reaction occurs at the electrode of interest.

    (a) Cl_2 from ClO_3^-
    (b) O_2 from H_2O
    (c) NO from NO_3^-

5. How many hours would a 3.00 ampere current have to flow to produce the following amounts of product?

    (a) 5.00 grams of Ca from molten $CaCl_2$
    (b) 125 grams of Cu from aqueous $CuSO_4$
    (c) 30.0 liters of N_2 at STP from N_2H_4
    (d) 60.0 liters of F_2 at STP from molten KF

(e) 85.0 grams of Ga from molten $GaCl_3$

(f) 46.0 grams of Te from molten TeO_2

6. What current in amperes is necessary to plate 27.5 grams of chromium from molten $Cr(OH)_3$ in 3.44 hours?

7. A 2.44 ampere current is passed for 45.0 minutes through a solution containing a copper salt. If a 2.17 gram sample of copper is plated out at the cathode, what is the charge on the copper ions in solution?

8. A solution of an acetate salt having a cation with a 1+ charge is electrolyzed with a 4.50 ampere current for 1.24 hours. As a result, 13.2 grams of metal plate out at the cathode. What is the metal?

9. An aqueous solution containing a vanadium salt is electrolyzed for 2.50 hours by a 3.00 ampere current. If a 2.84 gram sample of vanadium is obtained at the cathode, what is the charge on the vanadium ion in this salt?

10. A 8.65 ampere current applied for 14.00 minutes produces 3.49 grams of a trivalent metal. What is the metal?

Voltaic Cells

In Exercises 11-13, voltaic cells are described and observations after circuit completion are given. In each case, diagram the cell completely from the observations, as in the illustrative examples, and write balanced equations for half-reactions and the overall cell reaction.

11. One electrode consists of a strip of lead metal immersed in 1.0 M $Co(NO_3)_2$ solution. The other consists of a strip of copper metal immersed in 1.0 M $Cu(NO_3)_2$ solution. The cell is completed by a wire and a salt bridge. The observations are:

    (a) the cobalt electrode loses mass and $[Co^{2+}]$ increases in the solution around the electrode.

    (b) the copper electrode gains mass (due only to copper deposits) and the surrounding solution becomes lighter blue as $[Cu^{2+}]$ decreases.

12. One electrode is the standard hydrogen electrode and the other consists of an inert platinum electrode in contact with an oxygen-free solution that is 1.0 M in both $FeCl_2$ and $FeCl_3$. A wire and a salt bridge complete the circuit. The observations are:

    (a) the pH decreases at the SHE.

    (b) the $[Fe^{3+}]$ decreases at the platinum electrode.

One electrode consists of a strip of metallic silver immersed in 1.0 M NaCl and also in contact with solid AgCl. The other electrode is a strip of platinum immersed into a solution which is 1.0 M in Cr^{3+}, 1.0 M in $Cr_2O_7^{2-}$ and 1.0 M in H^+. A wire and salt bridge complete the circuit. The observations are:

(a) the $[Cr^{3+}]$ increases and the $[Cr_2O_7^{2-}]$ decreases, while the pH increases at the platinum electrode.

(b) the silver electrode loses mass and the amount of AgCl increases.

Prediction of Reaction Spontaneity from $E°$ Values

14. Calculate $E°$ for the observed reactions in Exercises 11-13.

15. Calculate the standard cell potentials for the following reactions at 25°C. Which ones are spontaneous under standard conditions?

    (a) $5(COOH)_2(aq) + 2ClO_3^-(aq) + 2H^+(aq) \rightarrow 10CO_2(g) + Cl_2(g) + 6H_2O(\ell)$

    (b) $Zr^{4+}(aq) + 2H_2Se(aq) \rightarrow Zr(s) + 2 Se(s) + 4H^+(aq)$

    (c) $2ClO^-(aq) + Si(s) + 2OH^-(aq) \rightarrow 2Cl^-(aq) + SiO_3^{2-}(aq) + H_2O(\ell)$

    (d) $Br^-(aq) + 3H_2O(\ell) + 6Cr^{3+}(aq) \rightarrow BrO_3^-(aq) + 6H^+(aq) + 6Cr^{2+}(aq)$

    (e) $3I_2(aq) + 10NO_3^-(aq) + 4H^+(aq) \rightarrow 6IO_3^-(aq) + 10NO(g) + 2H_2O(\ell)$

    (f) $3Si(s) + 4Cr(OH)_3(s) + 6OH^-(aq) \rightarrow 4 Cr(s) + 3SiO_3^{2-} + 9H_2O(\ell)$

16. Answer the following questions. Assume 1.0 M concentrations of all ions and one atmosphere partial pressure of all gases.

    (a) Will $TeO_2(s)$ oxidize chromium(III) ions to $Cr_2O_7^{2-}$ in aqueous solution or will $Cr_2O_7^{2-}$ oxidize tellurium to TeO_2?

    (b) Will hydrogen gas reduce zinc sulfide to metallic zinc and sulfide ions and be oxidized to hydrogen ions in aqueous solution?

    (c) Will hexaminecobalt(III) ion, $[Co(NH_3)_6]^{3+}$, oxidize zinc in basic solution to the tetrahydroxyzincate(II) ion, $[Zn(OH)_4]^{2-}$, and be reduced to the hexamine cobalt(II) ion, $[Co(NH_3)_6]^{2+}$?

    (d) Will chromium reduce tin(II) ions to tin in acidic solution and be oxidized to chromium(III) ions?

17. The standard cell potential, E°, for the reaction

$$3ClO_4^-(aq) + 6H^+(aq) + 2Y(s) \rightarrow 3ClO_3^-(aq) + 2Y^{3+}(aq) + 3H_2O(\ell)$$

is +3.56 volts. E° for the reduction of ClO_4^- to ClO_3^- is 1.19 volt. What is the standard reduction potential for the reduction of Y^{3+} to Y?

The Nernst Equation

18. Calculate reduction potentials for the following electrodes under the conditions stated at 25°C. (You may have to balance the appropriate half-reactions.)

   (a) SO_4^{2-}(aq) (2.78 M), H^+(aq) (pH = 3.215)/SO_2(g) (0.00624 atm)

   (b) NO_3^-(aq) (0.010 M), H^+(aq) (7.1 x 10^{-5} M)/NO(g) (2.50 atm)

   (c) HClO(aq) (3.00 M), H^+(aq) (pH = 4.25)/Cl_2(g) (6.00 atm)

   (d) $[RhCl_6]^{3-}$(aq) (1.65 x 10^{-6} M)/Rh(s), Cl^-(aq) (0.00100 M)

   (e) N_2(g) (0.0200 atm)/N_2H_4(aq) (0.50 M), OH^-(aq) (1.8 x 10^{-3} M)

19. Calculate the initial cell potentials, E_{cell}, for voltaic cells constructed from the following combinations of electrodes listed in Exercise 18. Write the balanced cell reaction.

   (a) 18(a) and 18(b)
   (b) 18(b) and 18(c)
   (c) 18(d) and 18(e)
   (d) 18(a) and 18(d)

Relationship of E° to ΔG° and K

20. Calculate the thermodynamic equilibrium constant at 25°C for each of the reactions below from standard potentials.

   (a) $Zr(s) + 2Se(s) + 4H^+(aq) \rightarrow Zr^{4+}(aq) + 2H_2Se(aq)$

   (b) $2H_2O_2(aq) \rightarrow 2H_2O(\ell) + O_2(g)$

   (c) $3NiO_2(s) + 2Cr(OH)_3(s) + 4OH^-(aq) \rightarrow 3Ni(OH)_2(s) + 2CrO_4^{2-}(aq) + 2H_2O(\ell)$

   (d) $2Cr(OH)_3(s) + 3SO_4^{2-}(aq) + 4OH^- \rightarrow 2CrO_4^{2-}(aq) + 5H_2O(\ell) + 3SO_3^{2-}(aq)$

21. Calculate ΔG° at 25°C for each of the reactions of Exercise 20 from $E°_{cell}$.

22. Calculate ΔG (nonstandard conditions) at 25°C for the reactions below under the conditions stated. Also give the thermodynamic equilibrium constant for each at 25°C.

(a) $O_2(g)$ (2.4 atm) + $4H^+$(aq) (pH 3.65) + 2Co(s) → $2H_2O(\ell)$ + $2Co^{2+}$(aq) (0.025 M)

(b) $3IO_3^-$(aq)(0.350 M) + 5NO(g)(7.50 atm) + $H_2O(\ell)$ →
$3/2 I_2$(aq)(0.040 M) + $5NO_3^-$(aq)(2.0 M) + $2H^+$(aq)(pH = 3.00)

(c) $2Cr^{3+}$(aq)(0.30 M) + $3N_2O_4$(g)(10.0 atm) + $12OH^-$(0.10 M) →
2Cr(s) + $6NO_3^-$(aq)(0.050 M) + $6H_2O(\ell)$

23. Use the following information to calculate K_{sp} for Hg_2Cl_2.

$Hg_2Cl_2(s) + 2e^- → 2Hg(\ell) + 2Cl^-(aq)$ $E° = +0.27$ V

$Hg_2^{2+}(aq) + 2e^- → 2Hg(\ell)$ $E° = +0.789$ V

24. Given the following information, calculate K_d for $[Zn(CN)_4]^{2-}$.

$[Zn(CN)_4]^{2-}(aq) + 2e^- → Zn(s) + 4CN^-(aq)$ $E° = -1.26$ V

$Zn^{2+}(aq) + 2e^- → Zn(s)$ $E° = -0.763$ V

25. Use the following information to find the K_{sp} of $PbSO_4$.

$PbSO_4(s) + 2e^- → Pb(s) + SO_4^{2-}(aq)$ $E° = -0.356$ V

$Pb^{2+}(aq) + 2e^- → Pb(s)$ $E° = -0.126$ V

Miscellaneous Exercises

26. How long must a 4.85 ampere current be applied to produce 35.8 liters of N_2 at STP? The applicable reaction is:

$2NH_4^+(aq) → N_2(g) + 8H^+(aq) + 6e^-$

27. Determine E°, ΔG°, and K for the process below at 25°C.

$3HgO(s) + 2Cr(OH)_3(s) + 4OH^-(aq) → 3Hg(\ell) + 2CrO_4^{2-}(aq) + 5H_2O(\ell)$

28. How many grams of MnO_2 can be produced in 15.4 minutes by a 2.75 ampere current applied to an aqueous basic $KMnO_4$ solution? The reaction is:

$MnO_4^-(aq) + 2H_2O(\ell) + 3e^- → MnO_2(s) + 4OH^-(aq)$

29. Find E_{cell} at 25°C for this reaction if: $[HClO] = 0.020\ M$, $pH = 3.75$, and $P_{Cl_2} = 0.64$ atm.

$$2HClO(aq) + 2H^+(aq) + 2e^- \rightarrow Cl_2(g) + 2H_2O(\ell)$$

30. $E°_{cell}$ for the process below is 1.94 V.

$$Cr_2O_7^{2-}(aq) + 8H^+(aq) + 2AsH_3(g) \rightarrow 2Cr^{3+}(aq) + 7H_2O(\ell) + 2As(s)$$

Use the text's tabulated value for the appropriate dichromate half-reaction to determine $E°_{red}$ for:

$$AsH_3(g) \rightarrow As(s) + 3H^+(aq) + 3e^-$$

31. Use the two half-reactions and $E°$ values below for this exercise.

$S_4O_6(aq) + 2e^- \rightarrow 2S_2O_3^{2-}(aq)$ $E° = +0.08$ V

$2HNO_2(aq) + 4H^+(aq) + 4e^- \rightarrow N_2O(g) + 3H_2O(\ell)$ $E° = 1.297$ V

(a) Write the equation for the spontaneous process and compute $E°_{cell}$.

(b) What is $\Delta G°$ for this process?

(c) What is K for this process?

(d) Calculate E_{cell} when $P_{N_2O} = 0.675$ atm, $[S_4O_6^{2-}] = 0.200\ M$, $[S_2O_3^{2-}] = 0.760\ M$, and the $pH = 2.171$.

ANSWERS TO EXERCISES

1.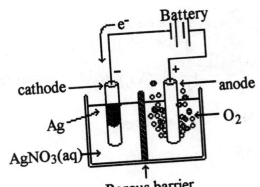

   Cathode (−): $4(Ag^+ + e^- \rightarrow Ag(s))$
   Anode (+): $2H_2O(\ell) \rightarrow O_2(g) + 4H^+(aq) + 4e^-$
   Overall: $4\,Ag^+(aq) + 2H_2O(\ell) \rightarrow 4Ag(s) + O_2(g) + 4H^+(aq)$

2.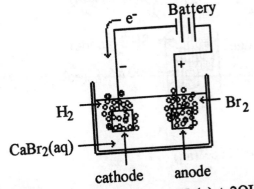

   Cathode (−): $2H_2O(\ell) + 2e^- \rightarrow H_2(g) + 2OH^-(aq)$
   Anode (+): $2Br^-(aq) \rightarrow Br_2(g) + 2e^-$
   Overall: $2H_2O(\ell) + 2Br^-(aq) \rightarrow Br_2(g) + H_2(g) + 2OH^-(aq)$

3. (a) 6.94 g Fe   (b) 14.5 g K   (c) 40.2 g Ag
   (d) 10.9 g Ni (e) 12.2 g Zn   (f) 8.51 g Zr

4. (a) 1.76 L Cl_2 (b) 4.39 L O_2 (c) 5.85 L NO

5. (a) 2.23 hr   (b) 35.1 hr   (c) 47.8 hr
   (d) 47.8 hr (e) 32.7 hr   (f) 12.9 hr

6. 12.3 A   7. Cu^{2+} 8. Cu   9. 5+   10. La

11.

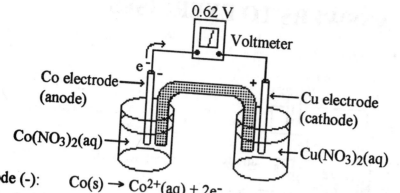

Anode (−): $Co(s) \rightarrow Co^{2+}(aq) + 2e^-$
Cathode (+): $Cu^{2+}(aq) + 2e^- \rightarrow Cu(s)$
Overall: $Pb(s) + Cu^{2+}(aq) \rightarrow Pb^{2+}(aq) + Cu(s)$

12.

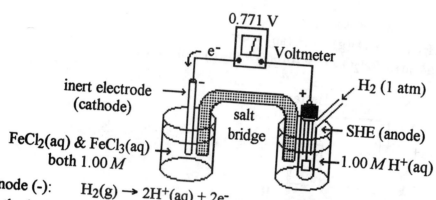

Anode (−): $H_2(g) \rightarrow 2H^+(aq) + 2e^-$
Cathode (+): $2(Fe^{3+}(aq) + e^- \rightarrow Fe^{2+}(aq))$
Overall: $H_2(g) + 2Fe^{3+}(aq) \rightarrow 2H^+(aq) + 2Fe^{2+}(aq)$

13.

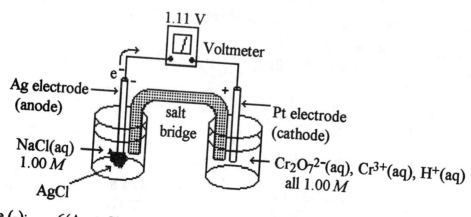

Anode (−): $6(Ag + Cl^- \rightarrow AgCl + e^-)$
Cathode (+): $Cr_2O_7^{2-} + 14H^+ + 6e^- \rightarrow 2Cr^{3+} + 7H_2O$
Overall: $6Ag + 6Cl^- + Cr_2O_7^{2-} + 14H^+ \rightarrow 6AgCl + 2Cr^{3+} + 7H_2O$

14. #11 = +0.62 V; #12 = +0.771 V; #13 = +1.11 V

15. (a) $E°_{cell}$ = +1.96 V, spontaneous
    (b) $E°_{cell}$ = -1.13 V, nonspontaneous
    (c) $E°_{cell}$ = +2.59 V, spontaneous
    (d) $E°_{cell}$ = +1.85 V, spontaneous
    (e) $E°_{cell}$ = +0.235 V, spontaneous
    (f) $E°_{cell}$ = +0.40 V, spontaneous

16. (a) No, $Cr_2O_7^{2-}$ oxdizes Te to TeO_2 with $E°_{cell}$ = +0.80 V
    (b) No H^+ ions oxidize Zn to Zn^{2+} in ZnS and are reduced to $H_2(g)$ with $E°_{cell}$ = 1.44 V
    (c) Yes, $E°_{cell}$ = +1.32 V   (d) Yes, $E°_{cell}$ = +0.60 V

17. $E°$ = -2.37 V

18. (a) E = -0.10 V   (b) E = +0.59 V   (c) E = +1.38 V
    (d) E = +0.68 V   (e) E = -1.02 V

19. (a) E_{cell} = +0.69 V
    $2NO_3^-(aq) + 3SO_2(g) + 2H_2O(\ell) \rightarrow 3SO_4^{2-}(aq) + 2NO(g) + 4H^+(aq)$

    (b) E_{cell} = +0.79 V
    $6HClO(aq) + 2NO(g) \rightarrow 3Cl_2(g) + 2NO_3^-(aq) + 2H^+(aq) + 2H_2O(\ell)$

    (c) E_{cell} = +1.70 V
    $4[RhCl_6]^{3-}(aq) + 3N_2H_4(aq) + 12OH^-(aq) \rightarrow$
    $4Rh(s) + 3N_2(g) + 12H_2O(\ell) + 24Cl^-(aq)$

    (d) E_{cell} = +0.58 V
    $2[RhCl_6]^{3-}(aq) + 3SO_2(g) \rightarrow 2Rh(s) + 12Cl^-(aq) + 3SO_4^{2-}(aq) + 12H^+(aq)$

20. (a) K = 2.7×10^{76} (b) K = 8×10^{36}
    (c) K = 8×10^{61} (d) K = 6.3×10^{-83}

21. $\Delta G°$ = -436 kJ for 20(a) $\Delta G°$ = -2.1×10^2 kJ for 20(b)
    $\Delta G°$ = -3.5×10^2 kJ for 20(c) $\Delta G°$ = 4.7×10^2 kJ for 20(d)

22. (a) ΔG = -521 kJ, K = 2×10^{235}
    (b) ΔG = -395 kJ, K = 4×10^{59}
    (c) ΔG = -5×10^1 kJ, K = 2×10^{11}

23. K_{sp} = 2.6×10^{-18} 24. K_d = 10^{-17} 25. K_{sp} = 1.7×10^{-8}

26. 53.0 hr

27. $E° = 0.22$ V, $\Delta G° = -1.3 \times 10^2$ kJ, $K = 2.0 \times 10^{22}$

28. 0.765 g MnO_2 29. 1.96 V   30. $E°_{red} = -0.61$ V

31. (a) $2HNO_2(aq) + 4H^+(aq) + 4S_2O_3^{2-}(aq) \rightarrow N_2O(g) + 3H_2O(\ell) + 2S_4O_6^{2-}(aq)$
     $E°_{cell} = 1.22$ V

   (b) $\Delta G° = -471$ kJ/mol   (c) $K = 4 \times 10^{82}$ (d) $E_{cell} = 1.12$ V